AF358731

Zoonoses and Wildlife: One Health Approach

Zoonoses and Wildlife: One Health Approach

Editor

David González-Barrio

MDPI • Basel • Beijing • Wuhan • Barcelona • Belgrade • Manchester • Tokyo • Cluj • Tianjin

Editor
David González-Barrio
Parasitology Reference and
Research Laboratory
Spanish National Centre
for Microbiology
Health Institute Carlos III
Majadahonda
Spain

Editorial Office
MDPI
St. Alban-Anlage 66
4052 Basel, Switzerland

This is a reprint of articles from the Special Issue published online in the open access journal *Animals* (ISSN 2076-2615) (available at: www.mdpi.com/journal/animals/special_issues/Zoonoses_Wildlife_One_Health_Approach).

For citation purposes, cite each article independently as indicated on the article page online and as indicated below:

LastName, A.A.; LastName, B.B.; LastName, C.C. Article Title. *Journal Name* **Year**, *Volume Number*, Page Range.

ISBN 978-3-0365-3736-8 (Hbk)
ISBN 978-3-0365-3735-1 (PDF)

Contents

About the Editor

David González-Barrio

David González-Barrio is a graduate in Veterinary Medicine by the University of Córdoba (Spain) in 2009, and Master in wildlife health in 2012 by Spanish Wildlife Research Institute (IREC-CSIC-UCLM). Dr. González-Barrio earned his European framework PhD in 2015 at the Spanish Wildlife Research Institute (IREC) of the University of Castilla-La Mancha (UCLM) and Superior Council of Scientific Research (CSIC) in Spain after research training periods in Italy and Netherlands. During this time, he aimed to develop a PhD Thesis on epidemiology and molecular typing of *Coxiella burnetii* from domestic and wild animals in Spain. His research experience is based in animal health and zoonosis with multidisciplinary skills in the fields of animal disease diagnosis, disease epidemiology, molecular epidemiology and animal disease prevention and control, and its public health involvement in zoonotic diseases. Currently, he is a Researcher at the Toxoplasmosis and Intestinal Protozoa Unit of Parasitology Reference and Research Laboratory, Spanish National Centre for Microbiology, Health Institute Carlos III from Spain.

Editorial

Zoonoses and Wildlife: One Health Approach

David González-Barrio 

Parasitology Reference and Research Laboratory, Spanish National Centre for Microbiology, Health Institute Carlos III, Ctra. Majadahonda-Pozuelo Km 2, Majadahonda, 28220 Madrid, Spain; dgonzalezbarrio@gmail.com or david.gonzalezb@isciii.es

Citation: González-Barrio, D. Zoonoses and Wildlife: One Health Approach. *Animals* **2022**, *12*, 480. https://doi.org/10.3390/ani12040480

Received: 25 January 2022
Accepted: 8 February 2022
Published: 15 February 2022

Publisher's Note: MDPI stays neutral with regard to jurisdictional claims in published maps and institutional affiliations.

Throughout history, wildlife has been an important source of infectious diseases transmissible to humans. Today, zoonoses with a wildlife reservoir constitute a major public health problem, affecting all continents. The importance of such zoonoses is increasingly recognized, and the need for more attention in this area is being addressed. The total number of zoonoses is unknown; some 1415 known human pathogens have been catalogued, and 62% are of zoonotic origin [1]. Over time, more and more human pathogens are found to be of animal origin. Moreover, most emerging infectious diseases in humans are zoonoses. Wild animals seem to be involved in the epidemiology of most zoonoses, and serve as major reservoirs for the transmission of zoonotic agents to domestic animals and humans [2]. The concept of the 'One Health' approach—involving collaboration between veterinary and medical scientists, policymakers, and public health officials—is necessary in order to foster joint cooperation and control of emerging zoonotic diseases [3]. Zoonotic diseases caused by a wide range of arthropods, bacteria, helminths, protozoans, and viruses can cause serious and even life-threatening clinical conditions in animals, with a number of them also affecting the human population due to their zoonotic potential.

The aim of the present Special Issue is to cover recent and novel research trends in zoonotic diseases in wildlife, including the relevant topics related to wildlife, zoonosis, public health, emerging diseases, infectious diseases, and parasitic diseases.

A total of 12 papers have been contributed by 96 authors from 14 countries to this issue, comprising 10 research articles, 1 communication, and 1 brief report (Figure 1). The number of specimens studied in this issue amounts to 5132, including wild animals, wild animals kept in captivity, domestic animals, and ticks; even human samples have been analyzed. More than 50 different species—including wild and domestic ungulates (e.g., red deer, roe deer, fallow deer, chamois, mouflon, European bison, wild boar, sheep, goat, cattle), wild carnivores (e.g., wolf, Eurasian lynx, Eurasian badger, coypu, beech marten, golden jackal), micromammals (e.g., yellow-necked field mouse, long-tailed field mouse, European water vole, white-toothed shrew, garden dormouse, common vole, house mouse, western Mediterranean mouse, black rat, Eurasian red squirrel), non-human primates (the genera *Cebuella, Cercocebus, Cercopithecus, Eulemur, Hylobates, Lemur, Macaca, Mandrillus, Saimiri,* and *Varecia*), turtles (e.g., *Testudo hermanni, T. h. boettgeri, T. graeca,* and *T. marginata*), bats (the families Pteropodidae, Emballonuridae, Rhinolophidae, Hipposideridae, and Vespertilionidae), and ticks (*Ixodes ricinus, Dermacentor marginatus, Hyalomma marginatum*)—are included. Regarding the zoonotic pathogens represented in this issue, the presence of or exposure to 17 different pathogens—including viruses [4] (West Nile virus), bacteria [5–13] (*Anaplasma phagocytophilum, Coxiella burnetii, Helicobacter pylori, H. suis, Mycobacterium tuberculosis* Complex, *Salmonella* sp., and *Leptospira interrogans* sensu stricto), and parasitic protists [14,15] (e.g., *Cryptosporidium* spp., *Giardia duodenalis, Blastocystis* sp., *Enterocytozoon bieneusi, Entamoeba histolytica, Entamoeba dispar, Balantioides coli, Troglodytella* spp., *Leishmania* spp.)—are presented.

Figure 1. A word cloud created from the titles of every article published in this Special Issue.

The study of zoonotic pathogens present in wildlife mainly involves serological and/or molecular analyses, among others, for their detection, which is somewhat costly due to the difficulty in obtaining the necessary samples for analysis and ensuring that they are of high quality [16]; therefore, samples are often obtained from wild animals kept in captivity or in rescue centers [7,14]. In addition, the study of parasites involves searching for them, or their DNA—mainly in the feces of animals. In remote areas or resource-poor settings where the cold chain cannot be maintained, preservation and conservation of biological specimens— including fecal samples—is a challenge; for this reason, Köster et al. [15] evaluated the suitability of filter cards for the long-term storage of fecal samples of animal and human origin that were positive for the diarrhea-causing protozoan parasites *Giardia duodenalis* and *Cryptosporidium hominis*. For this purpose, three commercially available Whatman® filter cards were comparatively evaluated: the FTA® Classic card, the FTA® Elute Micro card, and the 903 Protein Saver card. *Giardia duodenalis* ($n = 5$)- and *C. hominis* ($n = 5$)-positive human stool samples were used to impregnate the selected cards at selected storage times (1 month, 3 months, and 6 months) and temperatures ($-20\,°C$, $4\,°C$, and room temperature). Data presented by Köster et al. [15] demonstrate that Whatman® cards are a cost-effective option for the preservation and long-term storage (up to six months) of fecal samples under a wide range of temperatures (from $-20\,°C$ to room temperature), without compromising their biospecimen stability and suitability for molecular-based diagnostic methods. Indeed, Whatman® cards enable the molecular detection and genotyping of common diarrhea-causing enteric protozoan parasites, including *C. hominis* and *G. duodenalis*.

A significant proportion of wildlife studies are carried out in conservation centers— such as zoos—but also in wildlife rescue centers. Monitoring of infections that may be transmitted to humans by animals in wildlife rescue centers is very important in order to protect the staff engaged in rehabilitation practices. Casalino et al. [7] investigated the occurrence of non-typhoidal *Salmonella* in tortoises housed in a regional wildlife rescue center in Apulia, Southern Italy, to assess the presence of *Salmonella* serovars that may pose a risk to operators involved in wildlife management. *Salmonella* may be a natural inhabitant of the intestinal tracts of turtles, rarely causing disease in turtles. This may represent a potential risk for humans, increasing the sanitary risk for operators in wildlife rescue centers. Casalino et al. [7] tested 69 adult turtles (*Testudo hermanni*, *T. h. boettgeri*, *T. graeca*, and *T. marginata*); the distribution of *Salmonella* spp. was significantly higher in *T. hermanni* than in other species. Two different *Salmonella* species (*S. enterica* and *S. bongori*) three *S. enterica* subspecies (*enterica*, *diarizonae*, and *salamae*), and five different serovars (Hermannswerder, Abony, Ferruch, Richmond, and Vancouver) within the group *S. enterica* subspecies *enterica* were identified. Most of the detected *Salmonella* types may represent a potential risk to public health. Reducing turtles' stress in order to minimize *Salmonella* shedding, as well as adopting correct animal husbandry procedures and hygiene techniques, may be useful to minimize the risk of transmission of *Salmonella* to humans. In particular, the adoption of gloves to manage turtles is a relevant preventive measure. Nevertheless, the greater measure of prevention is information and education on the potential sanitary risks of each professional figure involved in wildlife management.

On the other hand, little information is currently available on the epidemiology and zoonotic potential of parasitic and commensal protist species in captive non-human primates (NHPs). Köster et al. [14] investigated the occurrence, molecular diversity, and potential transmission dynamics of parasitic and commensal protist species in a zoological garden in southern Spain. The prevalence and genotypes of the main enteric protist species were investigated in fecal samples from NHPs (n = 51), zookeepers (n = 19), and free-living rats (n = 64) via molecular (PCR and sequencing) methods between 2018 and 2019. The presence of *Leishmania* spp. was also investigated in tissues from sympatric rats using PCR. *Blastocystis* sp. (45.1%), *Entamoeba dispar* (27.5%), *Giardia duodenalis* (21.6%), *Balantioides coli* (3.9%), and *Enterocytozoon bieneusi* (2.0%) (but not *Troglodytella* spp.) were detected in NHPs. *Giardia duodenalis* (10.5%) and *Blastocystis* sp. (10.5%) were identified in zookeepers, while *Cryptosporidium* spp. (45.3%), *G. duodenalis* (14.1%), and *Blastocystis* sp. (6.25%) (but not *Leishmania* spp.) were detected in rats. *Blastocystis* ST1, ST3, and ST8, along with *G. duodenalis* sub-assemblage AII, were identified in NHPs, and *Blastocystis* ST1 was identified in zookeepers. In rats, four *Cryptosporidium* (*C. muris*, *C. ratti*, and rat genotypes IV and V), one *G. duodenalis* (assemblage G), and three *Blastocystis* (ST4) genetic variants were detected. These results indicate high exposure of NHPs to zoonotic protist species. In conclusion, strong evidence of the occurrence of zoonotic *Blastocystis* transmission between NHPs and their handlers was provided, despite the use of personal protective equipment and the implementation of strict health and safety protocols. Free-living sympatric rats are infected by host-specific species/genotypes of the investigated protists, and seem to play a limited role as a source of infections to NHPs or humans in this setting.

Interactions taking place between sympatric wildlife/livestock/humans may contribute to interspecies transmission of pathogens [17]—this is the case of the *Mycobacterium tuberculosis* complex [18]. Mycobacteria can cause medically and socioeconomically significant diseases, including several non-tuberculous infections and tuberculosis, and are considered a One Health challenge due to their impact on public and animal health. These microorganisms are maintained and shared between the environment, domestic and wild animals, and humans. In this Special Issue, two studies are related to the interaction between domestic and wild species and the detection of mycobacteria in wild species such as badgers. Varela-Castro et al. [6] characterized the interactions that take place between several wild mammals and cattle via camera-trapping in order to provide insights into the dynamics of mycobacterial transmission opportunities in the environment of cattle farms located in Atlantic habitats in the northern Iberian Peninsula. Camera traps were set during a one-year period in cattle farms with a history of tuberculosis and/or non-tuberculous mycobacteriosis. A total of 1293 visits were recorded during 2741 days of camera observation. Only 23 visits showed direct contacts with cattle, suggesting that mycobacterial transmission at the wildlife–livestock interface occurs mainly through indirect interactions.

Results showed that cattle pastures represented the most appropriate habitat for interspecies transmission of mycobacteria, and badgers' latrines appear to be a potential hotspot for mycobacterial circulation between badgers, wild boars, foxes, and cattle. According to both previous epidemiological information and the interaction patterns observed, wild boars, badgers, foxes, and small rodents are the species or groups most often in contact with livestock and, thus, may be the most involved in the epidemiology of mycobacteriosis in the wildlife–livestock interface in this area. As Valera-Castaro et al. [6] pointed out in their work, the badger and its latrines are a hotspot for interspecies transmission—both domestic and wild; more specifically, Blanco Vázquez et al. [9] investigated the prevalence, spatial distribution, and temporal distribution of tuberculosis in 673 free-ranging Eurasian badgers (*Meles meles*) in Asturias (Atlantic Spain) between 2008 and 2020. The study's objective was to assess the role of badgers as a tuberculosis reservoir for cattle and other sympatric wild species in the region. Serum samples were tested in an in-house indirect P22 ELISA to detect antibodies against the *Mycobacterium tuberculosis* complex (MTC). In parallel, data on MTC isolation and single intradermal tuberculin test results were extracted for cattle that were tested and culled as part of the Spanish National Program for the Eradication of Bovine Tuberculosis. A total of 27/639 badgers (4.23%) were positive for MTC based on bacterial isolation, while 160/673 badgers (23.77%) were found to be positive with the P22 ELISA. The rate of seropositivity was higher among adult badgers than sub-adults. The authors found that the tuberculosis status of badgers in Asturias during 2008–2020 was associated with the tuberculosis status of local cattle herds, and results could not determine the direction of possible interspecies transmission, but they were consistent with the idea that the two hosts may exert infection pressure on one another. Both studies highlight the importance of monitoring this multi-host infection and disease in wildlife during epidemiological interventions in order to optimize outcomes under the One Health concept.

Deadly emerging and re-emerging zoonotic pathogens are transmitted mostly from wildlife reservoirs to humans or other animals during spillover events, with or without a vector intervention. In this special issue, two papers are included in which vector-borne zoonotic pathogens were studied. Ain-Najwa et al. [4] highlight the first evidence of West Nile virus (WNV) infection—a mosquito-borne virus—in Malaysian macaques and bats. Of the 81 macaques from mangrove forests sampled, 24 of the long-tailed macaques were seropositive for WNV, indicating that they were exposed to the virus; meanwhile, 5 out of 41 bats that were found in the caves from northern Peninsular Malaysia showed susceptibility to WNV. The authors found a high WNV antibody prevalence in macaques and a moderate WNV RNA in various Malaysian bat species, suggesting that WNV circulates through Malaysian wild animals, and that Malaysian bat species may be susceptible to the WNV infection. On the other hand, Grassi et al. [12] researched the genetic variants of *Anaplasma phagocytophilum* (a tick-borne pathogen causing zoonotic disease) in wild ungulates (the leading reservoir species) and feeding ticks (the main vector of infection) from northeastern Italy. Using biomolecular tools and phylogenetic analysis, ecotypes I and II were detected in both ticks (*Ixodes ricinus* species) and wild ungulates. Specifically, ecotype II was mainly detected in roe deer and related ticks, while ecotype I—the potentially zoonotic variant—was detected in *Ixodes ricinus* ticks, and also in wild ungulates. These findings reveal not only the wide diffusion of *Anaplasma phagocytophilum*, but also the presence of zoonotic variants.

Žele-Vengušt et al. [5] analyzed the exposure of free-ranging wild animals to zoonotic *Leptospira interrogans* sensu stricto in Slovenia; for this, blood samples from 249 wild animals between 2019 and 2020 were tested using the microscopic agglutination test for specific antibodies against the *Leptospira* serovars Icterohaemorrhagiae, Bratislava, Pomona, Grippotyphosa, Hardjo, Sejroe, Australis, Autumnalis, Canicola, Saxkoebing, and Tarassovi. Antibodies to at least one of the pathogenic serovars were detected in 77 (30.9%; CI = 25–37%) sera. The proportion of positive samples varied intraspecifically, and was the greatest in large carnivores (86%), followed by mesopredators (50%) and large herbivores

(17%). Out of the 77 positive samples, 42 samples (53.8%) had positive titers against a single serovar, while 35 (45.4%) samples had positive titers against two or more serovars. The most frequently detected antibodies were those against the serovar Icterohaemorrhagiae. This study confirmed the presence of multiple pathogenic serovars in wildlife throughout Slovenia. It can be concluded that wild animals are reservoirs for at least some of the leptospiral serovars, and are a potential source of leptospirosis for other wild and domestic animals, as well as for humans.

In their study, Cortez Nunes et al. [13] investigated the presence of *Helicobacter pylori* and *H. suis* DNA in free-range wild boars. *Helicobacter pylori* and *H. suis* are associated with gastric pathologies in humans. Interactions between domestic animals, wildlife, and humans can increase the risk of bacterial transmission between species. Samples of the gastric tissue of 14 free range wild boars (*Sus scrofa*) were evaluated for the presence of *H. pylori* and *H. suis* using PCR. Two samples were PCR-positive for *H. pylori*, and another for *H. suis*. These findings indicate that these microorganisms were able to colonize the stomachs of wild boars, and raise awareness of their putative intervention in the transmission cycle of *Helicobacter* spp..

Finally, this Special Issue includes three articles dealing with the potential role of livestock and wildlife as potential sources of human Q fever. Q fever is a worldwide-distributed zoonosis caused by *Coxiella burnetii*—a small intracellular bacterium belonging to γ-Proteobacteria that infects a wide range of animal species, including mammals, birds, and arthropods. People are infected through inhalation of aerosols contaminated with the bacteria expelled by infected animals during abortion or normal deliveries. Domestic ruminants, sheep, and goats are considered the main reservoirs of the infection and the principal source of human outbreaks. *Coxiella burnetii* has a complex ecology that replicates in multiple host species; however, the role of wildlife in its transmission is poorly understood. Krzysiak et al. [11] examined 523 serum samples obtained from European bison for the presence of specific antibodies in order to assess whether infection occurs in this species, and whether European bison may be an important source of infection in the natural environment, as suggested by historical reports. Only one (0.19%) serum sample was positive in ELISA, and two other samples were doubtful; the only seropositive animal was a free-living bull. This suggests possible transmission from domestic cattle by sharing pastures. The transmission of *C. burnetii* into the European bison was rather accidental in the country, and its role as an important wild reservoir is unlikely. In their study, González-Barrio et al. [10] examined spleen samples from 816 micromammals of 10 species, and 130 vaginal swabs from *Microtus arvalis* by qPCR, to detect *C. burnetii* infection and shedding, respectively; 9.7% of the spleen samples were qPCR-positive. The highest infection prevalence (10.8%) was found in *Microtus arvalis*, in which *C. burnetii* DNA was also detected in 1 of the 130 vaginal swabs (0.8%) analyzed. Positive samples were also found in *Apodemus sylvaticus* (8.7%), *Crocidura russula* (7.7%), and *Rattus rattus* (6.4%). Positive samples were genotyped by coupling PCR with reverse line blotting, and a genotype II+ strain was identified for the first time in one of the positive samples from *M. arvalis*, whereas only partial results could be obtained for the rest of the samples. Acute Q fever was diagnosed in one of the researchers who participated in the study, and it was presumably linked to *M. arvalis* handling. The results of the study are consistent with previous findings suggesting that micromammals can be infected by *C. burnetii*. The authors additionally suggest that micromammals may be potential sources to trace back the origin of human Q fever and animal coxiellosis cases in Europe, and might be relevant in the maintenance of wild-type *C. burnetii* strains that can be a matter of concern for animal and human health authorities. Espí et al. [8] investigated the seroprevalence of *C. burnetii* in domestic ruminants and wild ungulates, as well as the current situation of Q fever in humans, in a small region in northwestern Spain, where close contact at the wildlife–livestock–human interface exists, and information on *C. burnetii* infection is scarce. Seroprevalence of *C. burnetii* was 8.4% in sheep, 18.4% in cattle, and 24.4% in goats. Real-time PCR analysis of environmental samples collected in 25 livestock farms detected *Coxiella* DNA in dust and/or aerosols collected in 20 of them. Analysis of sera from 327 wild ungulates revealed lower seroprevalence than that found in domestic ruminants. Exposure to the pathogen in humans was determined by IFAT analysis of 1312 blood samples collected from patients admitted to healthcare centers with Q-fever-compatible symptoms, such as fever and/or pneumonia. Results showed that 15.9% of the patients had IFAT titers $\geq 1/128$, suggestive of probable acute infection. This study is an example of a One Health approach with medical and veterinary institutions involved in investigating zoonotic diseases.

Overall, the papers in this Special Issue reveal different perspectives of current research on zoonotic disease and wildlife, from applied field studies to investigations into the intricate mechanisms involved in the interaction between pathogens, wildlife, livestock, and humans.

Funding: This research received no external funding.

Acknowledgments: I would like to thank all of the authors who contributed their papers to this Special Issue, and the reviewers for their helpful recommendations. I am also grateful to all members of the *Animals* Editorial Office for giving me this opportunity, and for their continuous support in managing and organizing this Special Issue. Finally, I would like to acknowledge the academic editors Julie Arsenault, Fulvio Marsilio, Scott C. Williams, Nicole Gottdenker, Stefania Perrucci, and Laila Darwich Soliva, who made decisions for certain papers.

Conflicts of Interest: The author declares no conflict of interest.

References

1. Billinis, C. Wildlife diseases that pose a risk to small ruminants and their farmers. *Small Rumin. Res.* **2013**, *110*, 67–70. [CrossRef]
2. González-Barrio, D.; Ruiz-Fons, F. *Coxiella burnetii* in wild mammals: A systematic review. *Transbound. Emerg. Dis.* **2019**, *66*, 662–671. [CrossRef] [PubMed]
3. Cunningham, A.A.; Daszak, P.; Wood, J.L.N. One Health, emerging infectious diseases and wildlife: Two decades of progress? *Philos. Trans. R. Soc. Lond. B Biol. Sci.* **2017**, *372*, 20160167. [CrossRef] [PubMed]
4. Ain-Najwa, M.Y.; Yasmin, A.R.; Arshad, S.S.; Omar, A.R.; Abu, J.; Kumar, K.; Mohammed, H.O.; Natasha, J.A.; Mohammed, M.N.; Bande, F.; et al. Exposure to Zoonotic West Nile Virus in Long-Tailed Macaques and Bats in Peninsular Malaysia. *Animals* **2020**, *10*, 2367. [CrossRef] [PubMed]
5. Žele-Vengušt, D.; Lindtner-Knific, R.; Mlakar-Hrženjak, N.; Jerina, K.; Vengušt, G. Exposure of Free-Ranging Wild Animals to Zoonotic *Leptospira interrogans Sensu Stricto* in Slovenia. *Animals* **2021**, *11*, 2722. [CrossRef] [PubMed]
6. Varela-Castro, L.; Sevilla, I.A.; Payne, A.; Gilot-Fromont, E.; Barral, M. Interaction Patterns between Wildlife and Cattle Reveal Opportunities for Mycobacteria Transmission in Farms from North-Eastern Atlantic Iberian Peninsula. *Animals* **2021**, *11*, 2364. [CrossRef] [PubMed]
7. Casalino, G.; Bellati, A.; Pugliese, N.; Camarda, A.; Faleo, S.; Lombardi, R.; Occhiochiuso, G.; D'Onghia, F.; Circella, E. *Salmonella* Infection in Turtles: A Risk for Staff Involved in Wildlife Management? *Animals* **2021**, *11*, 1529. [CrossRef] [PubMed]
8. Espí, A.; del Cerro, A.; Oleaga, Á.; Rodríguez-Pérez, M.; López, C.M.; Hurtado, A.; Rodríguez-Martínez, L.D.; Barandika, J.F.; García-Pérez, A.L. One Health Approach: An Overview of Q Fever in Livestock, Wildlife and Humans in Asturias (Northwestern Spain). *Animals* **2021**, *11*, 1395. [CrossRef] [PubMed]
9. Blanco Vázquez, C.; Barral, T.D.; Romero, B.; Queipo, M.; Merediz, I.; Quirós, P.; Armenteros, J.Á.; Juste, R.; Domínguez, L.; Domínguez, M.; et al. Spatial and Temporal Distribution of *Mycobacterium tuberculosis* Complex Infection in Eurasian Badger (*Meles meles*) and Cattle in Asturias, Spain. *Animals* **2021**, *11*, 1294. [CrossRef] [PubMed]
10. González-Barrio, D.; Jado, I.; Viñuela, J.; García, J.T.; Olea, P.P.; Arce, F.; Ruiz-Fons, F. Investigating the Role of Micromammals in the Ecology of *Coxiella burnetii* in Spain. *Animals* **2021**, *11*, 654. [CrossRef] [PubMed]
11. Krzysiak, M.K.; Puchalska, M.; Olech, W.; Anusz, K. A Freedom of *Coxiella burnetii* Infection Survey in European Bison (*Bison bonasus*) in Poland. *Animals* **2021**, *11*, 651. [CrossRef] [PubMed]
12. Grassi, L.; Franzo, G.; Martini, M.; Mondin, A.; Cassini, R.; Drigo, M.; Pasotto, D.; Vidorin, E.; Menandro, M.L. Ecotyping of *Anaplasma phagocytophilum* from Wild Ungulates and Ticks Shows Circulation of Zoonotic Strains in Northeastern Italy. *Animals* **2021**, *11*, 310. [CrossRef] [PubMed]
13. Cortez Nunes, F.; Letra Mateus, T.; Teixeira, S.; Barradas, P.; de Witte, C.; Haesebrouck, F.; Amorim, I.; Gärtner, F. Presence of *Helicobacter pylori* and *H. suis* DNA in Free-Range Wild Boars. *Animals* **2021**, *11*, 1269. [CrossRef] [PubMed]
14. Köster, P.C.; Dashti, A.; Bailo, B.; Muadica, A.S.; Maloney, J.G.; Santín, M.; Chicharro, C.; Migueláñez, S.; Nieto, F.J.; Cano-Terriza, D.; et al. Occurrence and Genetic Diversity of Protist Parasites in Captive Non-Human Primates, Zookeepers, and Free-Living Sympatric Rats in the Córdoba Zoo Conservation Centre, Southern Spain. *Animals* **2021**, *11*, 700. [CrossRef] [PubMed]

15. Köster, P.C.; Bailo, B.; Dashti, A.; Hernández-Castro, C.; Calero-Bernal, R.; Ponce-Gordo, F.; González-Barrio, D.; Carmena, D. Long-Term Preservation and Storage of Faecal Samples in Whatman ® Cards for PCR Detection and Genotyping of *Giardia duodenalis* and *Cryptosporidium hominis*. *Animals* **2021**, *11*, 1369. [CrossRef] [PubMed]

16. Kelly, T.R.; Karesh, W.B.; Johnson, C.K.; Gilardi, K.V.; Anthony, S.J.; Goldstein, T.; Olson, S.H.; Machalaba, C.; PREDICT Consortium; Mazet, J.A. One Health proof of concept: Bringing a transdisciplinary approach to surveillance for zoonotic viruses at the human-wild animal interface. *Prev. Vet. Med.* **2017**, *137*, 112–118. [CrossRef] [PubMed]

17. Barton Behravesh, C. Introduction. One Health: Over a decade of progress on the road to sustainability. *Rev. Sci. Tech.* **2019**, *38*, 21–50. [CrossRef] [PubMed]

18. Barroso, P.; Barasona, J.A.; Acevedo, P.; Palencia, P.; Carro, F.; Negro, J.J.; Torres, M.J.; Gortázar, C.; Soriguer, R.C.; Vicente, J. Long-Term Determinants of Tuberculosis in the Ungulate Host Community of Doñana National Park. *Pathogens* **2020**, *9*, 445. [CrossRef] [PubMed]

Article

Long-Term Preservation and Storage of Faecal Samples in Whatman® Cards for PCR Detection and Genotyping of *Giardia duodenalis* and *Cryptosporidium hominis*

Pamela Carolina Köster [1], Begoña Bailo [1], Alejandro Dashti [1], Carolina Hernández-Castro [1,2], Rafael Calero-Bernal [3], Francisco Ponce-Gordo [4], David González-Barrio [1,*] and David Carmena [1,*]

1   Parasitology Reference and Research Laboratory, Spanish National Centre for Microbiology, Majadahonda, 28220 Madrid, Spain; pamelakster@yahoo.com (P.C.K.); BEGOBB@isciii.es (B.B.); dashti.alejandro@gmail.com (A.D.); carolina.hernandez1@udea.edu.co (C.H.-C.)
2   Group of Parasitology, Faculty of Medicine, Corporation of Tropical Pathologies, University of Antioquia, Medellín 050010, Colombia
3   SALUVET, Department of Animal Health, Faculty of Veterinary, Complutense University of Madrid, 28040 Madrid, Spain; r.calero@ucm.es
4   Department of Microbiology and Parasitology, Faculty of Pharmacy, Complutense University of Madrid, 28040 Madrid, Spain; pponce@ucm.es
*   Correspondence: david.gonzalezb@isciii.es (D.G.-B.); dacarmena@isciii.es (D.C.); Tel.: +34-91-822-3641 (D.C.)

Citation: Köster, P.C.; Bailo, B.; Dashti, A.; Hernández-Castro, C.; Calero-Bernal, R.; Ponce-Gordo, F.; González-Barrio, D.; Carmena, D. Long-Term Preservation and Storage of Faecal Samples in Whatman® Cards for PCR Detection and Genotyping of *Giardia duodenalis* and *Cryptosporidium hominis*. *Animals* **2021**, *11*, 1369. https://doi.org/10.3390/ani11051369

Academic Editor: Stefania Perrucci

Received: 29 March 2021
Accepted: 10 May 2021
Published: 12 May 2021

Publisher's Note: MDPI stays neutral with regard to jurisdictional claims in published maps and institutional affiliations.

Simple Summary: Preservation and storage of biological samples prior to testing and analysis is a pressing issue in the epidemiological field studies conducted in remote or poor-resource areas with limited or no access to electricity where the cold chain cannot be maintained. This is particularly true for faecal specimens of human and animal origin exposed to high degradation rates under environmental conditions characterised by high temperatures and humidity, such as those present in tropical and subtropical regions. Under this scenario, simple, safe, and cost-effective methods are highly needed to allow the collection and transportation of well-preserved faecal samples intended for pathogen detection without compromising the performance, reliability, and accuracy of molecular procedures methods used for detection and genotyping purposes. This study assessed the suitability of three commercially available filter cards for the preservation of faecal samples containing common diarrhoea-causing enteric protozoan parasites at different storage periods and temperature conditions. Obtained results demonstrated that filter cards impregnated with faecal matrices containing these pathogens are fully compatible with downstream molecular methods for up to six months at room temperature. Therefore, filter cards can be used for the safe transportation, preservation, and storage of faecal samples without the need of the cold chain.

Abstract: Preservation and conservation of biological specimens, including faecal samples, is a challenge in remote areas or poor-resource settings where the cold chain cannot be maintained. This study aims at evaluating the suitability of filter cards for long-term storage of faecal samples of animal and human origin positive to the diarrhoea-causing protozoan parasites, *Giardia duodenalis* and *Cryptosporidium hominis*. Three commercially available Whatman® Filter Cards were comparatively assessed: the FTA® Classic Card, the FTA® Elute Micro Card, and the 903 Protein Saver Card. Human faecal samples positive to *G. duodenalis* (*n* = 5) and *C. hominis* (*n* = 5) were used to impregnate the selected cards at given storage (1 month, 3 months, and 6 months) periods and temperature (−20 °C, 4 °C, and room temperature) conditions. Parasite DNA was detected by PCR-based methods. Sensitivity assays and quality control procedures to assess suitability for genotyping purposes were conducted. Overall, all three Whatman® cards were proven useful for the detection and molecular characterisation of *G. duodenalis* and *C. hominis* under the evaluated conditions. Whatman® cards represent a simple, safe, and cost-effective option for the transportation, preservation, and storage of faecal samples without the need of the cold chain.

Keywords: filter card; faeces; transportation; storage; preservation; *Giardia duodenalis*; *Cryptosporidium hominis*; PCR

1. Introduction

Biological samples including blood, saliva, stools, urine, tissue, and cells have become increasingly valuable sources of genetic material for downstream DNA and RNA testing. Because PCR-based methods are extremely sensitive to the quality and purity of the starting nucleic acid material [1], the appropriate preservation of biospecimens during procurement, transportation, and storage is essential to maximise the success of laboratory analyses [2]. This is a pressing issue in field epidemiological studies conducted in non-clinical or remote locations where resources are scarce or lacking, or when the cold chain cannot be guaranteed. In these poor-resource settings, simple, safe, and cost-effective methods are highly needed to allow the collection and transportation of intact biospecimens without detrimental effect on their biophysical properties and diagnostic utility.

The use of solid supports, such as filter cards for the collection and analysis of biospecimens, began in 1961, when Robert Guthrie developed what is now known as the Guthrie Test by collecting drops of blood on filter paper for the detection of phenylketonuria in new-borns [3]. Since then, filter cards have been developed and commercialised by different companies [4], being the Whatman® FTA® card technology (Cytiva, Marlborough, MA, USA) one of the most frequently used. FTA® cards are cotton-based, cellulose paper containing chemicals that lyse cells on contact, denature proteins, remove contaminants, and protect DNA from degradation (including UV radiation) by immobilising it onto the card's matrix [5]. FTA® cards allow the collection, preservation, and shipment of biospecimens for subsequent DNA and RNA analysis in a small space and at room temperature reducing transportation cost. Because chemically-treated FTA® cards inactivate the pathogenic agents in the samples they are carrying, they do not usually require import/export permits. Some of them (e.g., the 903 Protein Saver Card) have been approved by the US Food and Drug Administration (FDA) as class 2 devices [6]. Whatman® cards have been successfully used for the transportation and storage of a wide range of biospecimens including blood and serum [7], saliva [8], tissue [9], urine [10], sperm [11], mucus [12], and cerebral spinal fluid [13], among others.

Additionally, Whatman® cards have been proven useful for the diagnosis of pathogenic blood parasites, including canine microfilariae [14] and malaria-causing *Plasmodium* species [15]. However, very few studies have attempted to evaluate their efficacy for the detection of enteric pathogens in stool samples [16–18]. The protozoan enteroparasites, *Giardia duodenalis* (syn. *G. intestinalis* and *G. lamblia*) and *Cryptosporidium* spp., are two of the major contributors to the global burden of diarrhoeal illness both in humans [19,20] and livestock [21,22] globally. In poor-resource settings, more than 200 million human cases of symptomatic giardiosis are reported annually [23], whereas cryptosporidiosis (primarily by *C. hominis* and *C. parvum*) is the second leading cause of diarrhoea and deaths (after rotavirus) in children younger than five years of age [24]. Remarkably, production (cattle, sheep, goats, horses, donkeys, Bactrian camels) and free-living (non-human primates, among others) animal species have all been demonstrated to be competent hosts for *C. hominis* globally, confirming that this *Cryptosporidium* species is indeed zoonotic [25].

This study aims at comparing the performance of three types of Whatman® cards for the medium-/long-term preservation and storage (up to six months) of faecal material at different temperatures for downstream PCR detection of *G. duodenalis* and *Cryptosporidium hominis*.

2. Materials and Methods

2.1. Selected Whatman® Cards

Three different commercially available Whatman® cards with specific properties for sample collection, storage capacity, and costs were selected for comparative performance purposes: Whatman® Classic Cards, FTA® Elute Micro Cards, and 903 Protein Saver Cards (GE Healthcare Ltd., Cardiff, UK) (Table 1).

Table 1. Main features of the Whatman® cards used in the present comparative study.

Card	Reference	Cards/Pack	Sample Areas/ Card	Total Volume/Card (μL)	Total Sample Area (cm²)/Card
FTA® Classic	WB120205	100	4	500	19.6
FTA® Elute Micro	WB120410	100	4	120	4.0
903 Protein Saver	10531018	100	5	390-400	6.1

Whatman® FTA® Classic Cards contains chemical denaturants and a free radical scavenger that have the ability to lyse cells on contact, denature proteins, and protect DNA from degradation. The extracted DNA remains tightly bound to the matrix while cell membranes and organelles are lysed and proteins and inhibitors are washed away [26]. FTA® Elute Micro Cards contains a chaotropic salt. Cells are lysed upon contact and proteins remain tightly bound while DNA is isolated from the matrix in a solution free of inhibitors with a simple water elution procedure [27]. Whatman® 903 Protein saver Card is an untreated cotton fibber-based matrix. It does not stabilise nor protect DNA from degradation.

2.2. Stool Samples

Five fresh, independent stool samples of human origin with a positive result for *G. duodenalis* by real-time PCR (qPCR, see Section 2.6) with cycle threshold (Ct) values ranging from 29.0 to 34.3 (median: 31.3; standard deviation: 2.0) and confirmed as assemblage B, sub-assemblage BIV at the *gdh* locus by Sanger sequencing were selected for this study. Five fresh, independent stool samples of human origin with a positive result for *C. hominis* by nested small subunit ribosomal RNA (*ssu* rRNA)-PCR (see Section 2.7) and confirmed as *C. hominis* (genotype IbA10G2) by Sanger sequencing were also included (Table S1). Human samples were chosen by mere convenience in terms of accessibility and quantity, but the faecal material from non-human animal sources is equally valid.

Initial diagnosis and subsequent genotyping of the *G. duodenalis*- and *C. hominis*-positive stool samples were conducted at the Parasitology Reference and Research Laboratory of the Spanish National Centre for Microbiology (Majadahonda, Spain). After faecal sample homogenisation, 200 mg aliquots (*n* = 27, enough to cover all the experimental conditions considered in the study, see below) were weighed and stored at 4 °C in clean 1.5 mL Eppendorf tubes. Therefore, 270 stool sample aliquots were prepared.

2.3. Impregnation of Whatman® Cards

To standardise the experimental conditions of the study, total sample areas of the three Whatman® cards compared in the present study were normalised taking into consideration the sample area for each card specified in Table 1. A single sampling area of the FTA® Classic Card (enough for impregnating 200 mg of faecal material, (Figure 1a) equalled to four sampling areas of the FTA® Elute Micro Card (Figure 1b) and to three sampling areas of the 903 Protein Saver Card (Figure 1c). Under this premise, original Whatman® cards were cut and rearranged, as shown in Figure 1, to allow the coverage of three (1 month, 3 months, and 6 months) storage periods. This arrangement was used in triplicate to test three (−20 °C, 4 °C, and room temperature) storage conditions. Room temperature was considered that in the range of 15 to 25 °C.

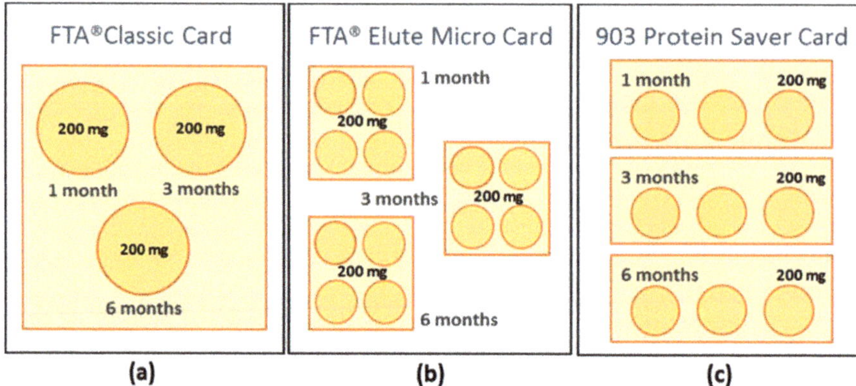

Figure 1. Standardisation of sampling areas for each Whatman® card used in the present study to assess storage periods and conditions of impregnated stool samples (**a**) FTA® Classic Card; (**b**) FTA® Elute Micro Card; (**c**) 903 Protein Saver Card.

Normalised sampling units for each Whatman® card were impregnated with 200 mg of each aliquoted stool sample described above using cotton swabs embedded in phosphate-buffered saline (PBS) to soften the faecal material and facilitate the impregnation process. Impregnated Whatman® cards were allowed to dry at room temperature and stored in individual zip-lock plastic bags containing silica desiccant to keep moisture level low at the periods and storage conditions evaluated.

2.4. Sensitivity Assay

To estimate the minimum amount of *G. duodenalis* DNA detectable by qPCR in positive stool samples impregnated in Whatman® cards, one stool sample positive to this pathogen was selected. Serial-halved amounts (200 mg, 100 mg, 50 mg, 25 mg, 12.5 mg, and 6.25 mg) of faecal material were weighed and aliquoted in clean 1.5 mL Eppendorf tubes and subsequently used to impregnate single sampling areas of the Whatman® FTA® Classic Card. Impregnated cards were allowed to dry at room temperature and stored in individual zip-lock plastic bags containing silica desiccant for 1 month at room temperature. This experiment was not conducted with *Cryptosporidium*-positive samples because no semi-quantitative qPCR method was available in our laboratory for the detection of this pathogen.

2.5. DNA Extraction and Purification

Genomic DNA was extracted and purified from impregnated Whatman® cards at each storage period and condition described in Sections 2.3 and 2.4 using the QIAamp DNA Stool Mini Kit (QIAGEN®, Hilden, Germany), following the manufacturer's instructions with minor modifications. Briefly, whole-normalised sampling surfaces of each compared Whatman® card were cut into small pieces using a sterilised scissor and transferred into clean 2 mL Eppendorf tubes containing 1 mL of Inhibitex buffer. After incubation at 95 °C for 10 min, the tubes were thoroughly vortexed and centrifuged at 13,000 rpm for 3 min. Then, 350 μL of the obtained supernatants were transferred to clean 1.5 mL Eppendorf tubes and the rest of the procedure was completed using the QIAcube (QIAGEN®) automated DNA extraction system. Purified genomic DNA (200 μL) was stored at 4 °C until downstream PCR testing.

2.6. Molecular Detection and Characterisation of Giardia duodenalis

Detection of *G. duodenalis* DNA was achieved using a real-time PCR (qPCR) method targeting a 62-bp region of the *ssu* rRNA) gene of the parasite, as described elsewhere [28]. Amplification reactions (25 μL) contained 3 μL of template DNA, 12.5 pmol of primers Gd-80F and Gd-127R, 10 pmol of probe, and 12.5 μL TaqMan® Gene Expression Master Mix

(Applied Biosystems, California, CA, USA). Detection of parasitic DNA was performed on a Corbett Rotor GeneTM 6000 real-time PCR system (QIAGEN®). Water (no template) and genomic DNA (positive) controls were included in each PCR run.

For genotyping purposes, a semi-nested PCR was used to amplify a 432-bp fragment of the glutamate dehydrogenase (*gdh*) of *G. duodenalis* [29]. Briefly, PCR reaction mixtures (25 µL) included 5 µL of template DNA and 0.5 µM of the primer pairs GDHeF/GDHiR in the primary reaction and GDHiF/GDHiR in the secondary reaction.

2.7. Molecular Detection and Characterisation of Cryptosporidium hominis

Detection of *C. hominis* DNA was achieved using a nested-PCR protocol to amplify a 587-bp fragment of the *ssu* rRNA gene of the parasite as described elsewhere [30]. Amplification reactions (50 µL) included 3 µL of DNA sample and 0.3 µM of the primer pairs CR-P1/CR-P2 in the primary reaction and CR-P3/CPB-DIAGR in the secondary reaction. Reaction mixes also contained 2.5 units of MyTAQ™ DNA polymerase (Bioline GmbH, Luckenwalde, Germany) and 5× MyTAQ™ Reaction Buffer containing 5 mM dNTPs and 15 mM $MgCl_2$.

For genotyping purposes, a nested PCR was used to amplify an 870-bp fragment of the 60 kDa glycoprotein (*gp60*) of *C. hominis* [31]. PCR reaction mixtures (50 µL) included 2-3 µL of template DNA and 0.3 µM of the primer pairs AL-3531/AL-3535 in the primary reaction and AL-3532/AL-3534 in the secondary reaction.

The semi-nested and nested PCR protocols described above were conducted on a 2720 Thermal Cycler (Applied Biosystems). Reaction mixes included 2.5 units of MyTAQ™ DNA polymerase (Bioline GmbH, Luckenwalde, Germany), and 5× MyTAQ™ Reaction Buffer containing 5 mM dNTPs and 15 mM $MgCl_2$. Amplicons were visualised under UV light after 2% agarose gel electrophoresis. Positive-PCR products were directly sequenced in both directions using inner primer sets. DNA sequencing was conducted by capillary electrophoresis using the BigDye® Terminator chemistry (Applied Biosystems) on an on ABI PRISM 3130 automated DNA sequencer.

2.8. Quality Control

To confirm the suitability of the three compared Whatman® cards for genotyping purposes, purified genomic DNAs from single *G. duodenalis*- and *C. hominis*-positive samples stored for six months (the maximum period covered in the present study) at 4 °C and room temperature (the most sensitive conditions to DNA damage evaluated here) were re-amplified by *gdh*-PCR (*G. duodenalis*) and *gp60*-PCR (*C. hominis*) and sequenced as described above. The quality of the obtained chromatograms was visually inspected, and the accuracy of the readings confirmed by alignment with appropriate reference sequences retrieved from GenBank.

2.9. Statistical Analyses

The Shapiro–Wilk's test was used to assess the normality of distribution of the Ct values obtained in *G. duodenalis*-positive samples during qPCR analyses at each period and storage condition evaluated. Once normality was demonstrated, an analysis of variance (ANOVA) for simultaneous comparison of conditions was conducted. A probability (*p*) value < 0.05 was considered evidence of statistical significance. Statistical analyses were performed using the R-software version 4.0.2 [32].

3. Results

3.1. Performance of Whatman® Cards for the Preservation and Storage of G. duodenalis-Positive Faecal Samples

All except two samples (an FTA® Elute Micro Card stored at room temperature for 1 month, and an FTA® Classic Card stored at -20 °C for three months) tested positive for *G. duodenalis* by qPCR, yielding Ct values similar to those generated at the time of initial diagnosis (Table S1). At 1 month-length storage, all three Whatman® cards performed equally well, with those kept at 4 °C yielding lower (but not statistically significant, $p = 0.96$) Ct values (Figure 2a). At 3 months-length storage, the FTA® Classic Card provided the best diagnostic values in terms of sensitivity (lower Ct values) and precision (lower standard deviation) regardless of the temperature. These results were statistically significant when compared with those obtained with the 903 Protein Saver Card ($p = 0.01$), but not with the FTA® Elute Micro Card ($p = 0.40$) (Figure 2b). No statistically significant differences ($p = 0.99$) were observed among the three compared Whatman® cards at 6 months-length storage (Figure 2c). It should be noted that the difference observed between the FTA® Classic Card and the 903 Protein Saver Card in Figure 2b is associated to the effect caused by an outlier data value generated with the latter at 4 °C storage conditions. Removal of this value resulted in the loss of statistical significance.

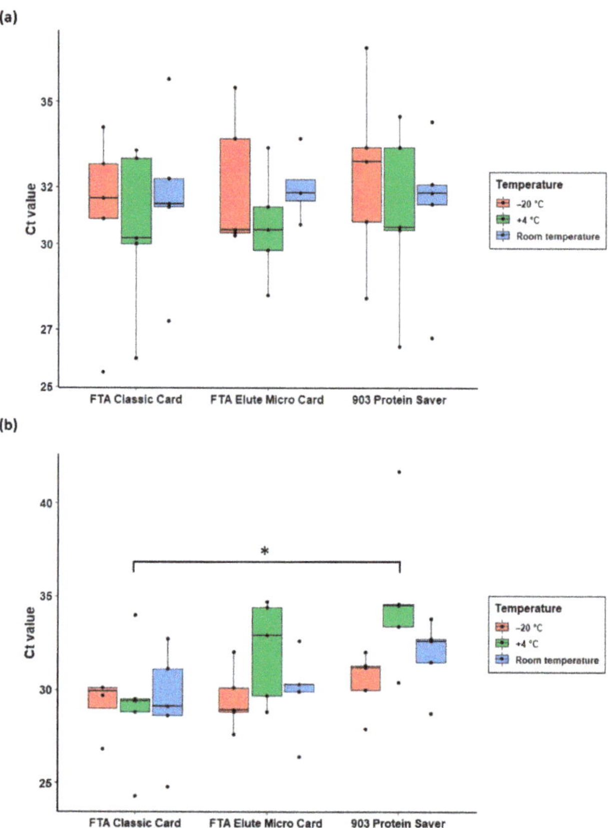

Figure 2. *Cont.*

(c)

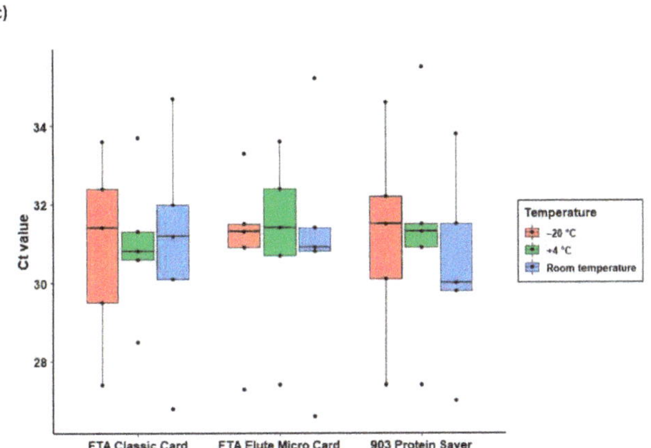

Figure 2. Box plot of cycle threshold (Ct) values generated from *Giardia duodenalis* isolates extracted from the three Whatman® cards evaluated in the present study at different storage conditions. (**a**): 1 month-length storage; (**b**): 3 months-length storage; (**c**): 6 month-length storage. Horizontal thick lines within boxes represent median values Upper and lower whiskers represent the data range. Plotted dots represent outliers. Using the Tukey's Honestly Significant Difference test as multiple post hoc comparison method, statistical significance is represented as * ($p < 0.05$).

3.2. Performance of Whatman® Cards for the Preservation and Storage of C. hominis-Positive Faecal Samples

The generated *ssu*-PCR results after the processing and testing of *C. hominis*-positive stool samples impregnated in the Whatman® cards at the storage periods and conditions assessed in the present study are shown in Figure 3. All tested samples stored for 1 month (Figure 3a), 3 months (Figure 3b), or 6 months (Figure 3c) yielded clear amplicons regardless of the storage temperature considered (Figure S1).

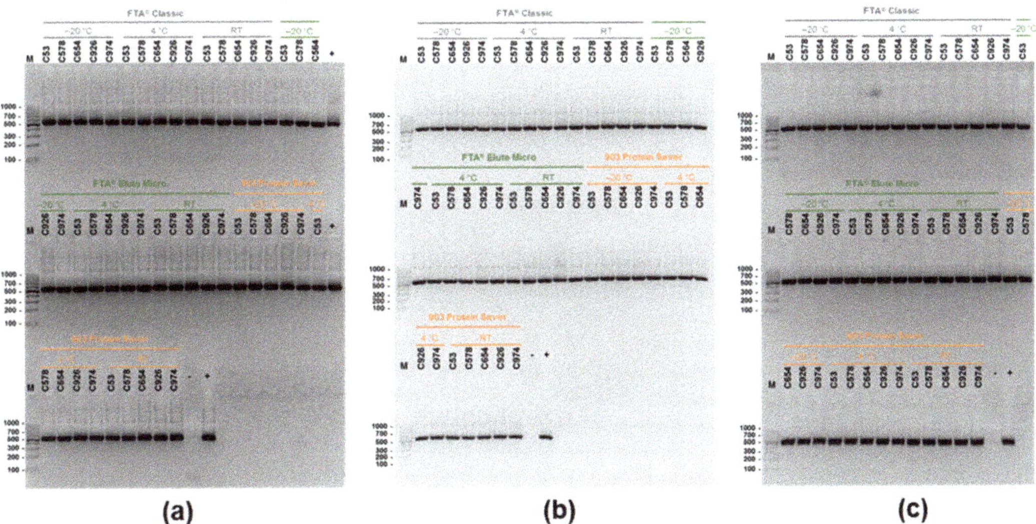

Figure 3. Agarose gel electrophoresis (2% *w/v*) detection of PCR products showing the presence of a 587-bp amplicon for the *Cryptosporidium hominis ssu* rRNA gene extracted from the three Whatman® cards evaluated in the present study at different storage conditions. (**a**): 1 month-length storage; (**b**): 3 months-length storage; (**c**): 6 month-length storage.

3.3. Sensitivity Assay

When tested by qPCR, serial-halved amounts of faecal material containing *G. duodenalis* cysts impregnated in Whatman® FTA® Classic Cards generated Ct values ranging from 25.6 (corresponding to 200 mg of faeces) to 33.1 (corresponding to 6.3 mg of faeces) (Table 2).

Table 2. Cycle threshold (Ct) values obtained by real-time PCR in serially-halved amounts of a faecal sample positive for *G. duodenalis* impregnated in Whatman® FTA® Classic Cards.

Faecal Material (mg)	Ct Value
200	25.6
100	27.3
50	27.1
25	28.7
12.5	30.2
6.25	33.1

3.4. Quality Control

Figure S1 shows the *gdh*-PCR and *gp60*-PCR amplification results obtained with the purified genomic DNAs from the two samples positive to *G. duodenalis* and *C. hominis*, respectively. Both samples were stored in all three compared Whatman® cards for 6 months (the maximum period covered in the present study) at 4 °C and room temperature (the most likely conditions to induce DNA damage evaluated here). Two of the six *G. duodenalis*-positive extracts were successfully amplified at the *gdh* locus (Figure S1a) and their associated chromatograms displayed good quality sequences confirming the identity (sub-assemblage BIV) of the parasite (Figure S2a). All *C. hominis*-positive extracts yielded clear amplicons at the *gp60* locus, irrespectively of the Whatman® card used or the temperature considered (Figure S1b). Sanger sequencing analysis revealed good quality sequence data confirming the identity (genotype IbA10G2) of the parasite (Figure S2b).

4. Discussion

This study evaluated the suitability of three commercially available Whatman® Filter Cards (the FTA® Classic Card, the FTA® Elute Micro Card, and the 903 Protein Saver Card) for the long-term storage of faecal material containing *G. duodenalis* cysts and *C. hominis* oocysts, two of the major contributors to the global burden of diarrhoeal illness both in humans [19,20] and livestock [21,22] globally. Of note, both protozoan parasites present aggregated distributions depending on the host species, genetic variants, or even geographical area considered [33–37]. Because both *G. duodenalis* and *Cryptosporidium* spp. are common findings in the faecal material of human and animal hosts, some of their species/genotypes have zoonotic potential, and are ubiquitous in the environment, research on the epidemiology and transmission of these pathogens should be always conducted under the One Health umbrella. This approach is particularly useful in those epidemiological scenarios where different epidemiological (e.g., domestic and sylvatic) cycles of the parasites overlap, allowing the occurrence of spillover events [33,34,36]. The three major contributions of this survey include the demonstration that (i) the three compared Whatman© cards performed near equally well in maintaining the stability of the faecal material for up to six months irrespectively of the storage temperature; (ii) the parasitic DNA extracted from impregnated Whatman® cards was suitable for subsequent molecular detection and genotyping purposes; and (iii) Whatman® cards represent simple, time- and cost-effective options for the safe storage and transportation of faecal samples of human and animal origin without the need of the cold chain.

FTA® card technology was originally designed as a matrix for blood storage and processing medium [5]. Because of their versatility and simplicity of use, FTA® cards were soon after tested for storing other biological samples including saliva [8], tissue [9], urine [10], sperm [11], mucus [12], and cerebral spinal fluid [13]. This tool has been also assessed for the molecular detection of gastrointestinal parasites (e.g., the coccidian *Cryptosporidium* spp. and *Cyclospora cayetanensis*, the flagellated *G. duodenalis*, and the microsporidia *Encephalitozoon intestinalis*) in matrices including clinical specimens and fresh produce [16–18]. In a seminal study, FTA® card templates prepared from purified *Cryptosporidium* spp. oocysts and *E. intestinalis* spores and subsequently assessed by PCR allowed the identification of as few as 10 oocysts/spores. Similar results were also observed with clinical samples including faeces, urine, sputum, and foods (berries) [16]. The authors concluded that PCR analysis using the FTA® card format for DNA template preparation was routinely unaffected by the matrix from which the sample was derived while still maintaining a high level of detection sensitivity [16]. In a subsequent survey, known concentrations of *G. duodenalis* cysts and *Cryptosporidium* spp. oocysts were serially diluted, spiked into a faecal suspension from a pathogen-negative stool, and smeared onto FTA® Elute Micro cards [17]. Stool cards were then stored at room temperature, DNA was extracted and purified using QIAGEN protocols and tested by multiplex PCR coupled with Luminex assay at 1 week, 1 month, and 3 months. A limited number of stool cards were also stored at 4 °C and in a humid incubator at 31 °C for 1 week to determine the impact of environmental conditions on detection. The authors detected *G. duodenalis* at 3 months with a 2-log reduction from the original concentration, whereas *Cryptosporidium* spp. was undetected after 1 month of storage. Failure to detect the presence of *Cryptosporidium* spp. for longer periods of time was attributed to suboptimal breakage of the parasite oocyst wall [17]. Finally, the 903 Protein Saver card has been evaluated for the detection of *G. duodenalis*, *Cryptosporidium* spp., and *Entamoeba histolytica* in either whole faecal samples or stool suspensions using QIAGEN procedures for DNA purification and qPCR for detection [18]. In this study the cards were stored for only 48 h before DNA purification and qPCR testing. Depending on the starting (whole or suspension) faecal material used for impregnation, obtained overall sensitivities were 85–95% for *G. duodenalis*, 60–85% for *E. histolytica*, and 35–40% for *Cryptosporidium* spp. In general, faecal suspensions yielded poorest qPCR amplification results than whole faecal samples. Parasite load was identified as a critical factor for qPCR success [18].

This study improves current knowledge on the practicality and performance of Whatman® cards for the molecular detection of diarrhoea-causing enteric protozoan parasites in several aspects. First, this is (to author´s knowledge) the first attempt conducted to date to compare simultaneously three different types of Whatman® cards including the FTA® Classic Card, the FTA® Elute Micro Card, and the 903 Protein Saver Card. Previous studies focused on a specific card type only [16–18]. Second, the evaluated storage period has been extended to 6 months, three more months that the maximum period covered in previous studies [17]. Third, this survey evaluated the effect of three different storage temperatures. Freezing (–20 °C) and refrigeration (4 °C) temperatures represented the most common conditions in routine laboratory practice, whereas the room temperature condition attempted to mimic those present in field work characterised by lack of electric supply where sample conservation is a pressing issue. Previous studies were conducted primarily at room temperature [16–18], with only few impregnated cards being tested at other temperatures [17]. Fourth, present results were obtained exclusively with true clinical faecal samples, whereas those from previous studies were mostly derived from purified parasitic material [16] or artificially spiked stools [17]. Fifth, quality control data presented here provided evidence demonstrating that Whatman® cards were suitable for genotyping (in addition to detection) purposes, including Sanger sequencing. None of the studies carried out before assessed this possibility.

A major contribution of this study was the finding that all three compared Whatman® cards yielded sensitivity values near 100%, irrespectively of the storage period, the temperature considered, or the parasite species investigated. These figures were considerably higher than those reported in similar surveys [16–18]. Several factors may account, at least partially, for the differences observed. For instance, we used high impregnation loads (200 mg) of faecal material, in line with the recommendations of the QIAGEN procedure used for DNA extraction and purification. In addition, our protocol included a modification (sample incubation with Inhibitex buffer at 95 °C for 10 min) specifically intended at improving the efficiency of cyst/oocyst breakage, an issue previously identified as a factor limiting the diagnostic sensitivity of PCR assays [17]. Variations in the diagnostic performance of the PCR methods used (conventional nested PCR, qPCR, multiplex qPCR) may also influence the amplification success rate obtained in these studies.

This study presents, however, some limitations that should be taken into consideration. For instance, a straightforward application of the data presented here is the potential usefulness of Whatman® cards as a convenient stool storage system for periods longer than 6 months (e.g., in biobanks), a possibility that should be conveniently evaluated in further studies. Of note, room temperature was considered here those in the range of 15 to 25 °C. More extreme temperature (and humidity) conditions, such as those typically present in tropical and sub-tropical regions, may affect the performance of the Whatman® cards. This possibility should be conveniently evaluated in future studies. Also, only faecal samples positive for *G. duodenalis* and *C. hominis* were investigated. Although we do not anticipate significant performance differences with other enteric protist species, this is also an issue that remains to be fully elucidated. Finally, our data can be used as proof of concept for the suitability of Whatman® technology for the safe storage and transportation of faecal material at room temperature without detrimental effects on stability and diagnostic features, although this fact should be demonstrated in *ad-hoc* studies.

5. Conclusions

Data presented here demonstrate that Whatman® cards are a cost-effective option for the preservation and long-term storage (up to six months) of faecal samples under a wide range of temperatures (from –20 °C to room temperature) without compromising their biospecimen stability and suitability for molecular-based diagnostic methods. Indeed, Whatman® cards enable the molecular detection and genotyping of common diarrhoea-causing enteric protozoan parasites, including *C. hominis* and *G. duodenalis*. Further research should be conducted to unambiguously demonstrate the usefulness of Whatman® cards in field epidemiological surveys involving larger number of faecal samples, wider ranges of temperature and humidity conditions, and storage periods longer than six months. In practical terms, Whatman® cards would allow the obtaining and safe transportation of faecal samples of human and animal origin from remote areas to clinical or research laboratories without the need of the cold chain.

Supplementary Materials: The following are available online at https://www.mdpi.com/article/10.3390/ani11051369/s1, Table S1: Real-time PCR cycle threshold values obtained after the amplification of *Giardia duodenalis*-positive stool samples impregnated in the Whatman® Cards at the storage periods and conditions assessed in the present study, Figure S1: Agarose gel electrophoresis (2% *w/v*) detection of PCR products used for evaluating the suitability of Whatman® cards for genotyping and Sanger sequencing purposes. (a) Results showing the presence of a 432-bp amplicon for the *Giardia duodenalis gdh* gene in sample G145. Some lanes have been cut and re-arranged to keep the same order in the whole figure; (b): Results showing the presence of an 870-bp amplicon for the *Cryptosporidium hominis gp60* gene in sample C578., Figure S2: Representative chromatograms showing Sanger sequencing results for PCR amplicons generated from genomic DNA extracted and purified from Whatman® Cards. (a): Results for the *Giardia duodenalis gdh* gene in sample G145; (b): Results for the *Cryptosporidium hominis gp60* gene in sample C578.

Author Contributions: Conceptualisation, P.C.K., R.C.-B., F.P.-G., and D.C.; methodology, P.C.K., B.B.; A.D., and C.H.-C.; validation, R.C.-B., F.P.-G., D.G.-B., and D.C.; formal analysis, P.C.K., A.D., R.C.-B., F.P.-G., D.G.-B., and D.C.; investigation, P.C.K.; resources, D.C.; writing—original draft preparation, P.C.K. and D.C.; writing—review and editing, P.C.K., R.C.-B., F.P.-G., D.G.-B., and D.C.; supervision, R.C.-B., F.P.-G., D.G.-B., and D.C.; project administration, D.C.; funding acquisition, D.C. All authors have read and agreed to the published version of the manuscript.

Funding: This research was funded by the Health Institute Carlos III (ISCIII), Ministry of Science, Innovation and Universities (Spain), grant number PI16CIII/00024.

Institutional Review Board Statement: The study was conducted according to the guidelines of the Declaration of Helsinki and approved by the Ethics Committee of the Health Institute Carlos III (CEI PI17_2017-v3; date of approval: 23 October 2017).

Data Availability Statement: All relevant data are within the article and its additional files.

Acknowledgments: David González-Barrio was recipient of a "Sara Borrell" postdoctoral fellow-ship (CD19CIII/00011) funded by the Spanish Ministry of Science, Innovation and Universities.

Conflicts of Interest: The authors declare no conflict of interest. The funders had no role in the design of the study; in the collection, analyses, or interpretation of data; in the writing of the manuscript, or in the decision to publish the results.

References

1. Paulos, S.; Mateo, M.; de Lucio, A.; Hernández-de Mingo, M.; Bailo, B.; Saugar, J.M.; Cardona, G.A.; Fuentes, I.; Mateo, M.; Carmena, D. Evaluation of five commercial methods for the extraction and purification of DNA from human faecal samples for downstream molecular detection of the enteric protozoan parasites *Cryptosporidium* spp., *Giardia duodenalis*, and *Entamoeba* spp. *J. Microbiol. Methods* **2016**, *127*, 68–73. [CrossRef] [PubMed]
2. Sánchez-Romero, M.I.; García-Lechuz Moya, J.M.; González López, J.J.; Orta Mira, N. Collection, transport and general processing of clinical specimens in microbiology laboratory. *Enferm. Infecc. Microbiol. Clin.* **2019**, *37*, 127–134. [CrossRef]
3. Guthrie, R.; Susi, A. A simple phenylalanine method for detecting phenylketonuria in large population of newborn infants. *Pediatrics* **1963**, *32*, 338–343. [PubMed]
4. Sharma, A.; Jaiswal, S.; Shukla, M.; Lal, J. Dried blood spots: Concepts, present status, and future perspectives in bioanalysis. *Drug. Test. Anal.* **2014**, *6*, 399–414. [CrossRef]
5. GE Healthcare Life Sciences. Your Forensic Samples, Our Experience. Available online: www.gelifesciences.com/forensics (accessed on 1 May 2021).
6. Min, K.L.; Ryu, J.Y.; Chang, M.J. Development and clinical applications of the dried blood spot method for therapeutic drug monitoring of anti-epileptic drugs. *Basic Clin. Pharmacol. Toxicol.* **2019**, *125*, 215–236. [CrossRef] [PubMed]
7. Choi, E.H.; Lee, S.K.; Ihm, C.; Sohn, Y.H. Rapid DNA extraction from dried blood spots on filter paper: Potential applications in biobanking. *Osong Public Health Res. Perspect.* **2014**, *5*, 351–357. [CrossRef]
8. Zheng, N.; Zeng, J.; Ji, Q.C.; Angeles, A.; Aubry, A.F.; Basdeo, S.; Buzescu, A.; Landry, I.S.; Jariwala, N.; Turley, W.; et al. Bioanalysis of dried saliva spot (DSS) samples using detergent-assisted sample extraction with UHPLC-MS/MS detection. *Anal. Chim. Acta* **2016**, *934*, 170–179. [CrossRef] [PubMed]
9. da Cunha Santos, G. FTA cards for preservation of nucleic acids for molecular assays: A review on the use of cytologic/tissue samples. *Arch. Pathol. Lab. Med.* **2018**, *142*, 308–312. [CrossRef]
10. Yan, X.; Yuan, S.; Yu, Z.; Zhao, Y.; Zhang, S.; Wu, H.; Yan, H.; Xiang, P. Development of an LC-MS/MS method for determining 5-MeO-DIPT in dried urine spots and application to forensic cases. *J. Forensic Leg. Med.* **2020**, *72*, 101963. [CrossRef] [PubMed]
11. Serra, O.; Frazzi, R.; Perotti, A.; Barusi, L.; Buschini, A. Use of FTA® classic cards for epigenetic analysis of sperm DNA. *Biotechniques* **2018**, *64*, 45–51. [CrossRef]
12. Kashiwagi, T.; Maxwell, E.A.; Marshall, A.D.; Christensen, A.B. Evaluating manta ray mucus as an alternative DNA source for population genetics study: Underwater-sampling, dry-storage and PCR success. *Peer J.* **2015**, *3*, e1188. [CrossRef] [PubMed]
13. Elliott, I.; Dittrich, S.; Paris, D.; Sengduanphachanh, A.; Phoumin, P.; Newton, P.N. The use of dried cerebrospinal fluid filter paper spots as a substrate for PCR diagnosis of the aetiology of bacterial meningitis in the Lao PDR. *Clin. Microbiol. Infect.* **2013**, *19*, E466–E472. [CrossRef]
14. Duscher, G.; Peschke, R.; Wille-Piazzai, W.; Joachim, A. Parasites on paper–The use of FTA Elute® for the detection of *Dirofilaria repens* microfilariae in canine blood. *Vet. Parasitol.* **2009**, *161*, 349–351. [CrossRef] [PubMed]
15. Hashimoto, M.; Bando, M.; Kido, J.I.; Yokota, K.; Mita, T.; Kajimoto, K.; Kataoka, M. Nucleic acid purification from dried blood spot on FTA Elute Card provides template for polymerase chain reaction for highly sensitive *Plasmodium* detection. *Parasitol. Int.* **2019**, *73*, 101941. [CrossRef] [PubMed]
16. Orlandi, P.A.; Lampel, K.A. Extraction-free, filter-based template preparation for rapid and sensitive PCR detection of pathogenic parasitic protozoa. *J. Clin. Microbiol.* **2000**, *38*, 2271–2277. [CrossRef] [PubMed]
17. Lalani, T.; Tisdale, M.D.; Maguire, J.D.; Wongsrichanalai, C.; Riddle, M.S.; Tribble, D.R. Detection of enteropathogens associated with travelers' diarrhea using a multiplex Luminex-based assay performed on stool samples smeared on Whatman FTA Elute cards. *Diagn. Microbiol. Infect. Dis.* **2015**, *83*, 18–20. [CrossRef] [PubMed]
18. Natarajan, G.; Kabir, M.; Perin, J.; Hossain, B.; Debes, A.; Haque, R.; George, C.M. Whatman protein saver cards for storage and detection of parasitic enteropathogens. *Am. J. Trop. Med. Hyg.* **2018**, *99*, 1613–1618. [CrossRef]
19. Checkley, W.; White, A.C., Jr.; Jaganath, D.; Arrowood, M.J.; Chalmers, R.M.; Chen, X.M.; Fayer, R.; Griffiths, J.K.; Guerrant, R.L.; Hedstrom, L.; et al. A review of the global burden, novel diagnostics, therapeutics, and vaccine targets for cryptosporidium. *Lancet Infect. Dis.* **2015**, *15*, 85–94. [CrossRef]

20. Escobedo, A.A.; Almirall, P.; Robertson, L.J.; Franco, R.M.; Hanevik, K.; Mørch, K.; Cimerman, S. Giardiasis: The ever-present threat of a neglected disease. *Infect. Disord. Drug Targets* **2010**, *10*, 329–348. [CrossRef] [PubMed]
21. Santin, M. *Cryptosporidium* and *Giardia* in ruminants. *Vet. Clin. N. Am. Food Anim. Pract.* **2020**, *36*, 223–238. [CrossRef] [PubMed]
22. Thomson, S.; Hamilton, C.A.; Hope, J.C.; Katzer, F.; Mabbott, N.A.; Morrison, L.J.; Innes, E.A. Bovine cryptosporidiosis: Impact, host-parasite interaction and control strategies. *Vet. Res.* **2017**, *48*, 42. [CrossRef] [PubMed]
23. Feng, Y.; Xiao, L. Zoonotic potential and molecular epidemiology of *Giardia* species and giardiasis. *Clin. Microbiol. Rev.* **2011**, *24*, 110–140. [CrossRef]
24. Kotloff, K.L.; Nataro, J.P.; Blackwelder, W.C.; Nasrin, D.; Farag, T.H.; Panchalingam, S.; Wu, Y.; Sow, S.O.; Sur, D.; Breiman, R.F.; et al. Burden and aetiology of diarrhoeal disease in infants and young children in developing countries (the Global Enteric Multicenter Study, GEMS): A prospective, case-control study. *Lancet* **2013**, *382*, 209–222. [CrossRef]
25. Widmer, G.; Köster, P.C.; Carmena, D. *Cryptosporidium hominis* infections in non-human animal species: Revisiting the concept of host specificity. *Int. J. Parasitol.* **2020**, *50*, 253–262. [CrossRef] [PubMed]
26. GE Healthcare. Helping You Build a Smarter Diagnostic Assay. Available online: www.gelifesciences.com (accessed on 1 May 2021).
27. Cytiva. Available online: https://www.cytivalifesciences.com/en/us/shop/whatman-laboratory-filtration/whatman-dx-components/blood-collection-cards-and-accessories/903-proteinsaver-card-p-01011#related-documents (accessed on 1 May 2021).
28. Verweij, J.J.; Shinkel, J.; Laeijendecker, D.; van Rooyen, M.A.; van Lieshout, L.; Polderman, A.M. Real-time PCR for the detection of *Giardia lamblia*. *Mol. Cell Probes* **2003**, *17*, 223–225. [CrossRef]
29. Read, C.M.; Monis, P.T.; Thompson, R.C. Discrimination of all genotypes of *Giardia duodenalis* at the glutamate dehydrogenase locus using PCR-RFLP. *Infect. Genet. Evol.* **2004**, *4*, 125–130. [CrossRef]
30. Tiangtip, R.; Jongwutiwes, S. Molecular analysis of *Cryptosporidium* species isolated from HIV-infected patients in Thailand. *Trop. Med. Int. Health* **2002**, *7*, 357–364. [CrossRef]
31. Feltus, D.C.; Giddings, C.W.; Schneck, B.L.; Monson, T.; Warshauer, D.; McEvoy, J.M. Evidence supporting zoonotic transmission of *Cryptosporidium* spp. in Wisconsin. *J. Clin. Microbiol.* **2006**, *44*, 4303–4308. [CrossRef] [PubMed]
32. R Core Team. A Language and Environment for Statistical Computing. R Foundation for Statistical Computing, Vienna, Austria 2020. Available online: https://www.R-project.org (accessed on 1 May 2021).
33. Ryan, U.; Zahedi, A. Molecular epidemiology of giardiasis from a veterinary perspective. *Adv. Parasitol.* **2019**, *106*, 209–254.
34. Ryan, U.; Cacciò, S.M. Zoonotic potential of *Giardia*. *Int. J. Parasitol.* **2013**, *43*, 943–956. [CrossRef] [PubMed]
35. Yang, X.; Guo, Y.; Xiao, L.; Feng, Y. Molecular epidemiology of human cryptosporidiosis in low- and middle-income countries. *Clin. Microbiol. Rev.* **2021**, *34*, e00087–e00119. [CrossRef] [PubMed]
36. Guo, Y.; Li, N.; Ryan, U.; Feng, Y.; Xiao, L. Small ruminants and zoonotic cryptosporidiosis. *Parasitol. Res.* **2021**, in press. [CrossRef] [PubMed]
37. Li, J.; Ryan, U.; Guo, Y.; Feng, Y.; Xiao, L. Advances in molecular epidemiology of cryptosporidiosis in dogs and cats. *Int. J. Parasitol.* **2021**, in press. [CrossRef] [PubMed]

Article

Salmonella Infection in Turtles: A Risk for Staff Involved in Wildlife Management?

Gaia Casalino [1,2], Adriana Bellati [3], Nicola Pugliese [1,2], Antonio Camarda [1,2], Simona Faleo [4], Roberto Lombardi [1,2], Gilda Occhiochiuso [4], Francesco D'Onghia [1,2] and Elena Circella [1,2,*]

[1]   Department of Veterinary Medicine, University of Bari "Aldo Moro", s.p. Casamassima km 3, 70010 Valenzano, Italy; gaia.casalino@uniba.it (G.C.); nicola.puglies@uniba.it (N.P.); antonio.camarda@uniba.it (A.C.); roberto.lombardi@uniba.it (R.L.); francesco.donghia@uniba.it (F.D.)
[2]   Regional Wildlife Rescue Centre, 70020 Bitetto, Italy
[3]   Department of Ecological and Biological Sciences, University of Tuscia, Largo dell'Università, 01100 Viterbo, Italy; adriana.bellati@unitus.it
[4]   Istituto Zooprofilattico Sperimentale della Puglia e della Basilicata, Via Manfredonia 20, 71121 Foggia, Italy; simona.faleo@izspb.it (S.F.); gilda.occhiochiuso@izspb.it (G.O.)
*   Correspondence: elena.circella@uniba.it; Tel./Fax: +39-080-467-9910

Citation: Casalino, G.; Bellati, A.; Pugliese, N.; Camarda, A.; Faleo, S.; Lombardi, R.; Occhiochiuso, G.; Circella, E.; D'Onghia, F. Salmonella Infection in Turtles: A Risk for Staff Involved in Wildlife Management? *Animals* **2021**, *11*, 1529. https://doi.org/10.3390/ani11061529

Academic Editor: David González-Barrio

Received: 16 April 2021
Accepted: 19 May 2021
Published: 24 May 2021

Publisher's Note: MDPI stays neutral with regard to jurisdictional claims in published maps and institutional affiliations.

Simple Summary: The aim of this study was to investigate the occurrence of non-typhoidal Salmonella in the turtles housed in a regional wildlife rescue centre of Apulia, in southern Italy, to assess the presence of *Salmonella* serovars that may represent a risk for operators involved in wildlife management. Sixty-nine tortoises, of which 36 were males and 33 were females, belonging to different species (*Testudo hermanni hermanni*, *T. h. boettgeri*, *T. graeca*, and *T. marginata*) were tested. All the turtles were adults (34 between 6 and 10 years of age and 35 more than 10 years of age). *Salmonella* was statistically detected more frequently in *T. hermanni hermanni*. No differences of the infection prevalence related to animal gender or age were found. Two different species, *S. enterica* and *S. bongori*, three *S. enterica* subspecies (*enterica*, *diarizonae*, *salamae*), and five different serovars (Hermannswerder, Abony, Ferruch, Richmond, Vancouver) within the group *S. enterica* subspecies *enterica* were identified. Two *Salmonella* types with different combinations were simultaneously found in specimens of *T. h. hermanni*. Most of the detected *Salmonella* types may represent a potential risk for operators in wildlife rescue centres.

Abstract: Monitoring of infections that may be transmitted to humans by animals in wildlife rescue centres is very important in order to protect the staff engaged in rehabilitation practices. *Salmonella* may be a natural inhabitant of the intestinal tract of turtles, rarely causing disease. This may represent a potential risk for humans, increasing the sanitary risk for operators in wildlife rescue centres. In this paper, the occurrence of non-typhoidal *Salmonella* among terrestrial turtles housed in a wildlife rescue centre in Southern Italy was investigated, in order to assess the serovars more frequently carried by turtles and identify those that may represent a risk for operators involved in wildlife management. Sixty-nine adult turtles (*Testudo hermanni hermanni*, *T. h. boettgeri*, *T. graeca*, and *T. marginata*) were tested. Detection and serotyping of *Salmonella* strains were performed according to ISO 6579-1 and ISO/TR 6579-3:2013, respectively. The distribution of *Salmonella* spp. was significantly higher in *T. hermanni hermanni* than in other species, independent of the age and gender of the animals. Two different *Salmonella* species, *S. enterica* and *S. bongori*, three *S. enterica* subspecies (*enterica*, *diarizonae*, *salamae*), and five different serovars (Hermannswerder, Abony, Ferruch, Richmond, Vancouver) within the group *S. enterica* subspecies *enterica* were identified. Different combinations of *Salmonella* types were simultaneously found in specimens of *T. h. hermanni*. Most of detected *Salmonella* types may represent a potential risk for public health. Adopting correct animal husbandry procedures and informing on potential sanitary risks may be useful for minimising the risk of transmission of *Salmonella* to workers involved in wildlife management.

Keywords: *Salmonella*; turtles; wildlife; zoonosis; wildlife rescue centres

1. Introduction

Biodiversity is being lost at an unprecedent rate, thought to be about 1000 times higher than before humans dominated the planet [1]. While about 90% of the extant species on Earth still remain to be discovered by scientists, our planet is now facing the so-called "sixth mass extinction" event [2]. Indeed, our species has the potential to disrupt natural ecosystems by impoverishing natural habitats, releasing alien species, accelerating climate change, and polluting the environment, with unpredictable consequences, even for the persistence of our own society [3–5]. In response to increasing awareness, conservation actions are presently funded by international organisations and national governments to reduce the human footprint on the planet, and to promote the restoration of the environment and natural populations. Protected areas are particularly crucial for preserving local variation from extinction, and indeed most of the conservation actions are realized at national or even finer spatial scales [6].

Reptiles have suffered dramatic population declines due to climate change and urbanisation, as well as landscape transformation, pollution, and illegal trade, particularly in recent times [7,8]. One of the most endangered reptiles in Europe nowadays is the western Hermann's tortoise *Testudo hermanni hermanni* (Gmelin, 1789), which started to be listed as "endangered" in the IUCN Red List more than 20 years ago due to its marked decline throughout most of its distribution range [9,10]. The nominal species (*Testudo hermanni*) is naturally spread along the coastal regions of the Mediterranean. According to morphology and coloration pattern, it comprises two well-differentiated subspecies [11], the western *T. hermanni hermanni*, spanning along the western coasts of Europe, and *T. h. boettgeri* (Mojsisovics 1889), showing more continuous distribution from northeastern Italy towards the Balkans. These two forms, which have differentiated in allopatric glacial refugia since the end of the Pliocene [12], are strongly supported from a molecular point of view [13,14]. Their contact zone occurs along the Po river in northeastern Italy [10,15]. Indeed, as a result of their phylogeographic history, private mitochondrial haplotypes and population-specific nuclear alleles at microsatellite loci allow for a clear distinction between the two subspecies. Nonetheless, hybridisation occasionally occurs, being rather common in captivity [16]. Although the hybrids show average phenotypic characteristics between the two forms, they are not easy to identify in the wild according to morphology alone, and it has been suggested that, irrespective of the parental cross, they mainly resemble the phenotype of the father [17].

Nowadays, *T. h. hermanni* shows a discontinuous distribution compared to the eastern form, mainly as a result of anthropic pressure like habitat destruction and overharvesting, which has amplified the effect of quaternary climatic warming.

To fight population and genetic erosion, many pristine areas have been raised to the status of protected reserves in the European Natura 2000 network. However, for species showing low dispersal ability like tortoises, specific conservation measures are needed to maintain populations, such as reintroduction projects, breeding programs, and reinforcement actions [9,18]. Following these initiatives, many individuals have been released in the wild according to morphology, avoiding a clear definition of their genetic composition. Moreover, individuals translocated for restocking purposes often come from wildlife rescue centres, where animals are maintained once seized from illegal captivity or found injured in the wild, and where they are aided and provided with safe housing and medical assistance in order to rehabilitate. In the absence of the preliminary genetic analysis, released individuals may therefore belong to cryptic lineages (like the oriental *T. h. boettgeri*, widely traded as a pet) and cause hybridization or outbreeding depression in target populations through the disruption of co-adapted gene complexes [19]. Similarly, appropriate sanitary control of the health status of captive animals before release is fundamental to enhance the survival of specimens in the natural environment.

Monitoring of wildlife health is also crucial to protect the staff engaged in rehabilitation practices and conservation projects. In particular, the detection of pathogens that may be transmitted to humans by animals is important.

Non-typhoidal (NT) *Salmonella* serovars are considered as potential zoonotic pathogens. Although salmonellosis in humans is usually associated with the ingestion of contaminated food of animal origin, the contact with infected animals, especially with poor hygienic practices, can provide an important source of *Salmonella* infection [20,21]. Several human salmonellosis cases have been associated to contact to reptiles [22–24]. In humans, non-typhoidal *Salmonella* may induce enteric forms characterized by fever, vomiting, and diarrhea [25], but also extraintestinal forms [26–29] depending on the serovar involved. Differently, *Salmonella* seems to be adapted to reptiles, causing prevalently asymptomatic infections [30], while disease and death occasionally occur [31]. Likewise, the infection rarely causes disease in turtles, and extraintestinal lesions may occur when some serotypes are involved [32]. This represents a potential risk for humans, because the apparent good state of health usually observed in the infected turtles induces operators in the management of these animals to handle them without adequate biosecurity measures.

The aims of the work illustrated in this paper were to (i) investigate the occurrence of NT *Salmonella* among *T. hermanni* specimens intended for a restocking project and other terrestrial turtles housed in a wildlife rescue centre, (ii) assess the serovars more frequently carried by animals, and (iii) identify the serovars that may represent a risk for operators involved in wildlife management.

2. Materials and Methods

The study was carried on 69 terrestrial turtles (57 *Testudo hermanni hermanni*, 8 *T. h. boettgeri*, 2 *T. graeca*, and 2 *T. marginata*) housed in the regional wildlife rescue centre of Apulia (Bitetto, BA) in southern Italy. All animals were adults aged 6 to 10 years (35 turtles) or over 10 years (34 turtles). Thirty-three were females and 36 were males (Table S1). All turtles appeared fully active, regularly ate, and showed no symptoms. Therefore, they were considered as clinically healthy. Faecal samples were collected from each turtle to perform bacteriological analyses for the identification of *Salmonella*. Sampling was carried out in June and July, in the morning, at average temperatures between 28 °C and 30 °C.

Moreover, in view of the possible hybridisation phenomena, 41 *T. h. hermanni* and 8 *T. h. boettgeri* were genetically tested in order to confirm their morphological identification, assess the compatibility of specimens of *T. h. hermanni* with their inclusion in a restocking plan, and evaluate the distribution of *Salmonella* isolates among different types of turtles. Therefore, oral swabs were set up for each turtle to perform molecular analysis.

2.1. Genotyping of Testudo h. hermanni and T. h. boettgeri Specimens

Genomic DNA was extracted from oral swabs using the GenElute Mammalian Genomic DNA Miniprep commercial kit (Sigma Aldrich, St. Louis, MO, USA), following the manufacturer's instructions, apart from initial digestion, which took 30 min instead of 3 h. Two swabs per individual were processed in order to maximize the recovery of DNA. The integrity of the extracted DNA was assessed on 1% agarose gel, returning high-molecular-weight bands for all the samples.

Captive individuals were genotyped by coupling the analysis of mtDNA and nuDNA variation. A portion of the *cytb* gene (372 bp) was initially amplified according to [14] (see Table S2 for details concerning primers and thermal profiles used in this study). PCR reactions were set up in 20 µL, containing 2 µL Buffer 10× (1.5 mM MgCl₂), 2 µL dNTPs (0.2 mM), 0.2 µL primer F and R (100 µM), 0.1 µL *Taq* (5 U; Biotech Rabbit, Berlin, Germany) and 1 µL DNA (50 ng/µL). PCR products were purified using the kit QIAquick PCR Purification (Qiagen, Hilden, Germany), and sequenced at GATC Biotech (Eurofins Genomics, Luxemburg).

Individuals were also genotyped at eight microsatellite loci available for the species (Ther 20, Ther 51, Ther 94, Test 71, Gal 263, Test 56, Test 10, and Test 76 [10,33,34]), setting the same PCR reactions (Table S2). Forward primers were fluorescently labelled using either HEX or FAM and sequenced at Applied Genetics (Eurofins Genomics). Three individuals previously genotyped as *T. h. boettgeri*, according to morphological and molecular evidence, were similarly genotyped as reference.

The obtained mtDNA sequences were checked by eye and translated into amino acids using MEGA 6.06 [35] to exclude the presence of premature stop codons indicative of NUMTs (nuclear DNA sequences of mitochondrial origin), then aligned using Geneious 11 (Biomatters Ltd., Auckland, New Zealand). Unique haplotypes were collapsed in FaBox 1.41 [36] and compared with public homologous sequences of *Testudo hermanni* in NCBI (National Center for Biotechnology Information) with the BLAST algorithm , accessed on 26 September 2020).

Microsatellite alleles were checked by eye and sized in Geneious 11. Individual assignment was performed by the Bayesian clustering algorithm implemented in Structure 3.2 [37], testing one to seven putative genetic clusters (K) for 10 million iterations with a burning of 300,000. The admixture model with correlated frequencies among populations was selected. The best K value was inferred according to [38], and the proportion of individual assignment to each cluster was visualized with CLUMPAK (http://clumpak. tau.ac.il, accessed on 31 October 2020).

2.2. Isolation of Salmonella spp.

Faecal samples were collected from each turtle and transported under cooling conditions to the laboratories of Avian Diseases Unit, Department of Veterinary Medicine, University of Bari, Italy.

Bacteriological analyses were performed according to ISO 6579-1, "Microbiology of the food chain—Horizontal method for the detection, enumeration and serotyping of *Salmonella*—Part 1: Detection of *Salmonella* spp." [39].

Briefly, each sample was inoculated in Buffered Peptone Water (Oxoid, Milan, Italy) in a ratio of 1:10. After the incubation at 37 °C for 24 h, each sample was plated into Modified Semi-solid Rappaport–Vassiliadis (MSRV) agar (Oxoid, Milan, Italy) with 10 mg/500 mL novobiocin (Oxoid, Milan, Italy) added, and incubated at 42 °C for 24 h.

Bacterial growth compatible with *Salmonella* spp. was plated into two selective media, Hektoen Enteric Agar and Xylose Lysine Deoxycholate (XLD) agar (Oxoid, Milan, Italy) and incubated at 37 °C for 24 h. All isolates were biochemically screened by using TSI (triple sugar iron) and a urea test.

The identification of *Salmonella* spp. was performed by colony PCR according to [40], with slight modifications. Briefly, a single, well-isolated colony was picked with a sterile stick and resuspended in 10 µL of sterile distilled water. Two µl of suspension were added to the reaction mixture. The reaction was carried out by the Platinum II Hot-Start Green PCR Master Mix (Thermo Scientific, Milan, Italy), and primers 139 and 141 were added at a final concentration of 15 µM. The thermal cycle was as follows: 94 °C for 10 min (cell lysis and initial denaturation of DNA), followed by 35 cycles consisting of 94 °C for 15 s, 60 °C for 15 s, 72 °C for 5 s, and a final elongation at 72 °C for 10 min. Each strain identified was stored at −20 °C until serotyping.

Two colonies identified as *Salmonella* spp. from each positive turtle were serotyped, in order to determine whether a turtle could carry two different types of *Salmonella*.

2.3. Salmonella Serotyping

Salmonella serotyping was performed according to ISO/TR 6579-3:2013 [41], based on the rapid slide agglutination method, to determine the antigenic formula of *Salmonella* spp. according to the White–Kauffmann–Le Minor scheme by means of specific antisera for the detection of O and H antigens.

Strains were inoculated into a trypticase soy agar (TSA) slant tube incubated for 24 h at 37 °C. Since auto-agglutinating strains cannot be investigated for serotyping, each bacterial culture was previously investigated for auto-agglutination by mixing 1 µL of bacterial culture with one drop of saline solution (0.9%). Subsequently, non-auto-agglutinant strains were tested with polyvalent and monovalent antisera against specific somatic (O) and flagellar (H) antigens.

Briefly, for detection of O antigens, one drop of antiserum was mixed with a small amount of bacterial culture on a slide, which was gently tilted for subsequent observation of agglutination. The positive reaction consisted of the presence of granules. Subsequently, the agglutination with polyvalent and monovalent H antisera was performed. Most of the *Salmonella* serovars possess two types of H-antigens (phase 1 and phase 2). When the two phases were not expressed simultaneously, the dominant H-phase was repressed so that the second H-phase could be expressed and identified. Phase inversion was performed using the Sven Gard method: specific phase inversion antiserum was added to Sven Gard swarming agar medium, and the *Salmonella* strain was spot-inoculated on the plate. After incubation, the isolate was tested with the specific antiserum against the previously unexpressed H phase.

2.4. Statistical Analysis

The association between testudo species and detection of *S. enterica* was evaluated by Fisher's exact test. The same test was used to establish the significance of the distribution of *S. enterica* serovars among isolates and *Testudo* species, and to verify the uniformity of such distribution by gender and age. In all cases, $p < 0.05$ was considered as a significance threshold. All tests were carried out in R v. 4.0.4 (R Foundation, Wien, Austria) [42].

3. Results

3.1. Genotyping of Testudo h. hermanni and T. h. boettgeri Specimens

T. hermanni hermanni specimens ($n = 41$) showed 99.7% to 100% identity to reference mtDNA haplotypes H1–H3–H4 of the same subspecies, which is widespread in the western part of *T. hermanni* distribution [21]. All tested *T. h. boettgeri* specimens ($n = 8$) showed 100% identity with reference mtDNA haplotypes B10–B11–B13 of the same subspecies, which occurs in Croatia and Epirus (Greece), according to [14]. Obtained mtDNA (cytb) sequences were submitted to GenBank (Accession numbers MZ197805-MZ197810).

Bayesian assignment of nuclear genotypes highlighted the occurrence of two distinct gene pools in the dataset, one that clustered together most of the turtles sharing the mtDNA of *T. hermanni hermanni* ($n = 34$), and the other which included the three reference genotypes of *T. h. boettgeri* and eight individuals assigned to the same subspecies, according to mtDNA analysis. The remaining seven turtles were identified as putative hybrids, according to cyto-nuclear mismatch or admixed nuclear genotype.

Therefore, they were considered as hybrids of *T. h. hermanni* × *T. h. boettgeri*.

3.2. Salmonella spp. Detection

Salmonella spp. was isolated from 42 out of 69 turtles (60.9%) (Table 1). The bacterium was particularly found in *T. h. hermanni*, where the positive rate (73.5%) was higher than in hybrids (57.1%). *Salmonella* was less frequently detected in *T. h. boettgeri*, and it was never identified in *T. marginata* and *T. graeca*.

The distribution of *Salmonella* spp. was significantly higher in *T. hermanni hermanni* than in *T. hermanni boettgeri* ($p = 0.019$), disregarding individual genotypes. Distribution of the pathogen within the *T. hermanni hermanni* groups was statistically uniform ($p = 0.650$). Despite the low number of isolates, the *Salmonella* spp. distribution in *T. graeca* and *T. marginata* was significantly lower than in *T. hermanni hermanni* ($p = 0.015$).

On the other hand, the distribution of *Salmonella* spp. was not biased by gender ($p = 1000$). The same result was obtained considering either *T. hermanni hermanni* individuals only, or the entire group of *Testudo* spp.

The distribution of *Salmonella* spp. in relation to the age of the turtles is reported in Table 2. No significant difference of distribution was found between younger (up to 10 years old) and older (over 10 years old) individuals ($p = 0.140$) (Table 2). However, considering the subspecies *T. hermanni hermanni* only, older animals were found significantly more prone to be infected ($p = 0.042$). Such a tendency was partially confirmed when the species *T. hermanni* was considered, but the results remained above the significance threshold ($p = 0.068$).

Table 1. Prevalence of *Salmonella* spp. in turtles housed in the rescue centre.

Turtles	Females		Males		Total	
	N° Pos/ N° Tested	% Positivity	N° Pos/ N° Tested	% Positivity	N° Pos/ N° Tested	% Positivity
T. h. hermanni	16/19	84.21	9/15	60.00	25/34	73.5
Hybrids of *T. h. hermanni*	0/2	0	4/5	80.00	4/7	57.1
N.G.A.* *T. h. hermanni*	4/8	50.00	7/8	87.50	11/16	68.8
T. h. boettgeri	0/2	0	2/6	33.33	2/8	25.0
T. graeca	0/1	0	0/1	0	0/2	0
T. marginata	0/1	0	0/1	0	0/2	0
Total	20/33	60.61	22/36	61.11	42/69	60.9

N.G.A.* = not genetically analysed.

Table 2. Prevalence of *Salmonella* spp. in relation to the age of the turtles.

Turtles	6 to 10 Years		More than 10 Years	
	N° Pos/ N° Tested	% Positivity	N° Pos/ N° Tested	% Positivity
T. h. hermanni	11/19	57.89	14/15	93.33
Hybrids of *T. h. hermanni*	2/3	66.67	2/4	50.00
N.G.A.* *T. h. hermanni*	5/9	55.56	6/7	85.71
T. h. boettgeri	0/3	0	2/5	40
T. graeca	0/1	0	0/1	0
T. marginata	0/0	0	0/2	0
Total	18/35	51.43	24/34	70.59

N.G.A.* = not genetically analysed.

3.3. Salmonella Serotyping

Salmonella isolates belonged to two different species: *S. enterica* and *S. bongori*. Five different serovars (Hermannswerder, Abony, Ferruch, Richmond, and Vancouver) with different prevalence were identified within the group *S. enterica* subsp. *enterica* (Table 3). Serovars Abony (43.5%) and Hermanswerder (28.3) were the most prevalent.

Table 3. Prevalence of serotypes among identified *Salmonella* strains.

Species	Subspecies	Serovar	Strains N°/Total	%
S. enterica	*enterica*	Hermannswerder	13/46	28.3
		Abony	20/46	43.5
		Ferruch	1/46	2.2
		Richmond	4/46	8.7
		Vancouver	1/45	2.2
	diarizonae		2/46	4.3
	salamae		4/46	8.7
S. bongori			1/46	2.2

One strain and four isolates belonged to *S. enterica* subsp. *diarizonae* and *S. enterica* subsp. *salamae*, respectively.

The frequency is significantly biased towards the serovars Abony and Hermannswerder (SD = 0.152, skewness = 1.600, kurtosis = 4.612).

The distribution of the different *Salmonella* types among the species of turtles is reported in Table 4.

Table 4. Distribution of different types of *Salmonella* among positive turtles.

Turtles	Hermannswerder (13) N° (%)	Abony (20) N° (%)	Ferruch (1) N° (%)	Richmond (4) N° (%)	Vancouver (1) N° (%)	*Bongori* (1) N° (%)	*Salamae* (4) N° (%)	*Diarizonae* (2) N° (%)
T. h. hermanni	7 (53.8)	12 (60)	0 (0)	4 (100)	1 (100)	0 (0)	2 (50)	0 (0)
Hybrids of *T. h. hermanni*	2 (15.4)	2 (10)	0 (0)	0 (0)	0 (0)	0 (0)	1 (25)	0 (0)
Not genotyped *T. h. hermanni*	4 (30.8)	5 (25)	1 (100)	0 (0)	0 (0)	1 (100)	1 (25)	1 (50)
T. h. boettgeri	0 (0)	1 (5)	0 (0)	0 (0)	0 (0)	0 (0)	0 (0)	1 (50)

All different serotypes of *S. enterica* subsp. *enterica* were found in *T. h. hermanni*, except for *Salmonella* ser Abony, which was also detected in *T. h. boettgeri*. Likewise, *S. enterica* subsp. *diarizonae* was identified in both *T. h. hermanni* and *T. h. boettgeri*.

The distribution of *Salmonella* serotypes within *Testudo* groups was statically homogeneous ($p = 0.474$ considering both all species and only the *Testudo hermanni* group).

Two different *Salmonella* serotypes with different combinations were simultaneously found in four specimens of *T. h. hermanni* (Table 5). *Salmonella* ser. Hermannswerder was the most identified in association with other types.

Table 5. Different combination of *Salmonella* serotypes in *T. h. hermanni*.

Turtles	Hermannswerder/ Ferruch	Hermannswerder/*Salamae*	Hermannswerder/Abony	Abony/ Richmond
T. h. hermanni	-	-	-	1
Hybrids of *T. h. hermanni*	-	1	-	-
N.G.A.* *T. h. hermanni*	1	-	1	-

N.G.A.* = not genetically analysed.

4. Discussion

Active conservation measures, like population reinforcement, reintroduction projects, or breeding programs are crucial to effectively protect species showing low dispersal ability, like *T. hermanni*. In turn, informative genetic analyses of the individuals to be released should be mandatory, in order to preserve both wildlife local gene pools and operators' safety. In the present study, 34 tortoises have been designated *T. h. hermanni*, which inhabit the most part of Italy, according both to mitochondrial and nuclear genetic screening. Therefore, they appeared eligible to be involved in restoking projects of wild Italian populations, pending the assessment of genetic compatibility—in term of allele frequencies—with the receiving local gene pool that is present in the protected area.

Salmonella was frequently identified in turtles housed in the rescue centre. All *Salmonella*-positive turtles appeared clinically healthy. This finding confirms the potential role of turtles as reservoirs for the bacterium. *Salmonella* may be a natural inhabitant of the intestinal tract of reptiles [43], and turtles in particular [25]. Although granulomatous hepatitis due to *Salmonella* typhimurium in a spur-thighed tortoise (*Testudo graeca*) has been reported [32], illness is usually not associated with non-typhoidal *Salmonella* infection in turtles, increasing the risk of transmission to humans. In humans, infections due to contact with turtles is mostly related to contact with pet animals and occurs more often in children than adults. Among reptiles, turtles are commonly housed as pets. Particularly, baby turtles are easy to handle, safe, inexpensive, and small enough to be kissed and held by children, increasing the likelihood of direct transmission of *Salmonella*. In addition, bacterium can be indirectly transmitted though cross-contamination by cleaning of turtle habitats in the kitchen sink and bathtub [25]. A prevalence of *Salmonella* infection of 49.1% was detected in terrestrial turtles reared in private farms of Italy [44]. The prevalence of infection found in the turtles tested in this study was higher (60.9%) and highlights the spread of the bacterium among the tortoises housed in the rescue centre. In previous studies involving turtles from wildlife rescue centres, the prevalence of infection ranged from 10% [30] to 79% [43], depending on the geographical area and the species of turtle involved in the study. A prevalence of 80% (*T. hermanni*) and 72% (*T. graeca*) has been reported in tortoises housed in wildlife rescue centres in Tuscany [43], while lower prevalence has been reported in *T. graeca* (36.8%) and in *T. hermanni* (25.4%) in Sicily [45], and in *T. graeca* (50%) and *T. marginata* (4.5%) in Sardinia [30]. In this study, *T. h. hermanni* tested positive (70.5%) more frequently than other species. *Salmonella* was never detected in *T. marginata* and *T. graeca*, but the datum is less relevant because of the small size of the sample.

Several factors may influence the prevalence of *Salmonella* infection in turtles. Aquatic turtles seem to host the germ less frequently than terrestrial species [46], probably because *Salmonella* spends less time on the skin and in the cloaca in the aquatic environment [47]. Moreover, terrestrial turtles usually practice geophagy and ingest faeces of other turtles or animals [48]. This behaviour also could explain why *Salmonella* is more frequently detected in terrestrial turtles [49].

Stressful conditions may interfere with the immune system and immunomodulation, potentially increasing susceptibility to pathogens [50] and the level of *Salmonella* shedding in the environment [25]. In a study carried out on specimens of *Testudo graeca* illegally introduced into Italy from Tunisia, the high prevalence of the infection was attributed to poor hygiene conditions, confined spaces, and high densities during transport [51]. Similarly, the relatively limited space, such as in the rescue centres in which the animals are located, may affect the shedding of the germ and the spread of the infection [52].

Usually, turtles are stressed during the mating season, because males continuously fight against each other and females are frequently chased and bitten by males [53]. This could explain the level of overlapping infection found in both sexes [43]. Also, high temperatures can lead to more intense bacterial replication. Therefore, sampling carried out during the warmer months, as in this and most studies, may influence the prevalence of infection [52].

Turtles can become infected with *Salmonella* throughout their lives through several routes, such as contaminated food, water, and soil [54]. In this study, no significant difference of *Salmonella* distribution was found between younger and older turtles. However, older animals were found significantly more prone to be infected when considering the subspecies *T. hermanni hermanni* only. Contaminated water can be an important source and method of spreading infection among specimens [55]. *Salmonella* strains display good resistance in water [56]. In captive reptiles, the prevalence of *Salmonella* spp. is several times higher when drinking water is not replaced regularly [57]. Water recirculation and oxygenation are reduced in artificial ponds, and this may amplify different possible kinds of contamination [52,58]. The incidence of *Salmonella* infection seems to be lower among free-living turtles, which are found in natural ponds [47]. *Salmonella* infection among turtles in rescue centres is an important issue also, considering that the germ has ability to survive and penetrate through turtle eggs [59,60].

In this study, a variability of NT *Salmonella* strains belonging to two different species, *S. enterica* and *S. bongori*, was found. The greatest number of isolates belonged to *S. enterica* subsp. *enterica* (*Salmonella* ser. Abony, *Salmonella* ser. Hermannswerder, *Salmonella* ser.

Richmond, *Salmonella* ser. Ferruch, and *Salmonella* ser. Vancouver), but *Salmonella enterica* subsp. *Salamae* and *Salmonella enterica* subsp. *Diarizonae* were also identified.

T. h. hermanni harboured different species and types of *Salmonella*, but *Salmonella* ser. Abony was the serovar most frequently identified. It was found prevalently in *T. h. hermanni*, but also in *T. h. boettgeri*. *Salmonella* ser. Abony, such as *Salmonella* ser. Richmond, has been previously detected in *T. graeca* [51,61], *T. hermanni* [45], and *T. marginata* [30] among both captive and free-living tortoises. The spread of this serovar among turtles exposes the personnel involved in their management to risk of infection. *Salmonella* ser. Abony is associated with human salmonellosis, particularly in infants and children, which also presents as more severe forms of disease. Gastroenteritis due to *Salmonella* ser. Abony occurred in a Japanese child infected by his turtle (*T. graeca*) bought in a pet shop [62]. *Salmonella* ser. Abony was associated with sepsis that occurred in an eight-month-old baby in Norway [26], and with sepsis and meningitis in a two-month-old baby in Belgium [27]. In adults, *Salmonella* ser. Abony usually causes severe pathological conditions in immunocompromised individuals [24], but a severe purulent pleuropneumonia associated to *Salmonella* ser. Abony was found in a fully immunocompetent woman [29].

In this study, *Salmonella* ser. Hermannswerder was found exclusively in *T. h. hermanni* and seemed to be linked to this turtle species. Previously, it has been detected in turtles, although the authors did not specify in what species [44]. *Salmonella* ser. Hermannswerder has never been associated with human salmonellosis, to our knowledge.

Salmonella ser. Richmond, which was found in *T. h. hermanni* in this study, was previously detected in captive and free-living tortoises [30,45,51,61]. *Salmonella* ser. Richmond has caused acute diarrhea in children [63]. Moreover, *Salmonella* ser. Richmond was responsible for an outbreak of acute gastroenteritis in a military detachment in Spain, and contaminated water was identified as the source of infection [64].

S. bongori is mostly associated with cold-blooded animals [65]. Also, *S. bongori* was responsible of diarrhea in a dog [66]. Human infections due to *S. bongori* have been frequently reported, although persistent endemicity of human cases due to this *Salmonella* species in southern Italy was observed [67]. In addition to reptiles, birds, such as pigeons and blackcaps, as well as healthy human carriers, urban sewerage plants, contaminated soft cheese, and eggs have been identified as other sources of infection for humans. The infection has particularly occurred with diarrhea and fever in children up to 3 years of age, although a case of *S. bongori* infection in a HIV-positive adult presenting diarrhea was reported [66]. More recently, human infections due to *S. bongori* have been reported in children from northern Italy and Switzerland [68,69].

Salmonella ser. Ferruch, which was occasionally found among the tested samples, has been previously isolated from faecal samples of pet chelonians in France [70], large intestine of wild boars in northern Italy [71], broiler giblets and skin [72], and cloacal samples of ducks farmed where chicken litter was used as fertilizer in Egypt [73]. *Salmonella* ser. Ferruch was also isolated from faecal swabs of diarrheic sheep in Egypt [74]. The serovar has not been identified as causing disease in humans, to our knowledge.

Both *S. enterica* subsp. *diarizonae* and *S. enterica* subsp *salamae* were previously identified in aquatic turtles [61,75,76], terrestrial turtles (*T. graeca*) [47], snakes [77], and reptiles of Norwegian zoos [78], but also in mammals like the quokka in Australia [79], wild boars in Italy [80], and sheep in Spain [81]. Overall, animals were asymptomatic. Among the aquatic turtles, *S. enterica* subsp *salamae* was previously found in specimens of *Emys orbicularis* [47] which is considered the only native aquatic turtle in Italy. Although rarely, *Emys* and *Testudo* specimens may share some natural habitats with potential risks of *Salmonella* cross-infection in free-living turtles.

While *S. enterica* subsp *salamae* is not associated with human infections, *S. enterica* subsp. *diarizonae* was responsible for acute gastroenteritis in a 77-year-old man affected by advanced rectal cancer, who later died due to multiple morbidities, including *Salmonella* infection [82], as well as for a case of gastroenteritis in a 10-day-old female infant with bloody mucous stools [83]. Furthermore, *S. enterica* subsp. *diarizonae* was isolated from maxillary sinusitis in a 29-year-old snake handler who suffered from intermittent fever, nasal discharge, and pain into the maxillary and frontal sinus area [28]. Finally, *S. enterica* subsp. *diarizonae* was isolated during an episode of extraintestinal infection in a eight-year-old African American boy who developed fever and a cervical mass on his neck after direct contact with snakes during an educational event at his school [84]. Although occasionally detected in this study, the finding of *Salmonella* ser. Vancouver in a hybrid of *Testudo hermanni* seems of particular interest. We found a single report about infection in humans due to *Salmonella* ser. Vancouver [85]. The serovar was responsible for fever, cramps, and abdominal pains, as well as bile-stained vomiting, nausea, anorexia, and diarrhea for several weeks in a woman in Canada. To our knowledge, there are no other reports about the detection of *Salmonella* ser. Vancouver in humans or animals.

In conclusion, terrestrial turtles may harbour a wide variety of *Salmonella* types in their gut, even simultaneously, as observed for *Salmonella* ser. Abony and S *Salmonella* ser. Richmond in this study. Among the detected strains, some serovars, such as *Salmonella* ser Abony, *Salmonella* ser Richmond, and *Salmonella* ser Vancouver, as well as the species *S. bongori*, may represent a potential risk, often underestimated, for public health, especially for workers in contact with turtles. The absence of clinical signs in infected tortoises and the lack of routine bacteriological testing for *Salmonella* in wildlife rescue centres increase the sanitary risk for operators.

Reducing turtles' stress to minimise *Salmonella* shedding [54], as well as adopting correct animal husbandry procedures and hygiene techniques, may be useful to minimise the risk of transmission of *Salmonella* to humans [86]. In particular, the adoption of gloves to manage turtles is a preventive measure of relevance. Nevertheless, the greater measure of prevention is information and education on potential sanitary risks of each professional figure involved in wildlife management.

Supplementary Materials: The following are available online at https://www.mdpi.com/article/10.3390/ani11061529/s1: Table S1: Gender, presumed age and morphometric measures of turtles tested in the study. Table S2: List of molecular markers used to genotype captive *T. hermanni* in this study. For each locus, the target DNA analyzed (Type), the length of the fragments amplified (bp), the primer pairs used (including sequences and labels) and their reference (Ref), and the thermal PCR profile have been reported.

Author Contributions: Conceptualization, E.C.; sampling, G.C. and F.D.; turtles genotyping, A.B.; Salmonella isolation and identification, E.C. and G.C.; Salmonella serotyping, S.F. and G.O.; turtle morphological data, G.C.; data curation E.C. and R.L.; statistical analysis N.P.; writing—original draft preparation, E.C. and G.C. with the contribution of A.B.; supervision, E.C., A.C., project administration, E.C.; funding acquisition, A.C. writing—review and editing, E.C. All authors have read and agreed to the published version of the manuscript.

Funding: This study received no external funding.

Institutional Review Board Statement: The study was conducted according to the guidelines of the Ethics Committee of Department of Veterinary Medicine of University of Bari "Aldo Moro", Italy (Approval number n. 14/2021).

Conflicts of Interest: The authors declare no conflict of interest.

References

1. Ceballos, G.; Ehrlich, P.R.; Barnosky, A.D.; Garcìa, A.; Pringle, R.M.; Palmer, T.M. Accelerated modern human-induced species losses: Entering the sixth mass extinction. *Sci. Adv.* **2015**, *1*, 1–5. [CrossRef] [PubMed]
2. McCallum, M.L. Vertebrate biodiversity losses point to a sixth mass extinction. *Biodivers. Conserv.* **2015**, *24*, 2497–2519. [CrossRef]
3. Harley, C.D. Climate change, keystone predation, and biodiversity loss. *Science* **2011**, *334*, 1124–1127. [CrossRef] [PubMed]
4. Bellard, C.; Cassey, P.; Blackburn, T.M. Alien species as a driver of recent extinctions. *Biol. Lett.* **2016**, *12*, 1–4. [CrossRef]
5. Supple, M.A.; Shapiro, B. Conservation of biodiversity in the genomics era. *Genome Biol.* **2018**, *19*, 1–12. [CrossRef]
6. Jenkins, C.N.; Pimm, S.L.; Joppa, L.N. Global patterns of terrestrial vertebrate diversity and conservation. *Proc. Natl. Acad. Sci. USA* **2013**, *110*, 2602–2610. [CrossRef]

7. Bouzat, J.L. Conservation genetics of population bottlenecks: The role of chance, selection, and history. *Conserv. Genet.* **2010**, *11*, 463–478. [CrossRef]
8. Böhm, M.; Collen, B.; Baillie, J.E.; Bowles, P.; Chanson, J.; Cox, N.; Hammerson, G.; Hoffman, M.; Livingstone, S.R.; Ram, M.; et al. The conservation status of the world's reptiles. *Biol. Conserv.* **2013**, *157*, 372–385. [CrossRef]
9. Bertolero, A.; Cheylan, M.; Hailey, A.; Livoreil, B.; Willemsen, R.E. *Testudo hermanni* (Gmelin 1789)—Hermann's tortoise. Conservation biology of freshwater turtles and tortoises: A compilation project of the IUCN/SSC Tortoise and Freshwater Turtle Specialist Group. *Chelonian Res. Monogr.* **2011**, *5*, 059.1–059.20.
10. Zenboudji, S.; Cheylan, M.; Arnal, V.; Bertolero, A.; Leblois, R.; Astruc, G.; Montgelard, C. Conservation of the endangered Mediterranean tortoise *Testudo hermanni hermanni*: The contribution of population genetics and historical demography. *Biol. Conserv.* **2016**, *195*, 279–291. [CrossRef]
11. Wermuth, H. *Testudo hermanni robertmertensi* n. subsp. und ihr Vorkommen in Spanien. *Senckenbergiana* **1952**, *33*, 157–164.
12. Morale-Perez, J.V.; Sanctis Serra, A. The Quaternary fossil record of the genus Testudo in the Iberian Peninsula. Archaeological implications and diachronic distribution in the western Mediterranean. *J. Archaeol. Sci.* **2009**, *36*, 1152–1162. [CrossRef]
13. Pérez, I.; Giménez, A.; Sánchez-Zapata, J.A.; Anadón, J.D.; Martínez, M.; Esteve, M.Á. Non-commercial collection of spur-thighed tortoises (*Testudo graeca graeca*): A cultural problem in southeast Spain. *Biol. Conserv.* **2004**, *118*, 175–181. [CrossRef]
14. Fritz, U.; Auer, M.; Bertolero, A.; Cheylan, M.; Fattizzo, T.; Hundsdorfer, A.K.; Marcos Martín Sampayo, M.M.; Pretus, J.L.; ŠIrokÝ, P.; Wink, M. A rangewide phylogeography of Hermann's tortoise, *Testudo hermanni* (Reptilia: Testudines: Testudinidae): Implications for taxonomy. *Zool. Scr.* **2006**, *35*, 531–543. [CrossRef]
15. Bertorelle, G. Il ruolo e i risultati delle analisi genetiche nella conservazione e nella gestione della testuggine di Hermann. *Quad. Stn. Ecol. Civ. Mus. Stor. Nat. Ferrara* **2007**, *17*, 107–110.
16. Kirsche, W. An F2-generation of *Testudo hermanni hermanni* Gmelin bred in captivity with remarks on the breeding of Mediterranean tortoises 1976–1981. *Amphib. Reptil.* **1984**, *5*, 31–35. [CrossRef]
17. Soler, J.; Pfau, B.; Martinez-Silvestre, A. Detectiong intraspecific hybrids in *Testudo hermanni*. *Radiata* **2012**, *21*, 4–30.
18. Devaux, B.; Stubbs, D. *Proceedings Conservation, Restoration, and Management of Tortoises and Turtles*, 1st ed.; Van Abbema, J., Ed.; Joint Publication New York Turtle and Tortoise Society & WCS Turtle Recovery Program: New York, NY, USA, 1997; pp. 330–332. ISBN 096-590-500-4.
19. Moritz, C. Defining 'evolutionarily significant units' for conservation. *Trends Ecol. Evol.* **1994**, *9*, 373–375. [CrossRef]
20. Hoelzer, K.; Moreno-Switt, A.; Wiedmann, M. Animal contact as a source of human non-typhoidal salmonellosis. *Vet. Res.* **2011**, *42*, 1–27. [CrossRef]
21. Hale, C.R.; Scallan, E.; Cronquist, A.B.; Dunn, J.; Smith, K.; Robinson, T.; Lathrop, S.; Tobin-D'Angelo, M.; Clogher, P. Estimates of Enteric Illness Attributable to Contact with Animals and Their Environments in the United States. *Clin. Infect. Dis.* **2012**, *54*, 472–479. [CrossRef]
22. Corrente, M.; Totaro, M.; Martella, V.; Campolo, M.; Lorusso, A.; Ricci, M.; Buonavoglia, C. Reptile-associated Salmonellosis in Man, Italy. *Emerg. Infect. Dis.* **2006**, *12*, 358–359. [CrossRef] [PubMed]
23. Lafuente, S.; Bellido, J.B.; Moraga, F.A.; Herrera, S.; Yagüe, A.; Montalvo, T.; de Simó, M.; Simón, P.; Caylà, J.A. *Salmonella paratyphi* B and *Salmonella litchfield* outbreaks associated with pet turtle exposure in Spain. *Enferm. Infecc. Microbiol. Clín.* **2013**, *31*, 32–35. [CrossRef] [PubMed]
24. Gambino-Shirley, K.; Stevenson, L.; Concepción-Acevedo, J.; Trees, E.; Wagner, D.; Whitlock, L.; Roberts, J.; Garret, N.; van Duyne, S.; McAllister, G.; et al. Flea market finds and global exports: Four multistate outbreaks of human Salmonella infections linked to small turtles, United States-2015. *Zoonoses Public Health* **2018**, *65*, 560–568. [CrossRef] [PubMed]
25. Sodagari, H.R.; Habib, I.; Shahabi, M.P.; Dybing, N.A.; Wang, P.; Bruce, M. A Review of the Public Health Challenges of Salmonella and Turtles. *Vet. Sci.* **2020**, *7*, 56. [CrossRef]
26. Torfoss, D.; Abrahamsen, T.G. *Salmonella smitte* fra skilpadder. *Tidsskr. Nor. Læge* **2000**, *120*, 3670–3671.
27. Van Meervenne, E.; Botteldoorn, N.; Lokietek, S.; Vatlet, M.; Cupa, A.; Naranjo, M.; Katelijne, D.; Bertrand, S. Turtle-associated *Salmonella septicaemia* and meningitis in a 2-month-old baby. *J. Med. Microbiol.* **2009**, *58*, 1379–1381. [CrossRef]
28. Horvath, L.; Kraft, M.; Fostiropoulos, K.; Falkowski, A.; Tarr, P.E. *Salmonella enterica* Subspecies *diarizonae* Maxillary Sinusitis in a Snake Handler: First Report. *Open Forum. Infect. Dis.* **2016**, *3*, 1–2. [CrossRef]
29. Pitiriga, V.; Dendrinos, J.; Nikitiadis, E.; Vrioni, G.; Tsakris, A. First Case of Lung Abscess due to *Salmonella enterica* Serovar Abony in an Immunocompetent Adult Patient. *Case Rep. Infect. Dis.* **2016**, 1–4. [CrossRef]
30. Lecis, R.; Paglietti, B.; Rubino, S.; Are, B.M.; Muzzeddu, M.; Berlinguer, F.; Chessa, B.; Pittau, M.; Alberti, A. Detection and Characterization of *Mycoplasma* spp. and *Salmonella* spp. in Free-living European Tortoises (*Testudo hermanni*, *Testudo graeca*, and *Testudo marginata*). *J. Wildl. Dis.* **2011**, *47*, 717–724. [CrossRef]
31. Le Souëf, A.; Barry, M.; Brunton, D.; Jakob-Hoff, R.; Jackson, B. Ovariectomy as treatment for ovarian bacterial granulomas in a Duvaucel's gecko (*Hoplodactylus duvaucelii*). *N. Z. Vet. J.* **2015**, *63*, 340–344. [CrossRef]
32. Candela, M.G.; Atance, P.M.; Seva, J.; Pallarés, F.J.; Vizcaíno, L.L. Granulamatous hepatitis caused by *Salmonella* Typhimurium in a spur-thighed tortoise (*Testudo graeca*). *Vet. Rec.* **2005**, *157*, 236–237. [CrossRef]
33. Ciofi, C.; Milinkovitch, M.C.; Gibbs, J.P.; Caccone, A.; Powell, J.R. Microsatellite analysis of genetic divergence among populations of giant Galápagos tortoises. *Mol. Ecol.* **2008**, *11*, 2265–2283. [CrossRef]

34. Forlani, A.; Crestanello, B.; Mantovani, S.; Livoreil, B.; Zane, L.; Bertorelle, G.; Congiu, L. Identification and characterization of microsatellite markers in Hermann's tortoise (*Testudo hermanni*, Testudinidae). *Mol. Ecol. Notes* **2005**, *5*, 228–230. [CrossRef]
35. Tamura, K.; Stecher, G.; Peterson, D.; Filipski, A.; Kumar, S. MEGA6: Molecular Evolutionary Genetics Analysis Version 6.0. *Mol. Biol. Evol.* **2013**, *30*, 2725–2729. [CrossRef]
36. Villensen, P. FaBox: An online toolbox for fasta sequences. *Mol. Ecol. Notes* **2007**, *7*, 965–968. [CrossRef]
37. Pritchard, J.K.; Stephens, M.; Donnelly, P. Inference of population structure using multilocus genotype data. *Genetics* **2000**, *155*, 945–959. [CrossRef]
38. Evanno, G.; Regnaut, S.; Goudet, J. Detecting the number of clusters of individuals using the software structure: A simulation study. *Mol. Ecol.* **2005**, *14*, 2611–2620. [CrossRef]
39. Mooijman, K.A.; Pielaat, A.; Kuijpers, A.F. Validation of EN ISO 6579-1—Microbiology of the food chain—Horizontal method for the detection, enumeration and serotyping of Salmonella—Part 1 detection of *Salmonella* spp. *Int. J. Food Microbiol.* **2018**, *288*, 3–12. [CrossRef]
40. Rahn, K.; de Grandis, S.A.; Clarke, R.C.; McEwen, S.A.; Galán, J.E.; Ginocchio, C.; Curtiss, R., III; Gyles, C.L. Amplification of an invA gene sequence of *Salmonella typhimurium* by polymerase chain reaction as a specific method of detection of Salmonella. *Mol. Cell. Probes* **1992**, *6*, 271–279. [CrossRef]
41. ISO/TR 6579-3:2013 "Microbiology of the Food Chain—Horizontal Method for the Detection, Enumeration and Serotyping of Salmonella—Part 3: Guidelines for Serotyping of *Salmonella* spp. Available online: https://www.iso.org/obp/ui/#iso:std:iso:tr:6579:-3:ed-1:v1:en (accessed on 15 July 2014).
42. R Core Team. A Language and Environment for Statistical Computing. Available online: https://www.Rproject.org (accessed on 25 April 2014).
43. Pasmans, F.; Herdt, P.; Chasseur-Libotte, M.L.; Ballasina, D.L.; Haesebrouck, F. Occurrence of Salmonella in tortoises in a rescue centre in Italy. *Vet. Rec.* **2000**, *146*, 256–258. [CrossRef]
44. Dipineto, L.; Capasso, M.; Maurelli, M.; Russo, T.; Pepe, P.; Capone, G.; Fioretti, A.; Cringoli, G.; Rinaldi, L. Survey of co-infection by Salmonella and oxyurids in tortoises. *BMC Vet. Res.* **2012**, *8*, 69. [CrossRef] [PubMed]
45. Percipalle, M.; Giardina, G.; Lipari, L.; Piraino, C.; Macrì, D.; Ferrantelli, V. Salmonella Infection in Illegally Imported Spur-Thighed Tortoises (*Testudo graeca*). *Zoonoses Public Health* **2010**, *58*, 262–269. [CrossRef] [PubMed]
46. Centeno, D.; Godínez, J.; Sandoval, C.V.; López, M.L.; Villatoro, F.; Escobar, J.; Rodríguez, M.D.; Ríos, L. Antibiotic-Resistant Salmonella, isolated from cloacal swab samples from turtles in Guatemala. *Tecnol. Salud* **2020**, *7*, 196–204.
47. Hidalgo-Vila, J.; Díaz-Paniagua, C.; de Frutos-Escobar, C.; Jiménez-Martínez, C.; Pérez-Santigosa, N. Salmonella in free living terrestrial and aquatic turtles. *Vet. Microbiol.* **2007**, *119*, 311–315. [CrossRef] [PubMed]
48. Terebiznik, M.; Moldowan, P.D.; Leivesley, J.A.; Massey, M.D.; Lacroix, C.; Connoy, J.W.; Rollinson, N. Hatchling turtles ingest natural and artificial incubation substrates at high frequency. *Behav. Ecol. Sociobiol.* **2020**, *74*, 1–12. [CrossRef]
49. Boycott, J. Salmonella Species in Turtles. *Science* **1962**, *137*, 761–762. [CrossRef]
50. Verbrugghe, E.; Boyen, F.; Gaastra, W.; Bekhuis, L.; Leyman, B.; van Parys, A.; Haesebrouck, F.; Pasmans, F. The complex interplay between stress and bacterial infections in animals. *Vet. Microbiol.* **2012**, *155*, 115–127. [CrossRef]
51. Giacopello, C.; Foti, M.; Passantino, A.; Fisichella, V.; Aleo, A.; Mammina, C. Serotypes and antibiotic susceptibility patterns of *Salmonella* spp. isolates from spur-thighed tortoise, *Testudo graeca* illegally introduced in Italy. *Hum. Vet. Med.* **2012**, *4*, 76–81.
52. Marenzoni, M.L.; Zicavo, A.; Veronesi, F.; Morganti, G.; Scuota, S.; Coletti, M.; Passamonti, F.; Santoni, L.; Natali, M.; Iolanda Moretta, I. Microbiological and parasitological investigation on chelonians reared in Italian facilities. *Vet. Ital.* **2015**, *51*, 173–178. [CrossRef]
53. Hailey, A.; Willemsen, R.E. Population density and adult sex ratio of the tortoise *Testudo hermanni* in Greece: Evidence for intrinsic population regulation. *J. Zool.* **2000**, *251*, 325–338. [CrossRef]
54. Bosch, S.; Tauxe, R.V.; Behravesh, C.B. Turtle-Associated Salmonellosis, United States, 2006–2014. *Emerg. Infect. Dis.* **2016**, *22*, 1149–1155. [CrossRef]
55. Chen, C.Y.; Chen, W.C.; Chin, S.C.; Lai, Y.H.; Tung, K.C.; Chiou, C.S.; Hsu, Y.M.; Chang, C.C. Prevalence and Antimicrobial Susceptibility of Salmonellae Isolates from Reptiles in Taiwan. *J. Vet. Diagn. Investig.* **2010**, *22*, 44–50. [CrossRef]
56. Cherry, W.B.; Hanks, J.B.; Thomason, B.M.; Murlin, A.M.; Biddle, J.W.; Croom, J.M. Salmonellae as an index of pollution of surface waters. *Appl. Microbiol.* **1972**, *24*, 334–340. [CrossRef]
57. Corrente, M.; Sangiorgio, G.; Grandolfo, E.; Bodnar, L.; Catella, C.; Trotta, A.; Martella, V.; Buonavoglia, D. Risk for zoonotic Salmonella transmission from pet reptiles: A survey on knowledge, attitudes and practices of reptile-owners related to reptile husbandry. *Prev. Vet. Med.* **2017**, *146*, 73–78. [CrossRef]
58. Circella, E.; Camarda, A.; Bano, L.; Marzano, G.; Lombardi, R.; D'Onghia, F.; Greco, G. Botulism in Wild Birds and Changes in Environmental Habitat: A Relationship to be Considered. *Animals* **2019**, *9*, 1034. [CrossRef]
59. Feeley, J.C.; Treger, M.D. Penetration of Turtle Eggs by *Salmonella braenderup*. *Public Health Rep.* **1969**, *84*, 156. [CrossRef]
60. Harris, J.R.; Neil, K.P.; Behravesh, C.B.; Sotir, M.J.; Angulo, F.J. Recent Multistate Outbreaks of Human Salmonella Infections Acquired from Turtles: A Continuing Public Health Challenge. *Clin. Infect. Dis.* **2010**, *50*, 554–559. [CrossRef]
61. Hidalgo-Vila, J.; Diaz-Paniagua, C.; Ruiz, X.; Portheault, A.; Lmouden, H.E.; Slimani, T.; de Frutos, C.; de Caso, M.S. Salmonella species in free-living spur-thighed tortoises (*Testudo graeca*) in central western Morocco. *Vet. Rec.* **2008**, *162*, 218–219. [CrossRef] [PubMed]

62. Kuroki, T.; Ito, K.; Ishihara, T.; Furukawa, I.; Kaneko, A.; Suzuki, Y.; Seto, J.; Kamiyama, T. Turtle-Associated Salmonella Infections in Kanagawa, Japan. *Jpn. J. Infect. Dis.* **2015**, *68*, 333–337. [CrossRef]
63. Shobirin, M.H.; Shuhaimi, M.; Abu-Bakar, F.; Ali, A.M.; Ariff, A.; Nur-Atiqah, N.A.; Yazid, A.M. Characterization of *Salmonella* spp. Isolated from patients belows 3 years old with acute diarrohea. *J. Microbiol. Biotechnol.* **2003**, *19*, 751–755. [CrossRef]
64. Pac Sa, M.R.; Arnedo, A.; Benedicto, J.; Arranz, A.; Aguilar, V.; Guillén, F. Epidemic outbreak of Salmonella richmond infection in Castellón, Spain. *Rev. Panam. Salud Pública* **1998**, *3*, 96–101. [CrossRef] [PubMed]
65. Lamas, A.; Miranda, J.M.; Regal, P.; Vázquez, B.; Franco, C.M.; Cepeda, A. A comprehensive review of non- enterica subspecies of *Salmonella enterica*. *Microbiol. Res.* **2018**, *206*, 60–73. [CrossRef] [PubMed]
66. Giammanco, G.M.; Pignato, S.; Mammina, C.; Grimont, F.; Grimont, P.A.; Nastasi, A.; Giammanco, G. Persistent Endemicity of *Salmonella bongori* 48:z35:- in Southern Italy: Molecular Characterization of Human, Animal, and Environmental Isolates. *J. Clin. Microbiol.* **2002**, *40*, 3502–3505. [CrossRef]
67. Bellio, A.; Bianchi, D.M.; Acutis, P.; Biolatti, C.; Luzzi, I.; Modesto, P.; Rocchetti, A.; Decastelli, L.; Gallina, S. *Salmonella bongori* 48:z35:—The first Italian case of human infection outside Sicily. *Microbiol. Med.* **2016**, *31*, 95–96. [CrossRef]
68. Romano, A.; Bellio, A.; Macori, G.; Cotter, P.D.; Bianchi, D.M.; Gallina, S.; Decastelli, L. Whole-Genome Shotgun Sequence of *Salmonella bongori*, First Isolated in Northwestern Italy. *Genome Announc.* **2017**, *5*. [CrossRef]
69. Stevens, M.J.; Cernela, N.; Müller, A.; Stephan, R.; Bloemberg, G. Draft Genome Sequence of *Salmonella bongori* N19-781, a Clinical Strain from a Patient with Diarrhea. *Microbiol. Resour. Announc.* **2019**, *8*. [CrossRef]
70. Kodjo, A.; Villard, L.; Prave, M.; Ray, S.; Grezel, D.; Lacheretz, A.; Bonnuau, M.; Richard, Y. Isolation and Identification of Salmonella Species from Chelonians Using Combined Selective Media, Serotyping and Ribotyping. *J. Vet. Med.* **1997**, *44*, 625–629. [CrossRef]
71. Chiari, M.; Zanoni, M.; Tagliabue, S.; Lavazza, A.; Alborali, L.G. Salmonella serotypes in wild boars (*Sus scrofa*) hunted in northern Italy. *Acta Vet. Scand.* **2013**, *55*, 1–4. [CrossRef]
72. Hassan, A.R.; Salam, H.S.; Abdel-Latef, G.K. Serological identification and antimicrobial resistance of Salmonella isolates from broiler carcasses and human stools in Beni-Suef, Egypt. *Beni-Suef Univ. J. Appl. Sci.* **2016**, *5*, 202–207. [CrossRef]
73. El-Nabarawy, A.M.; Shakal, M.A.; Hegazy, A.H.; Batikh, M.M. Comparative Clinicopathological study of Salmonellosis in integrated fish-duck farming. *J. World Poult. Res.* **2020**, *10*, 184–194. [CrossRef]
74. Farouk, M.M.; El-Molla, A.; Salib, F.A.; Soliman, Y.A.; Shaalan, M. The Role of Silver Nanoparticles in a Treatment Approach for Multidrug-Resistant Salmonella Species Isolates. *Int. J. Nanomed.* **2020**, *15*, 6933–7011. [CrossRef]
75. Marin, C.; Ingresa-Capaccioni, S.; González-Bodi, S.; Marco-Jiménez, F.; Vega, S. Free-Living Turtles Are a Reservoir for Salmonella but Not for Campylobacter. *PLoS ONE* **2013**, *8*, 1–6. [CrossRef]
76. Hossain, S.; de Silva, B.C.; Dahanayake, P.S.; Shin, G.W.; Heo, G.J. Molecular characterization of virulence, antimicrobial resistance genes and class 1 integron gene cassettes in *salmonella enterica* subsp. enterica isolated from pet turtles in Seoul, Korea. *J. Exot. Pet Med.* **2019**, 209–217. [CrossRef]
77. Pedersen, K.; Lassen-Nielsen, A.M.; Nordentoft, S.; Hammer, A.S. Serovars of Salmonella from captive reptiles. *Zoonoses Public Health* **2009**, *56*, 238–242. [CrossRef]
78. Bjelland, A.M.; Sandvik, L.M.; Skarstein, M.M.; Svendal, L.; Debenham, J.J. Prevalence of Salmonella serovars isolated from reptiles in Norwegian zoos. *Acta Vet. Scand.* **2020**, *62*, 1–9. [CrossRef]
79. Martínez-Pérez, P.; Hyndman, T.H.; Fleming, P.A. *Salmonella* in Free-Ranging Quokkas (*Setonix brachyurus*) from Rottnest Island and the Mainland of Western Australia. *Animals* **2020**, *10*, 585. [CrossRef]
80. Zottola, T.; Montagnaro, S.; Magnapera, C.; Sasso, S.; de Martino, L.; Bragagnolo, A.; D'Amici, L.; Condoleo, R.; Pisanelli, G.; Pagnini, U.; et al. Prevalence and antimicrobial susceptibility of Salmonella in European wild boar (*Sus scrofa*); Latium Region—Italy. *Comp. Immunol. Microbiol. Infect. Dis.* **2013**, *36*, 161–168. [CrossRef]
81. Davies, R.H.; Evans, S.J.; Chappell, S.; Kidd, S.; Jones, Y.E.; Preece, B.E. Increase in *Salmonella enterica* subspecies diarizonae serovar 61:k:l,5,(7) in sheep. *Vet. Rec.* **2001**, *149*, 555–557. [CrossRef]
82. Gerlach, R.G.; Walter, S.; McClelland, M.; Schmidt, C.; Steglich, M.; Prager, R.; Bender, J.K.; Fuchs, S.; Schoerner, C.; Rabsch, W.; et al. Comparative whole genome analysis of three consecutive *Salmonella diarizonae* isolates. *Int. J. Med. Microbiol.* **2017**, *307*, 542–551. [CrossRef]
83. Hervás, J.A.; Rosell, A.; Hervás, D.; Rubio, R.; Dueñas, J.; Mena, A. Reptile Pets–associated *Salmonella enterica* Subspecies diarizonae Gastroenteritis in a Neonate. *Pediatr. Infect. Dis. J.* **2012**, *31*, 1102–1103. [CrossRef] [PubMed]
84. Fermaglich, L.J.; Routes, J.M.; Lye, P.S.; Kehl, S.C.; Havens, P.L. Salmonella Cervical Lymphadenitis in an Immunocompetent Child Exposed to a Snake at an Educational Exhibit. *Infect. Dis. Clin. Pract.* **2012**, *20*, 289–290. [CrossRef]
85. Dolman, C.E.; Ranta, L.E.; Hudson, V.G.; Bynoe, E.T.; Bailey, W.R.; Laidley, R. A new salmonella type: Salmonella Vancouver. *Can. J. Public Health* **1950**, *41*, 23–26. [PubMed]
86. Koski, L.; Stevenson, L.; Huffman, J.; Robbins, A.; Latash, J.; Omoregie, E.; Kline, K.; Nichols, M. Notes from the Field: An Outbreak of *Salmonella* Agbeni Infections Linked to Turtle Exposure—United States, 2017. *Morb. Mortal. Wkly. Rep.* **2018**, *67*, 1350. [CrossRef]

animals

MDPI

Article

Occurrence and Genetic Diversity of Protist Parasites in Captive Non-Human Primates, Zookeepers, and Free-Living Sympatric Rats in the Córdoba Zoo Conservation Centre, Southern Spain

Pamela C. Köster [1], Alejandro Dashti [1], Begoña Bailo [1], Aly S. Muadica [1,2], Jenny G. Maloney [3], Mónica Santín [3], Carmen Chicharro [1], Silvia Miguéláñez [1], Francisco J. Nieto [1], David Cano-Terriza [4], Ignacio García-Bocanegra [4], Rafael Guerra [5], Francisco Ponce-Gordo [6], Rafael Calero-Bernal [7], David González-Barrio [1,7,*] and David Carmena [1,*]

[1] Parasitology Reference and Research Laboratory, Spanish National Centre for Microbiology, 28220 Madrid, Spain; pamelakster@yahoo.com (P.C.K.); dashti.alejandro@gmail.com (A.D.); BEGOBB@isciii.es (B.B.); muadica@gmail.com (A.S.M.); cchichar@isciii.es (C.C.); miguelanez170777@hotmail.com (S.M.); fjnieto@isciii.es (F.J.N.)

[2] Departamento de Ciências e Tecnologia, Universidade Licungo, Quelimane 106, Zambézia, Mozambique

[3] Environmental Microbial and Food Safety Laboratory, Agricultural Research Service, United States Department of Agriculture, Beltsville, MD 20705-2350, USA; jenny.maloney@usda.gov (J.G.M.); monica.santin-duran@usda.gov (M.S.)

[4] Animal Health and Zoonosis Research Group (GISAZ), Department of Animal Health, University of Córdoba, 14071 Córdoba, Spain; davidcanovet@gmail.com (D.C.-T.); v62garbo@uco.es (I.G.-B.)

[5] Veterinary Services, Córdoba Zoo Conservation Centre, 14071 Córdoba, Spain; veterinario.zoocordoba@gmail.com

[6] Department of Microbiology and Parasitology, Faculty of Pharmacy, Complutense University of Madrid, 28040 Madrid, Spain; pponce@ucm.es

[7] SALUVET, Department of Animal Health, Faculty of Veterinary, Complutense University of Madrid, 28040 Madrid, Spain; r.calero@ucm.es

* Correspondence: dgonzalezbarrio@gmail.com (D.G.-B.); dacarmena@isciii.es (D.C.)

Citation: Köster, P.C.; Dashti, A.; Bailo, B.; Muadica, A.S.; Maloney, J.G.; Santín, M.; Chicharro, C.; Miguéláñez, S.; Nieto, F.J.; Cano-Terriza, D.; et al. Occurrence and Genetic Diversity of Protist Parasites in Captive Non-Human Primates, Zookeepers, and Free-Living Sympatric Rats in the Córdoba Zoo Conservation Centre, Southern Spain. *Animals* **2021**, *11*, 700. https://doi.org/10.3390/ani11030700

Academic Editor: Scott C. Williams

Received: 9 February 2021
Accepted: 27 February 2021
Published: 5 March 2021

Publisher's Note: MDPI stays neutral with regard to jurisdictional claims in published maps and institutional affiliations.

Simple Summary: Little information is currently available on the epidemiology of parasitic and commensal protist species in captive non-human primates (NHP) and their zoonotic potential. This study investigates the occurrence, molecular diversity, and potential transmission dynamics of parasitic and commensal protist species in a zoological garden in southern Spain. The prevalence and genotypes of the main enteric protist species were investigated in faecal samples from NHP, zookeepers and free-living rats by molecular (PCR and sequencing) methods. A high prevalence of the diarrhoea-causing protists *Giardia duodenalis* and *Blastocystis* sp. (but not *Cryptosporidium* spp.) was observed in captive NHP at the Córdoba Zoo Conservation Centre. NHP can harbour zoonotic genotypes of *G. duodenalis*, *Blastocystis* sp., and *Enterocytozoon bieneusi*. Indeed, strong evidence of the occurrence of *Blastocystis* zoonotic transmission between NHP and their handlers was provided, despite the use of personal protective equipment and the implementation of strict health and safety protocols. Free-living sympatric rats are infected by host-specific species/genotypes of the investigated protists and seem to play a limited role as source of infections to NHP or humans in this setting. The extent of these findings should be confirmed in similar epidemiological surveys targeting other captive NHP populations.

Abstract: Little information is currently available on the epidemiology of parasitic and commensal protist species in captive non-human primates (NHP) and their zoonotic potential. This study investigates the occurrence, molecular diversity, and potential transmission dynamics of parasitic and commensal protist species in a zoological garden in southern Spain. The prevalence and genotypes of the main enteric protist species were investigated in faecal samples from NHP ($n = 51$), zookeepers ($n = 19$) and free-living rats ($n = 64$) by molecular (PCR and sequencing) methods between 2018 and 2019. The presence of *Leishmania* spp. was also investigated in tissues from sympatric rats using PCR. *Blastocystis* sp. (45.1%), *Entamoeba dispar* (27.5%), *Giardia duodenalis* (21.6%), *Balantioides coli* (3.9%), and *Enterocytozoon bieneusi* (2.0%) (but not *Troglodytella* spp.) were detected in NHP. *Giardia duodenalis*

(10.5%) and *Blastocystis* sp. (10.5%) were identified in zookeepers, while *Cryptosporidium* spp. (45.3%), *G. duodenalis* (14.1%), and *Blastocystis* sp. (6.25%) (but not *Leishmania* spp.) were detected in rats. *Blastocystis* ST1, ST3, and ST8 and *G. duodenalis* sub-assemblage AII were identified in NHP, and *Blastocystis* ST1 in zookeepers. *Giardia duodenalis* isolates failed to be genotyped in human samples. In rats, four *Cryptosporidium* (*C. muris*, *C. ratti*, and rat genotypes IV and V), one *G. duodenalis* (assemblage G), and three *Blastocystis* (ST4) genetic variants were detected. Our results indicate high exposure of NHP to zoonotic protist species. Zoonotic transmission of *Blastocysts* ST1 was highly suspected between captive NHP and zookeepers.

Keywords: *Cryptosporidium; Giardia; Blastocystis; Enterocytozoon bieneusi; Balantioides coli; Troglodytella;* non-human primates; rats; zoological garden

1. Introduction

Cryptosporidium spp., *Giardia duodenalis*, and *Entamoeba histolytica* are the most frequently identified protozoan parasites causing diarrhoeal disease in humans globally [1]. Clinical manifestations by these infections vary from self-limiting acute diarrhoea in immunocompetent individuals, to fatal chronic diarrhoea in immunocompromised patients [2]. In addition to these well-known enteric pathogens, other potential diarrhoea-causing protist species, including the Stramenopile *Blastocystis* sp. and the Microsporidia *Enterocytozoon bieneusi*, have gained wide clinical and scientific interest in recent years [3,4]. These parasites are transmitted via the faecal-oral route either directly (i.e., person-to-person) or indirectly (i.e., waterborne or foodborne). Remarkably, most of the species/ genotypes of the above-mentioned protists can be zoonotically transmitted [5–8]. For this reason, assessing the occurrence and genetic diversity of enteric protists in domestic, captive, and free-living animal hosts is essential to ascertaining their transmission dynamics, including the occurrence and directionality of zoonotic events.

Cryptosporidium spp., *G. duodenalis*, *Blastocystis* sp., and *E. bieneusi* exhibit extensive intra-species genetic diversity leading to the identification of several genotypes/subtypes with marked differences in host and geographical range. At least 40 valid *Cryptosporidium* species and a similar number of genotypes of unknown species status are currently recognized, with *C. hominis* and *C. parvum* causing most of the infections documented in humans and non-human primate (NHP) species [6,9]. *Giardia duodenalis* is currently regarded as a multi-species complex comprising eight (A to H) distinct assemblages, of which assemblages A and B are frequently reported in humans and NHP [5]. At least 28 subtypes (ST) have been proposed within *Blastocystis* sp. with apparent loose host specificity. Of them, ST1–9 and ST12 have been documented in humans and/or NHP, among other vertebrates [8,10,11]. A recent evaluation of ST1–ST26 subtypes concluded that only 22 of those subtypes (ST1–ST17, ST21, ST23–ST26) should be acknowledged as legitimate subtypes [10], with the remaining six pending confirmation in future investigations. Finally, nearly 500 *E. bieneusi* genotypes have been reported and distributed in 11 genetic groups, of which Group 1 (e.g., Type IV, D, and EbpC) and Group 2 (e.g., BEB4, BEB6, I, and J) include most of the potentially zoonotic genotypes [12].

Little is known about the epidemiology of gastrointestinal protist parasites in captive non-human primates (NHP). In Spain, most of the few studies published to date were based on conventional (microscopy) methods and conducted mainly at the zoological gardens of Almuñecar (Granada) and Barcelona [13–17]. Only a single study attempted to characterize the genetic diversity of *G. duodenalis* in NHP at the Madrid and Valencia zoological gardens [18]. Besides *Cryptosporidium* spp., *G. duodenalis*, *Blastocystis* sp., and *E. bieneusi*, ciliated protists in NHP have been even poorly studied. This is the case of *Balantioides coli*, a zoonotic parasite that primarily infect domestic and wild swine, but has also been reported in NHP including gorillas, chimpanzees, bonobos, hamadryas baboons, and Rhesus macaques [19]. Moreover, the commensal ciliate *Troglodytella abrassarti* has been demonstrated a common finding in the faeces of captive and free-living great apes including eastern and western gorillas, chimpanzees, bonobos, and orangutans [20], but little information is available on its occurrence in captive and free-living lesser apes and monkeys.

This molecular-based epidemiological study aims primarily at assessing the frequency and genetic diversity of generalist and host-adapted enteric protist species in captive NHP and their caretakers at the Córdoba Zoo Conservation Centre (CZCC) in southern Spain, with a special focus on the investigation of potential zoonotic transmission events and their directionality. Secondarily, the same survey was conducted in a free-living sympatric rat population in the same enclosure in order to (i) investigate the role of rodents as transmitters of protist infections to NHP and humans and (ii) assess the suitability of rats as natural reservoirs of *Leishmania* spp.

2. Materials and Methods

2.1. Study Area

The CZCC extends over 4.5 hectares and include 437 specimens belonging to 102 mammalian, reptilian, and avian species. The CZCC has a small but diverse population of NHP species belonging to 10 genera including *Cebuella* ($n = 2$), *Cercocebus* ($n = 4$), *Cercopithecus* ($n = 3$), *Eulemur* ($n = 2$), *Hylobates* ($n = 3$), *Lemur* ($n = 5$), *Macaca* ($n = 8$), *Mandrillus* ($n = 4$), *Saimiri* ($n = 3$), and *Varecia* ($n = 2$). Individuals of the same species are kept in specific enclosures without contact with other NHP, except members of the Lemuridae family (genera *Eulemur*, *Lemur*, and *Varecia*) that share the same enclosure. All NHP are housed in facilities littered with natural materials such as ground bark or earth. The CZCC has strict health and safety protocols in place to ensure that animals, employees, and visitors have a reduced exposure to risk of infection or injury. Employees routinely use appropriate Personal Protective Equipment when in contact with animals or their faecal material.

2.2. Sampling

This cross-sectional study included two sampling periods carried out between December 2018 and January 2019, and between November and December 2019. Fresh faecal samples from NHP were directly collected from the ground at the time of routine cleaning and sanitation of enclosures. Information regarding sex, age, and enclosure sharing with other NHP species was recorded at the time of sampling. In parallel, fresh stool samples were also collected from zookeepers and veterinarians in close contact with NHP that volunteered to participate in the study. Human and NHP stool samples were stored at $-20\ ^\circ$C without preservatives at the CZCC Veterinary Laboratory until the end of each sampling campaign, when they were shipped to the Spanish National Centre for Microbiology for downstream molecular analyses.

Taking advantage of an ongoing rodent control campaign carried out at the same time as the present study within the CZCC premises undertaken by the local authorities following European guidelines [21], free-living sympatric rats (*Rattus* spp.) were captured using Tomahawk live traps with bait (chicken, dry dog food, fruit, or peanut butter) (Figure 1A). Most captured rats were identified as brown rats (*Rattus norvegicus*), but differential detection with black rats (*Rattus rattus*) was not possible for younger individuals. Traps were placed in the evening and checked for captures the next morning. Rats were anaesthetized with medetomidine (1 mg/kg) and ketamine (50 mg/kg) and then humanely euthanized by an intracardiac injection of sodium pentobarbitone (Dolethal®, Vetoquinol Laboratories, Lure, France) at a dose >150 mg/kg (Figure 1B) [22]. Carcases were frozen at $-20\ ^\circ$C and shipped to the Spanish National Centre for Microbiology for necropsy (Figure 1C). After thawing and dissection, the small and large intestine was removed, and the intestinal content extracted for further investigation of enteric protists by molecular methods. Additionally, liver, spleen, and ear skin samples were taken to assess the presence of amastigote forms of *Leishmania* spp.

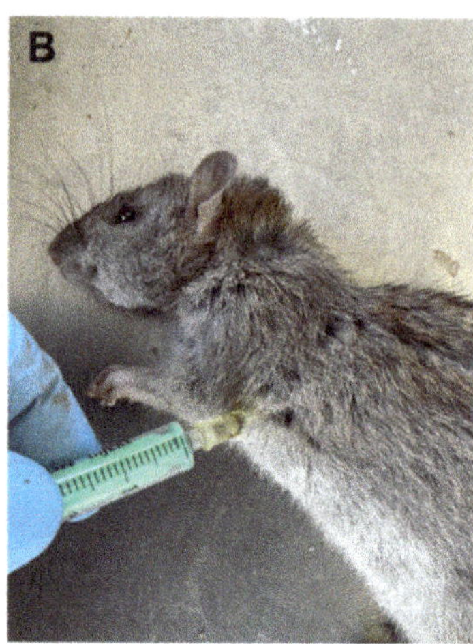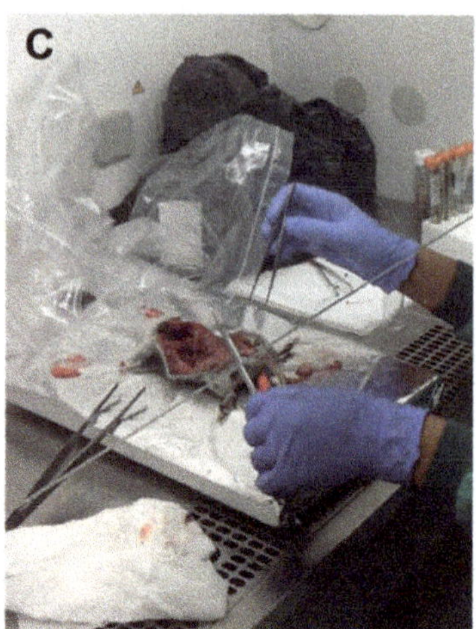

Figure 1. Sampling of rodent specimens within the premises of the Córdoba Zoo Conservation Centre. (**A**): Capture using live traps; (**B**): Humanely killing by intracardiac injection of sodium pentobarbitone; (**C**): Dissection of rat carcasses and organ removal.

2.3. Epidemiological Questionnaire

A standardized questionnaire (Table S1) and an informed consent was provided as part of the sampling kit to be completed by the CZCC personnel that volunteered to participate in the survey. Questions included: (i) demographic characteristics, e.g., age and sex, (ii) behavioural habits, e.g., hand and fruit/vegetable washing and whether there have been any occurrence of diarrhoea in the participant, their family members, and/or pets, (iii) work-related potential risk factors, e.g., contact with faecal material from NHP and/or other animal species, being a food handler, and (iv) additional questions on other risk factors, e.g., types of drinking water, use of recreational waters in the two weeks prior to sample collection, had any contact with pets and any recent travel abroad.

2.4. DNA Extraction and Purification

Genomic DNA was isolated from about 200 mg of each faecal specimen of human, NHP, or rodent origin by using the QIAamp DNA Stool Mini Kit (Qiagen, Hilden, Germany) according to the manufacturer's instructions, except that samples mixed with InhibitEX buffer were incubated for 10 min at 95 °C. Additionally, genomic DNA from murine tissues (liver, spleen, and ear skin) was isolated using the Speed Tools DNA Extraction Kit (Biotools, Madrid, Spain). To do so, 10–15 mg of each tissue was homogenized in 100 μl of NET-10 buffer (10 mM NaCl, 10 mM EDTA, 10 mM Tris-HCl pH 8) and digested overnight at 56 °C with 100 μL of BT1 buffer (Biotools) and 20 μL Proteinase K (20 mg/mL). After digestion, genomic DNA was extracted according to the manufacturer's instructions. In all cases, extracted and purified DNA samples were eluted in 200 μL of PCR-grade water and kept at 4 °C until further molecular analysis. A water extraction control was included in each sample batch processed.

2.5. Molecular Detection and Characterization of Giardia Duodenalis

Detection of *G. duodenalis* DNA was achieved using a real-time PCR (qPCR) method targeting a 62-bp region of the gene codifying the small subunit ribosomal RNA (*ssu* rRNA) of the parasite [23]. Amplification reactions (25 μL) consisted of 3 μL of template DNA, 0.5 μM of each primer Gd-80F and Gd-127R, 0.4 μM of probe (Table S2), and 12.5 μL TaqMan® Gene Expression Master Mix (Applied Biosystems, CA, USA). Detection of parasitic DNA was performed on a Corbett Rotor GeneTM 6000 real-time PCR system

(QIAGEN) using an amplification protocol consisting of an initial hold step of 2 min at 55 °C and 15 min at 95 °C followed by 45 cycles of 15 s at 95 °C and 1 min at 60 °C. Water (no-template) and genomic DNA (positive) controls were included in each PCR run.

Giardia duodenalis isolates that tested positive by qPCR were subsequently assessed by sequence-based multi-locus genotyping of the genes encoding for the glutamate dehydrogenase (*gdh*), β-giardin (*bg*), and triose phosphate (*tpi*) proteins of the parasite. A semi-nested PCR was used to amplify a 432-bp fragment of the *gdh* gene [24]. PCR reaction mixtures (25 µL) included 5 µL of template DNA and 0.5 µM of the primer pairs GDHeF/GDHiR in the primary reaction and GDHiF/GDHiR in the secondary reaction (Table S2). Both amplification protocols consisted of an initial denaturation step at 95 °C for 3 min, followed by 35 cycles of 95 °C for 30 s, 55 °C for 30 s, and 72 °C for 1 min, with a final extension of 72 °C for 7 min.

A nested PCR was used to amplify a 511 bp-fragment of the *bg* gene [25]. PCR reaction mixtures (25 µL) consisted of 3 µL of template DNA and 0.4 µM of the primers sets G7_F/G759_R in the primary reaction and G99_F/G609_R in the secondary reaction (Table S2). The primary PCR reaction was carried out with the following amplification conditions: one step of 95 °C for 7 min, followed by 35 cycles of 95 °C for 30 s, 65 °C for 30 s, and 72 °C for 1 min, with a final extension of 72 °C for 7 min. The conditions for the secondary PCR were identical to the primary PCR except that the annealing temperature was 55 °C.

A nested PCR was used to amplify a 530 bp-fragment of the *tpi* gene [26]. PCR reaction mixtures (50 µL) included 2-2.5 µL of template DNA and 0.2 µM of the primer pairs AL3543/AL3546 in the primary reaction and AL3544/AL3545 in the secondary reaction (Table S2). Both amplification protocols consisted of an initial denaturation step at 94 °C for 5 min, followed by 35 cycles of 94 °C for 45 s, 50 °C for 45 s, and 72 °C for 1 min, with a final extension of 72 °C for 10 min.

2.6. Molecular Detection and Characterization of Cryptosporidium spp.

The presence of *Cryptosporidium* spp. was assessed using a nested-PCR protocol to amplify a 587 bp fragment of the *ssu* rRNA gene of the parasite [27]. Amplification reactions (50 µL) included 3 µL of DNA sample and 0.3 µM of the primer pairs CR-P1/CR-P2 in the primary reaction and CR-P3/CPB-DIAGR in the secondary reaction (Table S2). Both PCR reactions were carried out as follows: one step of 94 °C for 3 min, followed by 35 cycles of 94 °C for 40 s, 50 °C for 40 s and 72 °C for 1 min, concluding with a final extension of 72 °C for 10 min.

2.7. Molecular Detection and Characterization of Blastocystis sp.

Identification of *Blastocystis* sp. was achieved by a direct PCR protocol targeting the *ssu* rRNA gene of the parasite [28]. The assay uses the pan-*Blastocystis*, barcode primer pair RD5/BhRDr to amplify a PCR product of ~600 bp. Amplification reactions (25 µL) included 5 µL of template DNA and 0.5 µM of each primer (Table S2). Amplification conditions consisted of one step of 95 °C for 3 min, followed by 30 cycles of 1 min each at 94, 59 and 72 °C, with an additional 2 min final extension at 72 °C.

In *Blastocystis*-positive samples from free-living sympatric rats for which Sanger sequencing data were of suboptimal quality, a next-generation amplicon sequencing strategy was used to identify *Blastocystis* sp. subtypes as previously described [29]. In brief, primers ILMN_Blast505_532F and ILMN_Blast998_1017R were used to generate amplicons. These primers amplify a fragment of the *ssu* rRNA gene of ~500 bp and are identical to Blast505_532F/Blast998_1017R [30] with the exception of containing the Illumina overhang adapter sequences on the 5′ end. Amplicons from two rats were used to prepare sequencing libraries, and final libraries were quantified by Qubit fluorometric quantitation (Invitrogen, Carlsbad, CA, USA) prior to normalization. A final pooled library concentration of 8 pM with 20% PhiX control was sequenced using Illumina MiSeq 600 cycle v3 chemistry (Illumina, San Diego, CA, USA). Paired end reads were processed and analyzed with an in-house pipeline that uses the BBTools package v38.82 [31], VSEARCH v2.15.1 [32], and BLAST + 2.10.1. Briefly, read pairs were merged, filtered for quality and length, denoised, and checked for chimeric sequences. Clustering and the assignment of centroid sequences to operational taxonomic units (OTUs) was performed within each sample at a 98% identity threshold. Only those OTUs with a minimum of 100 sequences were retained and then checked for chimeras once more. OTUs were then blasted against *Blastocystis* references from the National Center for Biotechnology Information (NCBI). Hits below an alignment length of 400 bp were removed.

2.8. Molecular Detection and Characterization of Enterocytozoon bieneusi

Detection of *E. bieneusi* was conducted by a nested PCR protocol to amplify the internal transcribed spacer (ITS) region as well as portions of the flanking large and small subunit of the ribosomal RNA gene as previously described [33]. The outer EBITS3/EBTIS4 and inner EBITS1/EBITS2.4 primer sets (Table S2) were used to generate a PCR product of 390 bp, respectively. Cycling conditions for the primary PCR consisted of one step of 94 °C for 3 min, followed by 35 cycles of amplification (denaturation at 94 °C for 30 s, annealing at 57 °C for 30 s, and elongation at 72 °C for 40 s), with a final extension at 72 °C for 10 min. Conditions for the secondary PCR were identical to the primary PCR except only 30 cycles were carried out with an annealing temperature of 55 °C.

2.9. Molecular Differential Detection of Entamoeba histolytica and Entamoeba dispar

Detection and differential diagnosis between pathogenic *E. histolytica* and non-pathogenic *E. dispar* were carried out by a qPCR method targeting a 172-bp fragment of the gene codifying the *ssu* rRNA gene of the *E. histolytica*/*E. dispar* complex [34,35]. Amplification reactions (25 μL) consisted of 3 μL template DNA, 12.5 pmol of the primer set Ehd-239F/Ehd-88R, 5 pmol of each TaqMan® probe (Table S2), and TaqMan® Gene Expression Master Mix (Applied Biosystems). Cycling conditions and data analysis were as described above for the detection of *G. duodenalis*.

2.10. Molecular Detection of Balantioides coli

Detection of *B. coli* was attempted by a direct PCR assay to amplify the complete ITS1–5.8s-rRNA–ITS2 region and the last 117 bp (3′ end) of the *ssu*-rRNA sequence of this ciliate using the primer set B5D/B5RC [36]. PCR reactions (25 μL) consisted of 2 μL of template DNA and 0.4 μM of each primer (Table S2). PCR conditions were as follows: 94 °C for 10 min; 30 cycles of 94 °C for 1 min, 60 °C for 1 min, 72 °C for 1 min, and a final extension for 5 min at 72 °C.

2.11. Molecular Detection of Troglodytella spp.

Detection of *Troglodytella* spp. was only attempted in captive NHP. Identification of this ciliate mutualist was carried out by a direct PCR method targeting a 401 bp fragment of the ITS region of the rDNA (ITS1-5.8S rDNA-ITS2) of the protist [37]. PCR reactions (25 μl) contained 2 μL of template DNA and 0.8 μM of each primer SSU-end/LSU-start (Table S2). Conditions of PCR for ITS amplification were initial denaturation for 2 min at 94 °C, 35 cycles of 45 s at 94 °C, 45 s at 50 °C, and 90 s at 72 °C, and terminal elongation for 5 min at 72 °C.

2.12. Molecular Detection of Leishmania spp.

Detection of *Leishmania* spp. was solely attempted in rodents, the only mammalian host for which tissue samples were available. Identification of this kinetoplastida parasite was carried out by a nested PCR protocol to amplify a partial fragment (358 bp) of the *ssu* rRNA gene of the parasite [38]. The primary PCR reaction (50 μL) contained 10 μL of template DNA and 15 pmol of the primer pair R221/R332 (Table S2). Conditions of PCR for *ssu* rRNA amplification were initial denaturation for 5 min at 94 °C, 35 cycles of 30 s at 94 °C, 30 s at 60 °C, and 30 s at 72 °C, and terminal elongation for 10 min at 72 °C. In the secondary PCR reaction (25 µL), 10 µL of a 1:40 dilution of the primary PCR product was re-amplified using 7.5 pmol of the primer pair R223/R333 (Table S2). Cycling conditions were as described above except that the annealing temperature was set at 65 °C.

All the direct, semi-nested, and nested PCR protocols described above were conducted on a 2720 Thermal Cycler (Applied Biosystems). Reaction mixes always included 2.5 units of MyTAQ^{TM} DNA polymerase (Bioline GmbH, Luckenwalde, Germany), and 5× MyTAQ^{TM} Reaction Buffer containing 5 mM dNTPs and 15 mM MgCl$_2$, except for the amplification of *Leishmania* spp., for which 0.7–1.4 units of Tth DNA polymerase (Biotools B&M Laboratories, S.A., Madrid, Spain) were used. Laboratory-confirmed positive and negative DNA samples of human and animal origin for each parasitic species investigated were routinely used as controls and included in each round of PCR. PCR amplicons were visualized on 1.5–2% D5 agarose gels (Conda, Madrid, Spain) stained with Pronasafe (Conda) or Gel Red (Biotium, Fremont, CA, USA) nucleic acid staining solutions. A 100 bp DNA ladder (Boehringer Mannheim GmbH, Baden-Wurttemberg, Germany) was used for the sizing of obtained amplicons. Positive-PCR products were directly sequenced in both directions using appropriate internal primer sets (Table S2). DNA sequencing was conducted by capillary electrophoresis using the BigDye® Terminator chemistry (Applied Biosystems) on an on ABI PRISM 3130 automated DNA sequencer.

The sequences obtained in this study have been deposited in GenBank under accession numbers MW417420–MW417422 (*G. duodenalis*), MW414634–MW414644 and MW581486 (*Blastocystis* sp.), MW406908–MW406921 (*Cryptosporidium* spp.) and MW414645 (*E. bieneusi*).

2.13. Statistical Analysis

Prevalence and 95% confidence intervals (95% CI) of any enteric protist infection/carriage, alone or in combination, in the study populations were calculated. Statistically significant differences between the prevalence of enteric protist species in NHP and sampling period were analyzed using the Pearson's chi-square or Fisher's exact test with crude odds ratios (OR) and 95% CI, as appropriate. A *p* value < 0.05 was considered evidence of statistical significance. Data were analyzed using R open-source software. Because of the relatively small sample size, limited number of positives obtained from human stool samples, and associated low statistical power, no attempts were conducted to investigate potential correlations between the occurrence of the detected protist species and the risk factors covered in the epidemiological questionnaire provided to volunteer zookeepers.

3. Results

3.1. Prevalence and Molecular Characterization of Enteroparasites in Captive Non-Human Primates

A total of 51 faecal samples from 10 different species of NHP hosted at the CZCC were collected during the period of study, 28 in the first sampling period and 23 in the second sampling period (Table 1). Members of all 10 NHP species were represented in the two sampling periods. All collected samples could be assigned to individual NHP, except those from the Lemuridae family sharing the same enclosure. Five protist species were detected, including *Blastocystis* sp. (45.1%, 23/51; 95% CI: 31.1–59.7), *E. dispar* (27.5%, 14/51; 95% CI: 15.9–41.7), *G. duodenalis* (21.6%, 11/51; 95% CI: 11.3–35.3), *B. coli* (3.9%, 2/51; 95% CI: 0.5–13.5), and *E. bieneusi* (2.0%, 1/51; 95% CI: 0.05–10.5). In contrast, *Cryptosporidium* spp., *E. histolytica*, and *Troglodytella* spp. were not detected in any of the NHP faecal samples analyzed (Table 1). *Blastocystis* sp. (39.1–50.0%), *E. dispar* (14.3–43.5%), *G. duodenalis* (17.9–26.1%), and *B. coli* (3.6–4.3%) were detected in both sampling campaigns, whereas the only sample that tested positive for *E. bieneusi* was obtained in the second sampling campaign. *Entamoeba dispar* was significantly more prevalent in the second sampling campaign than in the first sampling campaign (χ^2 = 5.4034, *p* = 0.0201).

Table 1. Frequency of enteric protists detected at each sampling campaign in faecal samples from captive non-human primates in the Córdoba Zoo Conservation Centre (Spain).

| | First Sampling Campaign | | | | | | Second Sampling Campaign | | | | | | All | | | | |
| | Frequency Positive Results (%) | | | | | | Frequency Positive Results (%) | | | | | | Frequency Positive Results (%) | | | | |
Species	No.	Bl	Ed	Gd	Bc	Eb	No.	Bl	Ed	Gd	Bc	Eb	No.	Bl	Ed	Gd	Bc	Eb
Cebuella pygmaea	1	0.0	0.0	0.0	0.0	0.0	2	0.0	0.0	50.0	0.0	0.0	3	0.0	0.0	33.3	0.0	0.0
Cercocebus lunulatus	3	100	0.0	100	0.0	0.0	3	33.3	100	66.7	0.0	0.0	6	66.7	50.0	83.3	0.0	0.0
Cercopithecus neglectus	2	50.0	0.0	50.0	0.0	0.0	3	66.7	66.7	0.0	0.0	0.0	5	60.0	40.0	20.0	0.0	0.0
Eulemur fulvus	2	0.0	0.0	0.0	0.0	0.0	0	0.0	0.0	0.0	0.0	0.0	2	0.0	0.0	0.0	0.0	0.0
Hylobates leucogenys Ogilby	4	75.0	100	25.0	0.0	0.0	3	33.3	33.3	0.0	0.0	33.3	7	57.1	71.4	14.3	0.0	14.3
Lemur catta	2	100	0.0	0.0	0.0	0.0	0	0.0	0.0	0.0	0.0	0.0	2	100	0.0	0.0	0.0	0.0
Macaca sylvanus	5	40.0	0.0	0.0	0.0	0.0	3	33.3	100	100	0.0	0.0	8	37.5	37.5	37.5	0.0	0.0
Mandrillus leucophaeus	5	40.0	0.0	0.0	20.0	0.0	3	33.3	33.3	0.0	33.3	0.0	8	37.5	12.5	0.0	25.0	0.0
Saimiri sciureus	2	0.0	0.0	0.0	0.0	0.0	3	0.0	0.0	0.0	0.0	0.0	5	0.0	0.0	0.0	0.0	0.0
Varecia variegata variegata	2	50.0	0.0	0.0	0.0	0.0	3	100	0.0	0.0	0.0	0.0	5	80.0	0.0	0.0	0.0	0.0
Total	28	50.0	14.3	17.9	3.6	0.0	23	39.1	43.5	26.1	4.3	4.3	51	45.1	27.5	21.6	3.9	2.0

Bc: *Balantioides coli*; Bl: *Blastocystis* sp., Eb: *Enterocytozoon bieneusi*; Ed: *Entamoeba dispar*; Gd: *Giardia duodenalis*.

Table 2 summarizes the occurrence of the enteric protist species detected in the present survey as single or multiple infections ($n = 32$). Multiple infections with two protist species were found in 14 samples (43.8%, 14/32) of which five (15.6%, 5/32) were co-infected with *Blastocystis* sp. and *E. dispar*, four (12.5%, 4/32) with *Blastocystis* sp. and *G. duodenalis* and three (9.4%, 3/32) with *G. duodenalis* and *E. dispar*. Three samples (9.4%, 3/32) were co-infected with *G. duodenalis*, *Blastocystis* sp., and *E. dispar*. No associations were demonstrated between *Blastocystis* sp. and *E. dispar* ($p = 0.630$; OR = 1.3, 95% CI: 0.41–4.37) or between *Blastocystis* sp. and *G. duodenalis* ($p = 0.190$; OR = 2.6, 95% CI: 0.54–14.1).

Table 2. Single and multiple enteric protist infections detected in faecal samples from captive non-human primates in the Córdoba Zoo Conservation Centre (Spain).

Species Combination	No. of Faecal Samples
Blastocystis sp. Only	10
E. dispar only	4
G. duodenalis only	1
Blastocystis sp. + *E. dispar*	5
Blastocystis sp. + *G. duodenalis*	4
G. duodenalis + *E. dispar*	3
Blastocystis sp. + *B. coli*	1
E. dispar + *E. bieneusi*	1
G. duodenalis + *Blastocystis* sp. + *E. dispar*	3
Total	32

Giardia duodenalis-positive results by qPCR ($n = 11$) generated cycle threshold (Ct) values ranging from 30.0 to 37.4 (median: 32.4; standard deviation: 2.4). Only a single sample could be genotyped at the *bg* locus, being identified as sub-assemblage AII (Table 3). Sequence alignment analysis revealed that this sequence was identical to its corresponding reference sequence (GenBank accession number: L40510). Out of the 23 *Blastocystis*-positive samples at the *ssu* rDNA (barcode region), the gene of the parasite confirmed by Sanger sequencing revealed the presence of three *Blastocystis* subtypes (STs), including zoonotic ST1 (39.1%, 9/23), ST3 (34.8%, 8/23), and ST8 (26.1%, 6/23) (Table 3). Additionally, 11 samples yielded amplicons of the expected size but in the form of faint bands on gel electrophoresis. Because their associated Sanger sequences were of poor quality (unreadable), these samples were conservatively considered as negative for *Blastocystis* sp. Neither mixed infection involving different STs of the parasite nor infections caused by animal-specific (ST10–ST17, ST21, ST23–ST28) subtypes were identified. A moderate genetic diversity was observed within ST1 (alleles 1 and 2, alone or in combination), and ST3 (alleles 34, 32 + 34), but not within ST8, where all isolates were assigned to allele 21. *Balantioides coli* was unmistakably identified in two isolates, but sequence data of insufficient quality precluded the possibility of determining the genotype of this parasite species. Finally, sequence analysis of the only sample positive to *E. bieneusi* revealed the presence of genotype D with 100% identity with reference sequence AF101200 (Table 3).

Table 3. Diversity, frequency, and molecular features of *Giardia duodenalis*, *Blastocystis* sp., *Balantioides coli*, and *Enterocytozoon bieneusi* in faecal samples from captive non-human primates in the Córdoba Zoo Conservation Centre (Spain). GenBank accession numbers are provided.

Species	Genotype	Sub-Genotype	Host Species	No. of Isolates	Locus	Reference Sequence	Stretch	Single Nucleotide Polymorphisms	GenBank ID
Giardia duodenalis	A	AII	C.t.	1	bg	AY072723	205–539	None	MW417420
Blastocystis sp.	ST1	Allele 1	M.l., H.l., C.t.	4	*ssu* rRNA	MK357786	4–602	None	MW414634
		Allele 2	C.t.	1	*ssu* rRNA	MT094302	36–539	None	MW414635
		Allele 2	C.n.	1	*ssu* rRNA	MT094302	32–539	C57A, 65InsG, A112G, C128A, C237T, C272T, A458C	MW414636
		Alleles 1 + 2	C.t., M.l.	3	*ssu* rRNA	MK357786	1–603	G128R, A474W	MW414637
	ST3	Allele 34	H.l., M.c., C.n.	7	*ssu* rRNA	MK801359	1–581	G114A, A115T, A116G, A159G, T160A, A161T	MW414638
		Alleles 32 + 34	C.n.	1	*ssu* rRNA	MK801359	1–586	G114A, A115T, A116G, A159K, T160R, A161K, A162R	MW414639
	ST8	Allele 21	V.v.v., L.c.	6	*ssu* rRNA	MT509451	1–525	None	MW414640
Balantioides coli	Unknown	-	M.l.	2	ITS	-	-	-	-
Enterocytozoon bieneusi	D	-	H.l.	1	ITS	AF101200	31–419	None	MW414645

bg: β-giardin; *C.n.*: *Cercopithecus neglectus*; *C.t.*: *Cercocebus torquatus*; *H.l.*: *Hylobates leucogenys*; ITS: Internal transcribed spacer; *L.c.*: *Lemur catta*; *M.c.*: *Macaca sylvanus*; *M.l.*: *Mandrillus leucophaeus*; *ssu* rRNA: Small subunit ribosomal RNA; *V.v.v.*: *Varecia variegata variegata*.

3.2. Prevalence and Molecular Characterization of Enteroparasites in Humans

A total of 19 members of the CZCC personnel, including zookeepers and veterinarians, participated in the study, 15 of them in the first sampling campaign and 11 in the second sampling campaign. Seven zookeepers participated in both sampling campaigns. The male/female ratio was 3.8, and the age range was 21 to 58 years (median: 49 years). Three individuals tested positive for at least one enteroparasite. Two enteric protist species were identified including *G. duodenalis* (10.5%, 2/19; 95% CI: 1.3–33.1) and *Blastocystis* sp. (10.5%, 2/19; 95% CI: 1.3–33.1). *Giardia duodenalis* was detected by qPCR (Ct values: 30.8 and 31.0) in a 58-year-old male and a 49-year-old female, respectively, participating in the second sampling campaign. Both samples failed to be amplified at the *gdh*, *bg*, and *tpi* loci, so the assemblages/sub-assemblages causing the infections were unknown. *Blastocystis* ST3 allele 34 (GenBank accession number: MW414642) was detected in a 56-year-old male participating in the first sampling campaign, whereas ST1 (GenBank accession number: MW414641) was identified in the same 49-year-old female co-infected with *G. duodenalis*. Sequence analysis of the later isolate revealed two clear double peaks (R and W) at positions 128 and 264, respectively, of reference sequence MK357786, compatible with mixed infections involving alleles 1, 2, 5 and/or 141. The variables potentially associated with *G. duodenalis* infections or *Blastocystis* sp. carriage are summarized in Table 4. The three individuals harbouring *G. duodenalis* and/or *Blastocystis* sp. declared no gastrointestinal symptoms at the moment of sampling. All three were food handlers and were regularly in contact with faecal material from NHP and other captive animal species at the CZCC. Other enteric protist species, including *Cryptosporidium* spp., *E. histolytica*, *E. dispar*, *E. bieneusi*, and *B. coli*, were apparently absent in the surveyed human population.

Table 4. Variables potentially associated to *G. duodenalis* infection and *Blastocystis* sp. carriage in staff at the Córdoba Zoo Conservation Centre (Spain).

Variable	Subject 38	Subject 79	Subject 86
Sociodemographic factors			
Sex	Male	Male	Female
Age (years)	56	58	49
Protist infection/carriage			
Giardia duodenalis	Negative	Positive	Positive
Blastocystis sp.	Positive	Negative	Positive
Clinical factors			
Diarrhoea in the last 7 days	No	No	No
Contact with children <5-years	No	No	No
Diarrhoea in family members/relatives	Yes	No	No
Work-related factors			
Activity	Veterinarian	Zookeeper	Zookeeper
Exposure to faeces from NHP	Yes	Yes	Yes
Exposure to faeces from animals other than NHP	Yes	Yes	Yes
Any of these animal species with diarrhoea	Yes	Yes	Yes
Food handler	Yes	Yes	Yes
Behavioural factors			
Recent travel	Yes	No	No
Contact with pet dogs	Yes	Yes	Yes
Contact with pet cats	Yes	No	Yes
Main drinking source—tap	Yes	Yes	Yes
Main drinking source—bottled	No	No	No
Swimming	No	No	No
Handwashing	Frequently	Always	Always
Vegetable washing	Always	Always	Always

3.3. Prevalence and Molecular Characterization of Enteroparasites and Leishmania spp. in Rats

A total of 64 faecal samples of free-living sympatric rats captured within the premises of the zoological garden were available for this study. Three enteric protist species were detected—*Cryptosporidium* spp. (45.3%, 29/64; 95% CI: 32.8–58.2), *G. duodenalis* (14.1%, 9/64; 95% CI: 6.6–25.0), and *Blastocystis* sp. (6.25%, 4/64; 95% CI: 0.4–10.8). None of the samples tested positive for *E. bieneusi* or *B. coli*. All the spleen, liver, and skin samples analyzed tested negative for *Leishmania* spp.

Sequence analyses of the murine *Cryptosporidium*-positive samples by *ssu*-PCR revealed the presence of *C. muris* (10.3%, 3/29), *C. ratti* (17.2%, 5/29,), rat genotype IV (69.0%, 20/29), and rat genotype V (3.5%, 1/29) (Table 5). All *C. muris* and *C. ratti* showed 100% identity with reference sequences AB089284 and MT504541, respectively. Conversely, a high genetic diversity was found within sequences belonging to rat genotype IV, with only four of them being identical to reference sequence JN172970. The remaining 16 sequences varied from JN172970 by 1–5 single nucleotide polymorphisms (SNPs), including a variety of mutations, insertions, deletions, and ambiguous (double peak) positions, the combination 448DelT + G493A being the most frequently detected (Table 5). The only sequence identified as rat genotype V varied from reference sequence MT504543 by a single (A667G) SNP.

Table 5. Diversity, frequency, and molecular features of *Cryptosporidium* spp. sequences at the *ssu* rRNA locus obtained in faecal samples from the rat population under study in Córdoba Zoo Conservation Centre (Spain). GenBank accession numbers are provided.

Species/Genotype	No. of Isolates	Reference Sequence	Stretch	Single Nucleotide Polymorphisms	GenBank ID
C. muris	3	AB089284	504–1012	None	MW406908
C. ratti [a]	6	MT504541	293–751	None	MW406909
Rat genotype IV	3	JN172970	377–775	None	MW406910
	1	JN172970	332–798	C342T, C410T, C423T, 448DelT, G493A	MW406911
	1	JN172970	332–752	C342T, C410T, C423T, 490_491DelTT, G493A	MW406912
	1	JN172970	328–814	A428G, 448DelT, G493A	MW406913
	1	JN172970	328–767	A445G, 448DelT, G493A, T541A, T542A	MW406914
	1	JN172970	348–776	A445G, 448DelT, G493A, T541W, T542W	MW406915
	5	JN172970	375–814	448DelT, G493A	MW406916
	1	JN172970	454–678	A459T, A475T, G493A	MW406917
	1	JN172970	332–788	A472G, 492InsT, G493A	MW406918
	3	JN172970	341–814	490_491DelTT, G493A	MW406919
	1	JN172970	481–814	490_491DelTT, G493A, G635A	MW406920
Rat genotype V	1	MT504543	306–699	A667G	MW406921

[a] Formerly known as Rat genotype I. See reference [39].

Rodent *G. duodenalis*-positive samples by qPCR generated Ct values ranging from 24.1 to 36.3 (median: 31.8). Of these, 55.6% (5/9) produced Ct values higher than 30. Two *G. duodenalis*-positive samples (Ct values: 24.1 and 28.2, respectively) were genotyped as assemblage G at the *gdh* locus. Both sequences were identical between them and showed an SNP (C262T) compared to reference sequence MF671912. Additionally, one of the sequences was confirmed as assemblage G at the *bg* locus and showed 100% identity with reference sequence MF671912. None of the two isolates could be amplified at the *tpi* locus.

The two *Blastocystis*-positive samples by *ssu*-PCR and Sanger sequencing were identified as ST4 alleles 92 and 94 (GenBank accession numbers: MW414643 and MW414644), and their sequences were identical with reference sequence MF186667 and MN526920, respectively. As in the case of NHP, three additional samples yielded amplicons of the expected size using barcoding primers but with faint bands on gel electrophoresis that did not produce readable sequences. Those three samples were subjected to a PCR to amplify a different region of the *ssu* rRNA gene, and two were found positive and subjected to next-generation amplicon sequencing. Those two samples were identified as ST4 and showed 100% identity with the reference sequence U26177. Because of lack of confirmation by Sanger sequencing or failing to amplify with an additional primer set, one sample was conservatively considered as negative for *Blastocystis* sp.

3.4. Molecular-Based Evidence of Zoonotic Transmission

Within NHP, a mangabey (*C. lunulatus*) investigated during the first sampling campaign was found infected with zoonotic *G. duodenalis* assemblage AII. Two zookeepers participating in the second sampling campaign were also positive for *G. duodenalis*, but lack of genotyping data and different sampling intervals precluded the unambiguous demonstration of zoonotic transmission of *G. duodenalis* infection between NHP and their zookeepers (Figure 2).

Blastocystis ST1 alleles 1 and 2 (alone or in combination) were consistently detected in mangabeys (*C. lunulatus*), drills (*M. leucophaeus*), and gibbons (*H. leucogenys* Ogilby) along the whole study period, and in a De Brazza´s monkey (*C. neglectus*) during the second sampling campaign. All these NHP species were housed in close proximity to each other within the CZCC premises. These data strongly suggest that these *Blastocystis* genetic variants were well established in the NHP population hosted at the CZCC. Interestingly, a zookeeper participating in the second sampling of the survey carried a genetic variant of *Blastocystis* ST1 compatible with a mixed infection by alleles 1 + 2 (Figure 2). This zookeeper was a food handler and declared regular contact with the faecal material of all NHP in the

CZCC. This finding suggests that NHP were acting as a source of *Blastocystis* infection to the zookeepers responsible for their wellbeing.

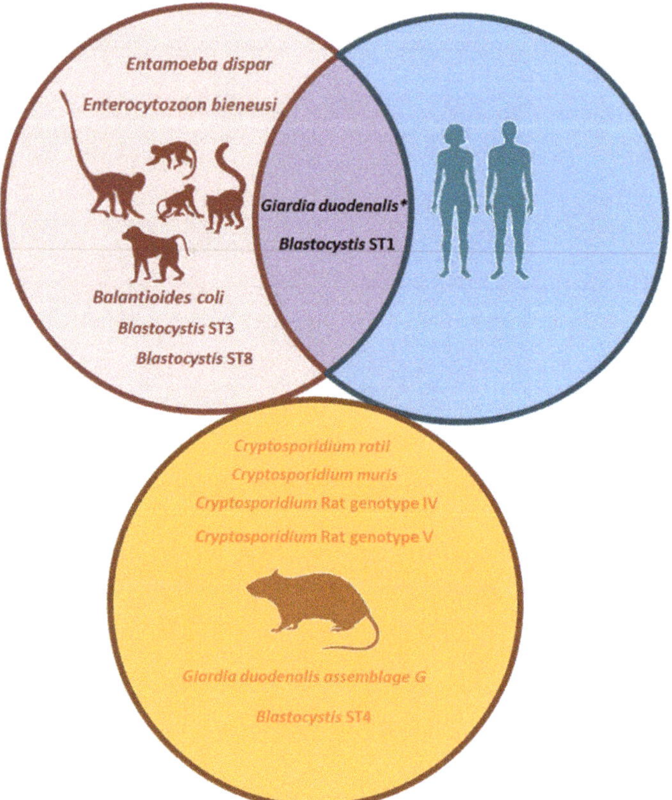

Figure 2. Molecular-based evidence of zoonotic transmission. Enteric protists detected at each species group (non-human primates (NHP), zookeepers and rats) in the Córdoba Zoo Conservation Centre (Spain). * Zoonotic *Giardia duodenalis* assemblage AII was detected in NHP; in addition, two zookeepers were positive for *G. duodenalis* of unknown assemblage, potentially making the assessment of zoonotic transmission for this protozoan parasite difficult.

Finally, the surveyed rat population was exclusively infected by rodent-specific species/genotypes of *Cryptosporidium*, *G. duodenalis*, and *Blastocystis* (Figure 2), indicating that this host species has a limited role as a source of potential infections for NHP and humans. Additionally, all tested rats were negative to zoonotic *Leishmania* spp.

4. Discussion

The epidemiology of pathogenic and commensal enteric protists in captive NHP is poorly understood. To fill this gap of knowledge, this survey provides new molecular-based data on the occurrence, transmission, genetic diversity, and zoonotic potential of the protist species that are most relevant from the public health point of view in NHP, their zookeepers/veterinarians, and free-living rats at the CZCC. In addition, the potential role of rodents as a natural reservoir of *Leishmania* spp. has been investigated.

Cryptosporidium spp. (particularly *C. hominis* and *C. parvum*) is, together with rotavirus, *Shigella*, and enterotoxigenic *Escherichia coli*, the major contributors to the global burden of diarrhoeal disease [40]. *Cryptosporidium* spp. is also a common diarrhoea-causing agent in livestock, companion species, and wildlife [6]. Interestingly, *Cryptosporidium* spp. was absent in the NHP population surveyed here and in the zookeepers/veterinarians that worked in their well care. This agrees with the previous findings observed in NHP ($n = 18$) from the Almuñecar zoological garden in southern Spain [17] but is in sharp contrast with those from the Barcelona zoological garden, where *Cryptosporidium* infections were consistently reported in 28–44% of the NHP investigated during a 10-year period [13,15,16]. In those studies, infected NHP were asymptomatic adults with intermittent (up to 10 months) shedding of oocysts irrespectively of the group or lone condition of the animals. This fact suggested that *Cryptosporidium* reinfection rather than continuous infection was taking part in that setting [17]. In contrast, *Cryptosporidium* spp. was found at a high prevalence rate (45%) in free-living rats captured within the CZCC enclosure. In the only two previous studies published in Spain, *Cryptosporidium* spp. infections have been reported in black rats from Catalonia (1/1) and the Canary Islands (14/101) [41,42]. Our sequence analyses revealed the presence of four distinct *Cryptosporidium* species/genotypes including rat genotype IV (69%), *C. ratti* (17%), *C. muris* (10%), and rat genotype V (4%). It should be noted that *C. ratti*, formerly known as rat genotype I, has been recently proposed as a valid *Cryptosporidium* species by Martin Kváč's laboratory [39]. Of these, *C. muris* and *C. ratti* (in addition to *C. meleagridis* and rat genotype II/III, not identified in the present survey) have been previously described in black rats from the Canary Islands [42]. Overall, our data indicate that rats captured at the CZCC were infected by murine-adapted *Cryptosporidium* species/genotypes and played a limited role as a source of cryptosporidiosis to NHP. Of interest, *C. muris* and rat genotype III have been sporadically reported in humans and/or companion animals, including dogs and cats [43–46].

In the present study, *G. duodenalis* infections were identified in 22% of NHP and 11% of zookeepers. Interestingly, all *Giardia*-positive cases by qPCR yielded Ct values >30, indicative of moderate-to-low parasite burdens. This agrees with the fact that all positive cases were asymptomatic and produced formed stools, also explaining the low genotyping success rate obtained (7.7%, 1/13). It should be noted that *gdh*, *bg*, and *tpi* are all single-copy genes with limited sensitivity compared with the multiple-copy *ssu* rRNA gene used in qPCR for detection purposes. An early epidemiological study detected the presence of *G. duodenalis* in 19.1% of NHP in the Barcelona zoological garden [14], but this parasite was absent in the NHP analyzed at the Almuñecar zoological garden [17]. Our sequence analyses identified the zoonotic sub-assemblage AII in a mangabey (*C. lunulatus*). In Spain, the *G. duodenalis* sub-assemblage AII has been found in 15–44% of documented clinical cases [47,48], and in 17–33% of children of paediatric age [49,50]. Although two of the CZCC zookeepers tested positive to this protozoan parasite, we were unable genotype these isolates, so their assemblage/sub-assemblage remained unknown and precluded us to propose a potential source of infection. Of interest, zoonotic AI and BIV have been previously identified in members of the Lemuridae family in the Valencia and Madrid zoological gardens [18].

Blastocystis infection/carriage was demonstrated in 45.1% of the NHP surveyed, a frequency rate considerably lower than those (67–95%) previously reported by conventional microscopy at the Barcelona and Almuñecar zoological gardens, respectively [14,17]. Sequence analyses of *Blastocystis* isolates revealed interesting data. ST1 was the most prevalent (39%) subtype circulating among captive NHP, being present in two genetic variants, allele 1 and allele 2, either alone or in combination. We have recently reported ST1 as the most common (82%) *Blastocystis* subtype in wild western chimpanzees (*Pan troglodytes verus*) in Senegal, although in that survey all the isolates characterized belonged to alleles 7 and 8 [51]. Moreover, we have also demonstrated that human cases of blastocystosis by ST1 in Spain are mainly due to allele 4 (and, to a much lesser extent, allele 77) both in asymptomatic [50,52] and clinical [53] individuals. The fact that one of the CZCC primate handlers carried a genetic variant of *Blastocystis* ST1 compatible with a mixed infection involving alleles 1, 2, 5 and/or 141 seems to indicate that this ST1 infection is most likely of primate origin and represents a zoonotic transmission event. Similarly, the vast majority of the ST3 isolates detected in NHP at the CZCC belonged to allele 32, a genetic variant not yet described in Spanish human populations [50,52,53]. This fact may indicate that ST3 allele 32 may be better adapted to infect NHP than humans. Finally, *Blastocystis* ST8 carriage was also a common finding (26%) in NHP at the CZCC. This result was highly expected as this *Blastocystis* subtype is well-known both in captive [54] and free-living [55] NHP globally. Although rarely reported in humans, the zoonotic potential of ST8 has been demonstrated in a zoological garden in the UK, where this subtype was responsible from one in four *Blastocystis* infections both in captive NHP and their handlers [56]. Also relevant was the finding of only identifying *Blastocystis* ST4 in rats. This finding supports that rodents appear to constitute the main animal reservoir of ST4 [56,57]. Additionally, the marked geographical distribution of ST4 in humans (commonly found in Europe but rarely or less frequently present in other geographical areas), together with its clonal structure strongly suggest that ST4 represents a lineage with a recent entry into the human population [58]. In Spain, all human cases carrying *Blastocystis* ST4 have been assigned to the allele 42 of the protist. As in the case of ST3 allele 32, this fact may indicate that ST4 alleles 92 and 94 (identified in rats in this study) may be particularly adapted to infect/colonize rodent species rather than humans.

Enterocytozoon bieneusi genotype D was detected in a single gibbon (*H. leucogenys* Ogilby). Genotype D has broad host and geographic ranges and belongs to Group 1 that includes zoonotic *E. bieneusi* genotypes most frequently found in humans, domestic and wild (including NHP) animal species worldwide [59,60]. In Spain, *E. bieneusi* genotype D has been described in renal transplant recipients [61], domestic rabbits [62] and cats [63], wild red foxes [62], and environmental (water) samples [64]. This result clearly indicates that NHP may act as suitable reservoirs for human microsporidiosis by *E. bieneusi*.

Regarding ciliate species, zoonotic *B. coli* was identified in a low (3.9%) proportion of the NHP investigated, but not in their handlers. At first sight, this result is much lower than those previously documented by microscopy examination in NHP at the Barcelona (38.1%) and Almuñecar (16.6%) zoological gardens [14,17]. However, when considering primate groups, *B. coli* has been found only in Catarrhini (*Mandrillus*, present results; *Cercocebus*, *Gorilla*, *Pan*, *Papio*, and *Pongo*) [14]. The data from Pérez-Cordón et al. [17] are aggregated and it is not possible to identify the positive primate genera. Our negative results in Strepsirrhini (*Eulemur*, *Lemur*, and *Varecia*) and Plathyrrhini (*Saimiri*) primates are in accordance with previous data [19] and suggest that these primates are uncommon or not valid hosts for this ciliate. It should be noted that, because *B. coli* cysts are morphologically indistinguishable from other ciliate species (i.e., *Buxtonella* spp.), it is possible that some of the microscopy-based prevalence rates described above do indeed represent an overestimation of the true occurrence of the parasite. While *B. coli*-like cysts are easily identifiable by microscopy, differential diagnosis based on molecular (PCR and Sanger sequencing) methods should be used for the correct identification of *B. coli*.

Finally, the absence of *T. abrassarti* in the investigated species is in accordance with previous surveys [65]. This ciliate is commonly reported in wild great apes [20] but there is no conclusive evidence in lesser apes and monkeys (e.g., red colobus, red-tailed monkeys, vervet monkeys, and yellow baboons). In captive chimpanzees, prevalence and infection intensities are influenced by the dietary starch concentration, suggesting a symbiotic function and participation in nourishment degradation [66].

Vector-borne *Leishmania infantum*, the causative agent of visceral and cutaneous leishmaniasis in Spain, is one of the most important neglected zoonosis in the Mediterranean region. In Spain, and in addition to domestic dogs, leporids such as rabbits and hares have been demonstrated as competent reservoirs of the infection [67]. Micromammals (e.g., mice, shrews) seem to play a limited role in the epidemiology of the parasite [68], although an unanticipated high prevalence rate (33%) of the parasite has been recently described in rats captured in sewers in the city of Barcelona [69]. To confirm the accuracy and extent of these previous findings, we investigated by PCR the occurrence of *Leishmania* spp. in rat tissues, including liver, spleen, and ear skin. In all cases, we failed to detect the presence of the parasite. It is possible that the limited number of rodent samples (n = 64) analyzed in the present study may have biased the obtained results. Further studies are warranted to assess the role of rodent on *Leishmania* spp. transmission in these epidemiological scenarios.

5. Conclusions

A high prevalence of the diarrhoea-causing protists *G. duodenalis* and *Blastocystis* sp. (but not *Cryptosporidium* spp.) was observed in captive NHP at the CZCC. NHP can harbour zoonotic genotypes of *G. duodenalis*, *Blastocystis* sp., and *E. bieneusi*. Indeed, strong evidence of the occurrence of *Blastocystis* zoonotic transmission between NHP and their handlers was provided, despite the use of personal protective equipment and the implementation of strict health and safety protocols. Free-living sympatric rats are infected by host-specific species/genotypes of the investigated protists and seem to play a limited role as a source of infections to NHP or humans in this setting. The extent of these findings should be confirmed in similar epidemiological surveys targeting other captive NHP populations.

Supplementary Materials: The following are available online at https://www.mdpi.com/2076-261 5/11/3/700/s1, Table S1: English version of the standardized epidemiological questionnaire used in this study, Table S2: Oligonucleotides used for the molecular identification and/or characterization of the intestinal parasitic and commensal protist species investigated in the present study.

Author Contributions: Conceptualization, P.C.K., M.S., C.C., F.J.N., I.G.-B., R.G., D.G.-B., F.P.-G., R.C.-B. and D.C.; methodology, P.C.K., J.G.M., M.S., R.G., I.G.-B., D.C.-T., F.P.-G., R.C.-B. and D.C.; software, P.C.K.; A.D.; J.G.M., D.C.; validation, P.C.K., M.S., F.P.-G., R.C.-B., D.G.-B. and D.C.; formal analysis, P.C.K., A.D., B.B., A.S.M., J.G.M., M.S.; C.C., S.M., D.C.-T., and R.G.; investigation, P.C.K., F.P.-G., R.C.-B. and D.C.; resources, M.S., D.C. and I.G.-B.; data curation, P.C.K., A.D., M.S., C.C. and D.C.; writing—original draft preparation, P.C.K., J.G.M., M.S., F.P.-G., R.C.-B. and D.C.; writing— review and editing, P.C.K., J.G.M., M.S., F.P.-G., R.C.-B., D.G.-B. and D.C.; supervision, F.P.-G., R.C.-B., D.G.-B., and D.C.; project administration, D.C.; funding acquisition, I.G.-B. and D.C. All authors have read and agreed to the published version of the manuscript.

Funding: This research was funded by the Health Institute Carlos III (ISCIII), Ministry of Economy and Competitiveness (Spain), under project PI16CIII/00024. Additional funding was obtained from the University of Córdoba (Spain), under the project UCO-FEDER-1264967.

Institutional Review Board Statement: This study was carried out in accordance with Spanish legislation guidelines (RD 8/2003) and with the International Guiding Principles for Biomedical Research Involving Animals issued by the Council for International Organization of Medical Sciences and the International Council for Laboratory Animal Science (RD 53/2013). This study has been approved by Ethics Committee of the Health Institute Carlos III on 17 December 2018 under the reference number CEI PI 90_2018-v2. Written informed consent was obtained from zookeepers that volunteered to participate in the survey.

Data Availability Statement: All relevant data are within the article and its additional files. The sequences data were submitted to the GenBank database under the accession numbers MW417420– MW417422 (*G. duodenalis*), MW414634–MW414644 (*Blastocystis* sp.), MW406908–MW406921 (*Cryptosporidium* spp.) and MW414645 (*E. bieneusi*).

Acknowledgments: David González-Barrio was recipient of a 'Sara Borrell' postdoctoral fellowship (CD19CIII/00011) funded by the Spanish Ministry of Science, Innovation and Universities.

Conflicts of Interest: The authors declare no conflict of interest. The funders had no role in the design of the study; in the collection, analyses, or interpretation of data; in the writing of the manuscript, or in the decision to publish the results.

References

1. DuPont, H.L. Persistent diarrhea: A clinical review. *JAMA* **2016**, *315*, 2712–2723. [CrossRef]
2. Hemphill, A.; Müller, N.; Müller, J. Comparative pathobiology of the intestinal protozoan parasites *Giardia lamblia*, *En-tamoeba histolytica*, and *Cryptosporidium parvum*. *Pathogens* **2019**, *8*, 116. [CrossRef]
3. Ajjampur, S.S.; Tan, K.S. Pathogenic mechanisms in *Blastocystis* spp.—Interpreting results from in vitro and in vivo studies. *Parasitol. Int.* **2016**, *65*, 772–779. [CrossRef]

4. Li, W.; Feng, Y.; Santin, M. Host Specificity of *Enterocytozoon bieneusi* and Public Health Implications. *Trends Parasitol.* **2019**, *35*, 436–451. [CrossRef] [PubMed]
5. Ryan, U.; Cacciò, S.M. Zoonotic potential of *Giardia*. *Int. J. Parasitol.* **2013**, *43*, 943–956. [CrossRef] [PubMed]
6. Ryan, U.; Fayer, R.; Xiao, L. *Cryptosporidium* species in humans and animals: Current understanding and research needs. *Parasitology* **2014**, *141*, 1667–1685. [CrossRef]
7. Li, W.; Feng, Y.; Zhang, L.; Xiao, L. Potential impacts of host specificity on zoonotic or interspecies transmission of *Enterocytozoon bieneusi*. *Infect. Genet. Evol.* **2019**, *75*, 104033. [CrossRef] [PubMed]
8. Hublin, J.S.Y.; Maloney, J.G.; Santin, M. *Blastocystis* in domesticated and wild mammals and birds. *Res. Vet. Sci.* **2020**. [CrossRef] [PubMed]
9. Widmer, G.; Köster, P.C.; Carmena, D. *Cryptosporidium hominis* infections in non-human animal species: Revisiting the concept of host specificity. *Int. J. Parasitol.* **2020**, *50*, 253–262. [CrossRef]
10. Stensvold, C.R.; Clark, C.G. Pre-empting Pandora's Box: *Blastocystis* Subtypes Revisited. *Trends Parasitol.* **2020**, *36*, 229–232. [CrossRef]
11. Maloney, J.G.; Molokin, A.; Da Cunha, M.J.R.; Cury, M.C.; Santin, M. *Blastocystis* subtype distribution in domestic and captive wild bird species from Brazil using next generation amplicon sequencing. *Parasite Epidemiol. Control* **2020**, *9*, e00138. [CrossRef] [PubMed]
12. Ou, Y.; Jiang, W.; Roellig, D.M.; Wan, Z.; Li, N.; Guo, Y.; Feng, Y.; Xiao, L. Characterizations of *Enterocytozoon bieneusi* at new genetic loci reveal a lack of strict host specificity among common genotypes and the existence of a canine-adapted *Enterocytozoon* species. *Int. J. Parasitol.* **2021**, *51*, 215–223. [CrossRef]
13. Gómez, M.S.; Gracenea, M.; Gosalbez, P.; Feliu, C.; Enseñat, C.; Hidalgo, R. Detection of oocysts of *Cryptosporidium* in several species of monkeys and in one prosimian species at the Barcelona zoo. *Parasitol. Res.* **1992**, *78*, 619–620. [CrossRef]
14. Soledad Gómez, M.; Gracenea, M.; Montoliu, I.; Feliu, C.; Monleon, A.; Fernandez, J.; Enseñat, C. Intestinal parasitism—protozoa and helminths—in primates at the Barcelona Zoo. *J. Med. Primatol.* **1996**, *25*, 419–423. [CrossRef] [PubMed]
15. Gómez, M.S.; Torres, J.; Gracenea, M.; Fernandez-Morán, J.; Gonzalez-Moreno, O. Further report on *Cryptosporidium* in Barcelona zoo mammals. *Parasitol. Res.* **2000**, *86*, 318–323. [CrossRef] [PubMed]
16. Gracenea, M.; Gómez, M.S.; Torres, J.; Carné, E.; Fernández-Morán, J. Transmission dynamics of *Cryptosporidium* in primates and herbivores at the Barcelona zoo: A long-term study. *Veter. Parasitol.* **2002**, *104*, 19–26. [CrossRef]
17. Pérez Cordón, G.; Hitos Prados, A.; Romero, D.; Sánchez Moreno, M.; Pontes, A.; Osuna, A.; Rosales, M.J. Intestinal parasitism in the animals of the zoological garden "Peña Escrita" (Almuñecar, Spain). *Vet. Parasitol.* **2008**, *156*, 302–309. [CrossRef] [PubMed]
18. Martínez-Díaz, R.A.; Sansano-Maestre, J.; Martínez-Herrero, M.D.C.; Ponce-Gordo, F.; Gómez-Muñoz, M.T. Occurrence and genetic characterization of *Giardia duodenalis* from captive nonhuman primates by multi-locus sequence analysis. *Parasitol. Res.* **2011**, *109*, 539–544. [CrossRef]
19. Ponce-Gordo, F.; García-Rodríguez, J.J. *Balantioides coli*. *Res. Veter. Sci.* **2020**, *S0034–5288*, 31066–31073. [CrossRef]
20. Pomajbíková, K.; Petrželková, K.J.; Profousová, I.; Petrášová, J.; Kišidayová, S.; Váradyová, Z.; Modrý, D. A survey of entodiniomorphid ciliates in chimpanzees and bonobos. *Am. J. Phys. Anthr.* **2009**, *142*, 42–48. [CrossRef]
21. UNE 16636:2015. Available online: https://www.aenor.com/normas-y-libros/buscador-de-normas/une/?Tipo=N&c=N0055762 (accessed on 6 February 2021).
22. Mayer, J.; Mans, C. Rodents. In *Exotic Animal Formulary*, 5th ed.; Carpenter, J.W., Marion, C.J., Eds.; Elsevier: St. Louis, MO, USA, 2018; pp. 673–675.
23. Verweij, J.J.; Schinkel, J.; Laeijendecker, D.; van Rooyen, M.A.; van Lieshout, L.; Polderman, A.M. Real-time PCR for the detection of *Giardia lamblia*. *Mol. Cell. Probes* **2003**, *17*, 223–225. [CrossRef]
24. Read, C.M.; Monis, P.T.; Thompson, R.C. Discrimination of all genotypes of *Giardia duodenalis* at the glutamate dehydro-genase locus using PCR-RFLP. *Infect. Genet. Evol.* **2004**, *4*, 125–130. [CrossRef]
25. Lalle, M.; Pozio, E.; Capelli, G.; Bruschi, F.; Crotti, D.; Cacciò, S.M. Genetic heterogeneity at the beta-giardin locus among human and animal isolates of *Giardia duodenalis* and identification of potentially zoonotic subgenotypes. *Int. J. Parasitol.* **2005**, *35*, 207–213. [CrossRef]
26. Sulaiman, I.M.; Fayer, R.; Bern, C.; Gilman, R.H.; Trout, J.M.; Schantz, P.M.; Das, P.; Lal, A.A.; Xiao, L. Triosephosphate isomerase gene characterization and potential zoonotic transmission of *Giardia duodenalis*. *Emerg. Infect. Dis.* **2003**, *9*, 1444–1452. [CrossRef] [PubMed]
27. Tiangtip, R.; Jongwutiwes, S. Molecular analysis of *Cryptosporidium* species isolated from HIV-infected patients in Thai-land. *Trop. Med. Int. Health* **2002**, *7*, 357–364. [CrossRef] [PubMed]
28. Scicluna, S.M.; Tawari, B.; Clark, C.G. DNA Barcoding of *Blastocystis*. *Protist* **2006**, *157*, 77–85. [CrossRef] [PubMed]
29. Maloney, J.G.; Molokin, A.; Santin, M. Next generation amplicon sequencing improves detection of *Blastocystis* mixed subtype infections. *Infect. Genet. Evol.* **2019**, *73*, 119–125. [CrossRef] [PubMed]
30. Santín, M.; Gómez-Muñoz, M.T.; Solano-Aguilar, G.; Fayer, R. Development of a new PCR protocol to detect and subtype *Blastocystis* spp. from humans and animals. *Parasitol. Res.* **2011**, *109*, 205–212. [CrossRef]
31. Bushnell, B. BBMap Download. 2014. Available online: SourceForge.net (accessed on 6 February 2021).
32. Rognes, T.; Flouri, T.; Nichols, B.; Quince, C.; Mahé, F. VSEARCH: A versatile open source tool for metagenomics. *PeerJ* **2016**, *4*, e2584. [CrossRef]

33. Buckholt, M.A.; Lee, J.H.; Tzipori, S. Prevalence of *Enterocytozoon bieneusi* in swine: An 18-month survey at a slaughter-house in Massachusetts. *Appl. Environ. Microbiol.* **2002**, *68*, 2595–2599. [CrossRef] [PubMed]

34. Verweij, J.J.; Oostvogel, F.; Brienen, E.A.; Nang-Beifubah, A.; Ziem, J.; Polderman, A.M. Prevalence of *Entamoeba histolytica* and *Entamoeba dispar* in northern Ghana. *Trop. Med. Int. Health* **2003**, *8*, 1153–1156. [CrossRef]

35. Gutiérrez-Cisneros, M.J.; Cogollos, R.; López-Vélez, R.; Martín-Rabadán, P.; Martínez-Ruiz, R.; Subirats, M.; Merino, F.J.; Fuentes, I. Application of real-time PCR for the differentiation of *Entamoeba histolytica* and *E. dispar* in cyst-positive faecal samples from 130 immigrants living in Spain. *Ann. Trop. Med. Parasitol.* **2010**, *104*, 145–149. [CrossRef]

36. Ponce-Gordo, F.; Fonseca-Salamanca, F.; Martínez-Díaz, R.A. Genetic heterogeneity in internal transcribed spacer genes of *Balantidium coli* (*Litostomatea, Ciliophora*). *Protist* **2011**, *162*, 774–794. [CrossRef]

37. Vallo, P.; Petrželková, K.J.; Profousová, I.; Petrášová, J.; Pomajbíková, K.; Leendertz, F.; Hashimoto, C.; Simmons, N.; Babweteera, F.; Machanda, Z.; et al. Molecular diversity of entodiniomorphid ciliate *Troglodytella abrassarti* and its coevolution with chimpanzees. *Am. J. Phys. Anthropol.* **2012**, *148*, 525–533. [CrossRef] [PubMed]

38. Cruz, I.; Chicharro, C.; Nieto, J.; Bailo, B.; Canavate, C.; Figueras, M.A.-C.; Alvar, J. Comparison of new diagnostic tools for management of pediatric Mediterranean visceral leishmaniasis. *J. Clin. Microbiol.* **2006**, *44*, 2343–2347. [CrossRef]

39. Ježková, J.; Prediger, J.; Holubová, N.; Sak, B.; Konečný, R.; Feng, Y.; Xiao, L.; Rost, M.; McEvoy, J.; Kváč, M. *Cryptosporidium ratti* n. sp. (Apicomplexa: Cryptosporidiidae) and genetic diversity of *Cryptosporidium* spp. in brown rats (*Rattus norvegicus*) in the Czech Republic. *Parasitology* **2021**, *148*, 84–97. [CrossRef] [PubMed]

40. Kotloff, K.L. The burden and etiology of diarrheal illness in developing countries. *Pediatr. Clin. North Am.* **2017**, *64*, 799–814. [CrossRef] [PubMed]

41. Torres, J.; Gracenea, M.; Gómez, M.S.; Arrizabalaga, A.; González-Moreno, O. The occurrence of *Cryptosporidium parvum* and *C. muris* in wild rodents and insectivores in Spain. *Vet. Parasitol.* **2000**, *92*, 253–260. [CrossRef]

42. García-Livia, K.; Martín-Alonso, A.; Foronda, P. Diversity of *Cryptosporidium* spp. in wild rodents from the Canary Islands, Spain. *Parasites Vectors* **2020**, *13*, 445. [CrossRef] [PubMed]

43. Guy, R.A.; Yanta, C.A.; Muchaal, P.K.; Rankin, M.A.; Thivierge, K.; Lau, R.; Boggild, A.K. Molecular characterization of *Cryptosporidium* isolates from humans in Ontario, Canada. *Parasites Vectors* **2021**, *14*, 1–14. [CrossRef]

44. Ayinmode, A.B.; Obebe, O.O.; Falohun, O.O. Molecular detection of *Cryptosporidium* species in street-sampled dog faeces in Ibadan, Nigeria. *Vet. Parasitol. Reg. Stud. Rep.* **2018**, *14*, 54–58. [CrossRef] [PubMed]

45. Yang, R.; Ying, J.L.J.; Monis, P.; Ryan, U. Molecular characterisation of *Cryptosporidium* and *Giardia* in cats (*Felis catus*) in Western Australia. *Exp. Parasitol.* **2015**, *155*, 13–18. [CrossRef]

46. Pavlásek, I.; Ryan, U. The first finding of a natural infection of *Cryptosporidium muris* in a cat. *Vet. Parasitol.* **2007**, *144*, 349–352. [CrossRef] [PubMed]

47. De Lucio, A.; Martínez-Ruiz, R.; Merino, F.J.; Bailo, B.; Aguilera, M.; Fuentes, I.; Carmena, D. Molecular genotyping of *Giardia duodenalis* isolates from symptomatic individuals attending two major public hospitals in Madrid, Spain. *PLoS ONE* **2015**, *10*, e0143981. [CrossRef] [PubMed]

48. Azcona-Gutiérrez, J.M.; De Lucio, A.; Hernández-De-Mingo, M.; García-García, C.; Soria-Blanco, L.M.; Morales, L.; Aguilera, M.; Fuentes, I.; Carmena, D. Molecular diversity and frequency of the diarrheagenic enteric protozoan *Giardia duodenalis* and *Cryptosporidium* spp. in a hospital setting in Northern Spain. *PLoS ONE* **2017**, *12*, e0178575. [CrossRef] [PubMed]

49. Cardona, G.A.; Carabin, H.; Goñi, P.; Arriola, L.; Robinson, G.; Fernández-Crespo, J.C.; Clavel, A.; Chalmers, R.M.; Carmena, D. Identification and molecular characterization of *Cryptosporidium* and *Giardia* in children and cattle populations from the province of Álava, North of Spain. *Sci. Total Environ.* **2011**, *412/413*, 101–108. [CrossRef]

50. Muadica, A.S.; Köster, P.C.; Dashti, A.; Bailo, B.; Hernández-de-Mingo, M.; Reh, L.; Balasegaram, S.; Verlander, N.Q.; Ruiz Chércoles, E.; Carmena, D. Molecular diversity of *Giardia duodenalis*, *Cryptosporidium* spp. and *Blastocystis* sp. in asymptomatic school children in Leganés, Madrid (Spain). *Microorganisms* **2020**, *8*, 466. [CrossRef]

51. Renelies-Hamilton, J.; Noguera-Julian, M.; Parera, M.; Paredes, R.; Pacheco, L.; Dacal, E.; Saugar, J.M.; Rubio, J.M.; Poulsen, M.; Köster, P.C.; et al. Exploring interactions between *Blastocystis* sp., *Strongyloides* spp. and the gut microbiomes of wild chimpanzees in Senegal. *Infect. Genet. Evol.* **2019**, *74*, 104010. [CrossRef]

52. Paulos, S.; Köster, P.C.; De Lucio, A.; Hernández-De-Mingo, M.; Cardona, G.A.; Fernández-Crespo, J.C.; Stensvold, C.R.; Carmena, D. Occurrence and subtype distribution of *Blastocystis* sp. in humans, dogs and cats sharing household in northern Spain and assessment of zoonotic transmission risk. *Zoonoses Public Health* **2018**, *65*, 993–1002. [CrossRef]

53. Köster, P.C.; Molina, A.; García, M.; Cifre, S.; Trelis, M.; Pérez de Ayala, A.; Azcona, J.M.; García, C.; Paulos, S.; Hernández-de-Mingo, M.; et al. Molecular diversity and frequency of Blastocystis sp. subtypes in Spanish clinical patients: A pilot multicentre study. In Proceedings of the 2nd International Blastocystis Conference, Bogotá, Colombia, 9–12 October 2018.

54. Oliveira-Arbex, A.P.; David, É.B.; Tenório, M.D.S.; Cicchi, P.J.P.; Patti, M.; Coradi, S.T.; Lucheis, S.B.; Jim, J.; Guimarães, S. Diversity of *Blastocystis* subtypes in wild mammals from a zoo and two conservation units in southeastern Brazil. *Infect. Genet. Evol.* **2020**, *78*, 104053. [CrossRef]

55. Helenbrook, W.D.; Shields, W.M.; Whipps, C.M. Characterization of *Blastocystis* species infection in humans and mantled howler monkeys, *Alouatta palliata aequatorialis*, living in close proximity to one another. *Parasitol. Res.* **2015**, *114*, 2517–2525. [CrossRef]

56. Stensvold, C.R.; Alfellani, M.A.; Nørskov-Lauritsen, S.; Prip, K.; Victory, E.L.; Maddox-Hyttel, C.; Nielsen, H.V.; Clark, C.G. Subtype distribution of *Blastocystis* isolates from synanthropic and zoo animals and identification of a new subtype. *Int. J. Parasitol.* **2009**, *39*, 473–479. [CrossRef] [PubMed]
57. Mohammadpour, I.; Bozorg-Ghalati, F.; Gazzonis, A.L.; Manfredi, M.T.; Motazedian, M.H.; Mohammadpour, N. First molecular subtyping and phylogeny of *Blastocystis* sp. isolated from domestic and synanthropic animals (dogs, cats and brown rats) in southern Iran. *Parasites Vectors* **2020**, *13*, 1–11. [CrossRef]
58. Stensvold, C.R.; Alfellani, M.; Clark, C.G. Levels of genetic diversity vary dramatically between *Blastocystis* subtypes. *Infect. Genet. Evol.* **2012**, *12*, 263–273. [CrossRef] [PubMed]
59. Li, W.; Zhong, Z.; Song, Y.; Gong, C.; Deng, L.; Cao, Y.; Zhou, Z.; Cao, X.; Tian, Y.; Li, H.; et al. Human-pathogenic *Enterocytozoon bieneusi* in captive giant pandas (*Ailuropoda melanoleuca*) in China. *Sci. Rep.* **2018**, *8*, 6590. [CrossRef]
60. Chen, L.; Zhao, J.; Li, N.; Guo, Y.; Feng, Y.; Feng, Y.; Xiao, L. Genotypes and public health potential of *Enterocytozoon bie-neusi* and *Giardia duodenalis* in crab-eating macaques. *Parasites Vectors* **2019**, *12*, 254. [CrossRef]
61. Galván, A.L.; Sánchez, A.M.; Valentín, M.A.; Henriques-Gil, N.; Izquierdo, F.; Fenoy, S.; del Aguila, C. First cases of microsporidiosis in transplant recipients in Spain and review of the literature. *J. Clin. Microbiol.* **2011**, *49*, 1301–1306. [CrossRef]
62. Galván-Díaz, A.L.; Magnet, A.; Fenoy, S.; Henriques-Gil, N.; Haro, M.; Gordo, F.P.; Millán, J.; Miró, G.; Del Águila, C.; Izquierdo, F. Microsporidia detection and genotyping study of human pathogenic *E. bieneusi* in animals from Spain. *PLoS ONE* **2014**, *9*, e92289. [CrossRef]
63. Dashti, A.; Santín, M.; Cano, L.; de Lucio, A.; Bailo, B.; de Mingo, M.H.; Köster, P.C.; Fernández-Basterra, J.A.; Aramburu-Aguirre, J.; López-Molina, N.; et al. Occurrence and genetic diversity of *Enterocytozoon bieneusi* (Microsporidia) in owned and sheltered dogs and cats in Northern Spain. *Parasitol. Res.* **2019**, *118*, 2979–2987. [CrossRef]
64. Galván, A.L.; Magnet, A.; Izquierdo, F.; Fenoy, S.; Rueda, C.; Fernández Vadillo, C.; Henriques-Gil, N.; Del Aguila, C. Molecular characterization of human-pathogenic Microsporidia and *Cyclospora cayetanensis* isolated from various water sources in Spain: A year-long longitudinal study. *Appl. Environ. Microbiol.* **2013**, *79*, 449–459. [CrossRef]
65. Kooriyama, T.; Hasegawa, H.; Shimozuru, M.; Tsubota, T.; Nishida, T.; Iwaki, T. Parasitology of five primates in Mahale Mountains National Park, Tanzania. *Primates* **2012**, *53*, 365–375. [CrossRef] [PubMed]
66. Petrželková, K.J.; Schovancová, K.; Profousová, I.; Kišidayová, S.; Váradyová, Z.; Pekár, S.; Kamler, J.; Modrý, D. The effect of low- and high-fiber diets on the population of entodiniomorphid ciliates *Troglodytella abrassarti* in captive chimpanzees (*Pan troglodytes*). *Am. J. Primatol.* **2012**, *74*, 669–675. [CrossRef] [PubMed]
67. Ortega-García, M.V.; Salguero, F.J.; Rodríguez-Bertos, A.; Moreno, I.; García, N.; García-Seco, T.; Luz Torre, G.; Domínguez, L.; Domínguez, M. A pathological study of *Leishmania infantum* natural infection in European rabbits (*Oryctolagus cuniculus*) and Iberian hares (*Lepus granatensis*). *Transbound. Emerg. Dis.* **2019**, *66*, 2474–2481. [CrossRef] [PubMed]
68. Millán, J. Molecular investigation of vector-borne parasites in wild micromammals, Barcelona (Spain). *Parasitol. Res.* **2018**, *117*, 3015–3018. [CrossRef]
69. Galán-Puchades, M.T.; Gómez-Samblás, M.; Suárez-Morán, J.M.; Osuna, A.; Sanxis-Furió, J.; Pascual, J.; Bueno-Marí, R.; Franco, S.; Peracho, V.; Montalvo, T.; et al. Leishmaniasis in Norway Rats in Sewers, Barcelona, Spain. *Emerg. Infect. Dis.* **2019**, *25*, 1222–1224. [CrossRef]

Article

Exposure to Zoonotic West Nile Virus in Long-Tailed Macaques and Bats in Peninsular Malaysia

Mohd Yuseri Ain-Najwa [1], Abd Rahaman Yasmin [1,2,*], Siti Suri Arshad [3], Abdul Rahman Omar [2,3], Jalila Abu [4], Kiven Kumar [5], Hussni Omar Mohammed [6], Jafar Ali Natasha [1], Mohammed Nma Mohammed [1], Faruku Bande [7], Mohd-Lutfi Abdullah [8] and Jeffrine J. Rovie-Ryan [8]

[1] Department of Veterinary Laboratory Diagnosis, Faculty of Veterinary Medicine, Universiti Putra Malaysia, UPM Serdang, Selangor 43400, Malaysia; ainnajwa2111@gmail.com (M.Y.A.-N.); natashali036@gmail.com (J.A.N.); mohdnma@live.com (M.N.M.)

[2] Laboratory of Vaccines and Biomolecules, Institute of Bioscience, UPM Serdang, Selangor 43400, Malaysia; aro@upm.edu.my

[3] Department of Veterinary Pathology and Microbiology, Faculty of Veterinary Medicine, Universiti Putra Malaysia, UPM Serdang, Selangor 43400, Malaysia; suri@upm.edu.my

[4] Department of Veterinary Clinical Studies, Faculty of Veterinary Medicine, Universiti Putra Malaysia, UPM Serdang, Selangor 43400, Malaysia; jalila@upm.edu.my

[5] Department of Pathology, Faculty of Medicine and Health Sciences, Universiti Putra Malaysia, UPM Serdang, Selangor 43400, Malaysia; kivenkumar@yahoo.com

[6] Department of Population Medicine and Diagnostic Sciences, Cornell University, Ithaca, NY 14853, USA; hussni.mohammed@cornell.edu

[7] Department of Veterinary Services, Ministry of Animal Health and Fisheries Development, Sokoto 840, Sokoto State, Nigeria; bandeyabo@gmail.com

[8] Department of Conservation of Biodiversity of Wildlife and National Park Malaysia, Ministry of Energy and Natural Resources, Kuala Lumpur 56000, Malaysia; lutfi@wildlife.gov.my (M.-L.A.); jeffrine@wildlife.gov.my (J.J.R.-R.)

* Correspondence: noryasmin@upm.edu.my

Received: 29 October 2020; Accepted: 5 December 2020; Published: 10 December 2020

Simple Summary: The role of wildlife animals, such as macaques and bats, in the spreading and maintenance of deadly zoonotic pathogens in nature are documented in several studies. The present study substantially highlights the first evidence of West Nile Virus (WNV) infection, a mosquito borne virus in the Malaysian macaques and bats. Of the 81 macaques sampled, 24 of the long-tailed macaques were seropositive to WNV, indicating that they were exposed to the virus in the past. The long-tailed macaques were found in the mangrove forests located in the Central, Southern, and West Peninsular Malaysia. Meanwhile, five out of 41 bats (Lesser Short-nosed Fruit Bats, Lesser Sheath-tailed Bats, and Thai Horseshoe Bats) that were found in the caves from Northern Peninsular Malaysia showed susceptibility to WNV. Therefore, a constant bio surveillance of WNV in the wildlife in Malaysia is a proactive attempt. This study was aligned with the Malaysian government's mission under the Malaysia Strategy for Emerging Diseases and Public Health Emergencies (MYSED) II (2017–2021) and the Ministry of Health priorities in order to enhance the regional capability to rapidly and accurately survey, detect, diagnose, and report outbreaks of pathogens and diseases of security concern.

Abstract: The role of wildlife such as wild birds, macaques, and bats in the spreading and maintenance of deadly zoonotic pathogens in nature have been well documented in many parts of the world. One such pathogen is the mosquitoes borne virus, namely the West Nile Virus (WNV). Previous research has shown that 1:7 and 1:6 Malaysian wild birds are WNV antibody and RNA positive, respectively, and bats in North America may not be susceptible to the WNV infection. This study was conducted to determine the status of WNV in Malaysian macaques and bats found in mangrove forests and caves, respectively. Archive sera and oropharyngeal swabs from long-tailed macaques were subjected to the antibody detection using WNV competitive enzyme-linked immunosorbent assay (c-ELISA) and WNV RNA using RT-PCR, respectively, while the archive oropharyngeal and rectal swabs from bats were subjected to RT-PCR without serological analysis due to the unavailability of serum samples. The analysis revealed a WNV seropositivity of 29.63% (24/81) and none of the macaques were positive for WNV RNA. Meanwhile, 12.2% (5/41) of the bats from Pteropodidae, Emballonuridae, and Rhinolophidae families tested positive for WNV RNA. Here, we show a high WNV antibody prevalence in macaques and a moderate WNV RNA in various Malaysian bat species, suggesting that WNV circulates through Malaysian wild animals and Malaysian bat species may be susceptible to the WNV infection.

Keywords: west nile virus; arbovirus; zoonotic; macaque; bats; c-ELISA; RT-PCR

---

1. Introduction

Deadly emerging and re-emerging zoonotic pathogens are transmitted mostly from wildlife reservoirs to humans or other animals during spillover events, with or without a vector intervention. Evidence has shown that some of the medically important mosquito borne illnesses causing West Nile fever, dengue, malaria, chikungunya, zika, and Japanese encephalitis were isolated from the wildlife [1–3]. The West Nile fever is distributed in Africa, USA, Europe, and Western Asia causing febrile illness and encephalitis in humans and animals [4]. The causative agent is an envelope RNA virus known as the West Nile Virus (WNV), which is classified in the genus of *Flavivirus* under the family of Flaviviridae [5]. Despite the role played by wild birds as amplifier hosts of WNV and mosquitoes as vectors, the role of wildlife such as macaques and bats in the WNV transmission cycle remains poorly understood [6].

Several epidemiological studies attested that the macaque species could become infected with WNV [6–8]. Nevertheless, the macaques developed a low level of viraemia based on the experimental study, making them unlikely to perpetuate the virus [8]. Meanwhile, bats have been demonstrated as a competent amplifying host of arthropod borne Flaviviruses transmission, but have not been proven to maintain WNV infections as seen in the North American bats, which were infrequently infected with WNV [9,10]. Due to the nature of wildlife habitats and constant exposure to mosquito bites during blood meals, there is an abundant considerable opportunity for WNV introduction to the wildlife. There is also the possibility of spillover events as well as the sylvatic transmission of the virus from the wildlife to humans due to human activities such as deforestation and urbanization, and also due to other climate-related factors that lead to a loss of wildlife habitat [11,12].

In Malaysia, some studies have reported evidence of exposure of humans, birds, and mosquitos to WNV. A study by Marlina et al. [13] reported a WNV seroprevalence of 1.21% (9/742) in the Orang Asli from several states in Peninsular Malaysia. Additionally, Rais et al. [14] and Ain-Najwa et al. [15] reported a WNV seroprevalence of 4.41% (3/68) in captive birds and 18.71% (29/155) in wild birds, respectively. In 1970, a sub-type of WNV designated as the Kunjin Virus (KUNV), which was originally endemic in Australia was detected in *Culex pseudovishnui* mosquitoes in Sarawak, a Malaysian state of Borneo [16]. More recently, WNV was also detected in pooled samples of *Culex* spp. mosquitoes, which were trapped close to the migratory birds landing areas in Malaysia (under review). These findings collectively demonstrated evidence of WNV exposure with asymptomatic infection in animals and humans in Malaysia [17,18]. Nevertheless, the status of WNV infection in Malaysian macaques and bats remains poorly understood. Therefore, this study was conducted to determine the serological and/or molecular prevalence of WNV in bats and macaques in selected areas of Malaysia.

2. Materials and Methods

2.1. Ethical Statement

All experimental procedures involving the archived samples that originated from the Department of Wildlife and National Parks (DWNP) were conducted in accordance with guidelines approved by DWNP, Malaysia with the research permit number JPHL&TN (IP):100-6/1/14. The no. IACUC approval was needed in this study, since the archive samples were used

2.2. Study Design

The archived macaque and bat samples originating from several states in Peninsular Malaysia were shared by the Department of Conservation of Biodiversity of Wildlife and National Park Malaysia. A total of 88 long-tailed macaques and 41 bats, which were sampled from the year 2014 to 2017, were included in this study. The samples obtained from macaques included 81 sera and 63 oropharyngeal swabs, while 38 rectal swabs and 34 oropharyngeal swabs were obtained from bats. Sera and swabs samples (oropharyngeal and rectal) were subjected to serological and molecular analysis, respectively. Since the archive samples were used, the samples obtained in this study were based on availability. Consequently, an inconsistent number of samples relative to the total number of animals were observed in this study. In addition, due to the unavailability of bats sera samples, no serological analysis was performed for bats.

Macaques were sampled from mangrove forests in Pahang state (central Peninsular Malaysia), Perak state (West Coast Peninsular Malaysia), and Johor state (Southern Peninsular Malaysia) (Figure 1). In Pahang state, the macaques were captured at (1) Kuala Lipis (4.1843° N, 102.0542° E) and (2) Temerloh (3.4486° N, 102.4163° E); in Perak state, they were captured at (3) Sungai Siput (4.8190° N, 101.0737° E) and (4) Kuala Gula (4.933° N, 100.467° E), while in Johor state, they were captured at (5) Ayer Hitam (1.9183° N, 103.1800° E) and (6) Batu Pahat (1.8469° N, 102.9352° E). Meanwhile, bats were sampled from a cave located in (7) Hutan Simpan Wang Mu in Perlis State Park (6.467° N, 100.250° E) from Perlis state (Northern Peninsular Malaysia) (Figure 1). Both sites are a natural habitat for macaques and bats found in Malaysia.

2.3. Serological Analysis

The status of WNV seropositivity in macaques was determined using a commercial WNV IgG antibody-based c-ELISA kit (ID Screen West Nile Competition Multi-species ELISA, ID VET, Montpellier, France) pre-coated with the WNV envelope protein (prE) from macaque's sera. However, the kit cross-reacted with other Flavivirus namely the Japanese Encephalitis Virus (JEV) and Yellow Fever. Since Yellow Fever is not endemic in Malaysia, sera were subjected to JEV screening using a specific double-antibody sandwich ELISA (DAS-ELISA) (Sun red, Shanghai, China) to rule out cross-reactivity.

For the WNV ELISA analysis, the positive and negative controls were run in duplicate experiments. If the sample showed a percentage of S/N (S: Sample optical density (OD); N: Negative control OD) less than or equal to 40%, the reaction was considered positive. In this study, the prevalence of WNV antibodies in macaques was calculated as a percentage of positive samples over the total number of examined samples at a 95% confidence interval. Data obtained in this study could not be analyzed for risk factors due to the uneven number of samples from different locations, which may have led to a potential bias.

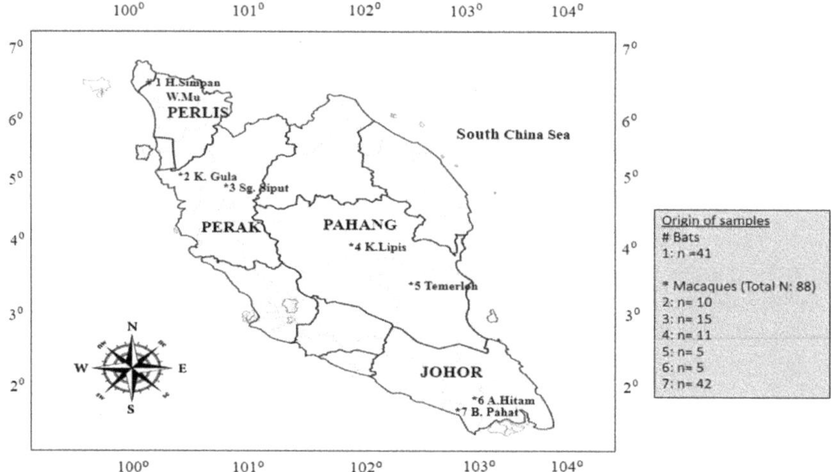

Figure 1. Map of Peninsular Malaysia showing the origin of bat and macaque samples. Note: The hash (#) symbol indicates the origin of bat samples, which were collected from (1) Hutan Simpan Wang Mu located in Perlis state (Northern Malaysia), while the asterisk (*) symbol indicates the origin of macaque samples from (2) Kuala Gula and (3) Sungai Siput located in Perak state (West Coast Malaysia); (4) Kuala Lipis and (5) Temerloh Pahang state (central Malaysia); and (6) Ayer Hitam and (7) Batu Pahat from Johor state (Southern Malaysia). N: Total animals; n: No. of animals.

2.4. Reverse Transcriptase—Polymerase Chain Reaction (RT-PCR) Assay

Oropharyngeal and rectal swabs were subjected to total RNA extraction using TRIsure (Bioline, London, UK) according to the manufacturer's instructions. The concentration and purity of the RNA extracted were determined using a BioPhotometer (Eppendorf, Hamburg, Germany). Synthetic plasmid was used as a positive control and for the RT-PCR primer set which targeted highly conserved regions between WNV Capsid (C) and Pre-Membrane (prM) proteins, as previously described by Ain-Najwa et al. [15] (Table S1). The one-step RT-PCR using MyTaq (Bioline, Memphis, TN, USA) in a total of 25 µL reaction was performed, as previously described by Ain-Najwa et al. [15]. Gel electrophoresis was conducted to view the amplification of positive reactions visualized by the presence of a 470-bp amplicon fragment between C and prM genes aligned with a positive control band.

2.5. DNA Sequencing and Bioinformatic Analysis

The DNA sequencing analysis was performed by purifying the band using a gel purification kit (Nucleospin Gel and PCR Clean-up kit (Macherey-Nagel, Duren, Germany) and sequenced in both directions using gene specific primers in the ABI PRISM 3730xl Genetic Analyzer (Applied Biosystems, Foster City, CA, USA). The sequences obtained were searched using the Basic Local Alignment Search Tool (BLAST) algorithm (https://blast.ncbi.nlm.nih.gov/Blast.cgi). A total of 63 sequences of WNV (Table S2) including 16 sequences based on a previous study [15] were included in the sequences alignment using Multiple Sequence Alignment (MAFFT) software version 7. The phylogenetic tree was constructed using neighbor-joining with the Maximum Composite Likelihood model in Molecular Evolutionary Genetics Analysis (MEGA) 7 [19]. A bootstrapped confidence interval with 1000 replicates was set. The Newick file was created and the tree was viewed and edited using the Interactive Tree Of Life (iTOL). The percentage identity of each nucleotide and amino acids sequences of WNV sequences used in the phylogenetic tree were subjected into a pairwise distance by MEGA-7.

3. Results

3.1. WNV Antibodies in Macaques

The serological analysis of WNV using c-ELISA revealed that 24 out of 81 macaques (29.63% (24/81) at 95% CI (0.203 to 0.410)) were seropositive with a S/N% (S: Sample OD; N: Negative control OD) value less than or equal to 40%. All of these serum samples were further analyzed using JEV DAS-ELISA and none of the samples showed a positive reaction towards JEV. The distribution of the WNV positive antibody in macaques according to the sample origin, age, and sex is provided in Table 1.

Among the three states sampled, 14 macaques from Johor showed the highest positive WNV followed by eight macaques from Perak and the remaining two from Pahang. Of the total WNV seropositive macaques, 16 were male and eight were female. Furthermore, adult macaques showed the highest WNV seropositive number with 17 followed by seven macaques from the juvenile group.

3.2. Molecular Analysis of West Nile Virus in Bats

The one-step RT-PCR revealed that 12.2% (5/41) of the bats tested positive for WNV RNA, while none of the macaques were positive. The distribution of positive WNV RNA in bats from the present study is shown in detail in Table 2. Of the five bat families studied, WNV positive bats were detected from Pteropodidae, Emballonuridae, and Rhinolophidae families. With regard to species, the highest number of positive bats was from *Emballonura monticola* (three bats positive) followed by *Cynopterus brachyotis* (one bat positive), and *Rhinolophus siamensis* (one bat positive). Both oropharyngeal and rectal swabs from one *R. siamensis* bat showed a positive WNV RNA in RT-PCR analysis, while the other bats were positive either from oropharyngeal or rectal swabs. Of the five positive bats, three were male, while two were female bats. Among them, three out of five bats were adult, while the remaining two were juvenile.

Six positive sequences were submitted to the GenBank under the following accession numbers: MK327803–MK327808. The sequencing analysis revealed that isolates from the present study showed a 99.25–100% similarity with the SPU116/89 strain. The pairwise percentage identity between the nucleotide and amino acid of selective WNV strains sequences with local isolates from bats showed a range of high, medial, and low similarity. The highest identical percentage was shown by the SPU116_89 strain (a human strain), while the Rabensburg isolate 97–103, Dak Ar D 5443, ArD96655/1993/SN, LEIV-Krnd88-190, and 101_5-06-Uu strains were less similar (Figure S1). A phylogenetic tree constructed using neighbor-joining methods, showing evolutionary relationships of taxa of WNV positive isolates from this study, were grouped together with WNV strains from the WNV lineage 2 (Figure 2).

Table 1. West Nile Virus (WNV) c-ELISA and RT-PCR results in long-tailed macaques according to states, sex, age, and type of sample.

Family	Species	Malaysian State	No. of Macaques	No. of Serum Sample	No. of Positive WNV Antibody (c-ELISA)	No. of Oropharyngeal Swabs	No. of Positive WNV RNA (RT-PCR)
Cercopithecidae	Long-tailed macaque (*Macaca Fascicularis*)	Pahang	8	6	1	8	0
			0	0	-	0	-
			4	2	1	4	0
			0	0	-	0	-
			0	0	-	0	-
			4	1	0	3	0
		Perak	5	5	3	0	-
			0	0	-	0	-
			11	11	2	1	0
			5	5	2	0	-
			0	0	-	0	-
			4	4	1	0	0
		Johor	5	5	1	5	0
			3	3	0	3	0
			21	21	8	21	0
			6	6	0	6	0
			2	2	0	2	0
			10	10	5	10	0
Total			88	81	24	63	0

M: Male; F: Female; c-ELISA: Competitive ELISA; RT-PCR: Reverse transcriptase-PCR. Adult ≥ 5 years; sub adult ≥ 3–5 years, and juvenile ≥ 1–3 years old.

Table 2. West Nile Virus RNA detection in bats according to families, species, sex, age, and type of sample.

Family	Species	No. of Bats	No. of Oropharyngeal Swabs	No. of Rectal Swabs	No. of Positive WNV RNA (RT-PCR)
	Horsfield's Fruit Bats (*Cynopterus horsfieldii*)	1	1	1	0
		5	4	5	0
		2	2	2	0
		1	1	1	0
Pteropodidae	Lesser Short-nosed Fruit Bats (*Cynopterus brachyotis*)	0	0	0	-
		0	0	0	-
		1	1	1	1
		0	0	0	-
	Long-tongued Fruit Bats (*Macroglossus sobrinus*)	0	0	0	-
		1	1	1	0
		0	0	0	-
		0	0	0	-
Emballonuridae	Lesser Sheath-tailed Bats (*Emballonura monticola*)	0	0	0	-
		3	3	3	2
		1	1	1	0
		2	2	2	1
	Blyth's Horseshoe Bats (*Rhinolophus lepidus*)	1	1	1	0
		2	1	2	0
		5	1	5	0
		2	1	2	0
Rhinolophidae	Malayan Horseshoe Bats (*Rhinolophus malayanus*)	0	0	0	0
		2	2	2	0
		0	0	0	0
		0	0	0	0
	Thai Horseshoe Bats (*Rhinolophus siamensis*)	1	1	1	2
		0	0	0	-
		0	0	0	-
		0	0	0	-

Table 2. *Cont.*

Family	Species	No. of Bats	No. of Oropharyngeal Swabs	No. of Rectal Swabs	No. of Positive WNV RNA (RT-PCR)
Rhinolophidae	Croslet Horseshoe Bats (*Rhinolophus coelophyllus*)	0	0	0	-
		1	1	1	0
		0	0	0	-
		0	0	0	-
	Bourret's Horseshoe Bats (*Rhinolophus paradoxolophus*)	0	0	0	-
		0	0	0	-
		0	0	0	-
		1	1	1	0
Hipposideridae	Great Roundleaf Bats (*Hipposideros armiger*)	1	1	1	0
		0	0	0	-
		0	0	0	-
		1	1	1	0
	Shield-faced Roundleaf Bats (*Hipposideros lylei*)	1	1	0	0
		1	1	1	0
		2	2	1	0
		1	1	0	0
	Diadem Leaf-nosed Bats (*Hipposideros diadema*)	0	0	0	-
		1	1	1	0
		0	0	0	-
		0	0	0	-
Vespertilionidae	Lesser Asiatic Yellow Bats (*Scotophilus kuhlii*)	0	0	0	-
		1	1	1	0
		0	0	0	-
		0	0	0	-
	TOTAL	41	34	38	Bats +ve: 5 RT-PCR +ve: 6

M: Male; F: Female; c-ELISA: Competitive ELISA; RT-PCR: Reverse transcriptase-PCR; O: Orophryngeal swab; R: Rectal. Juvenile < 9 months; Adult > 9 months.

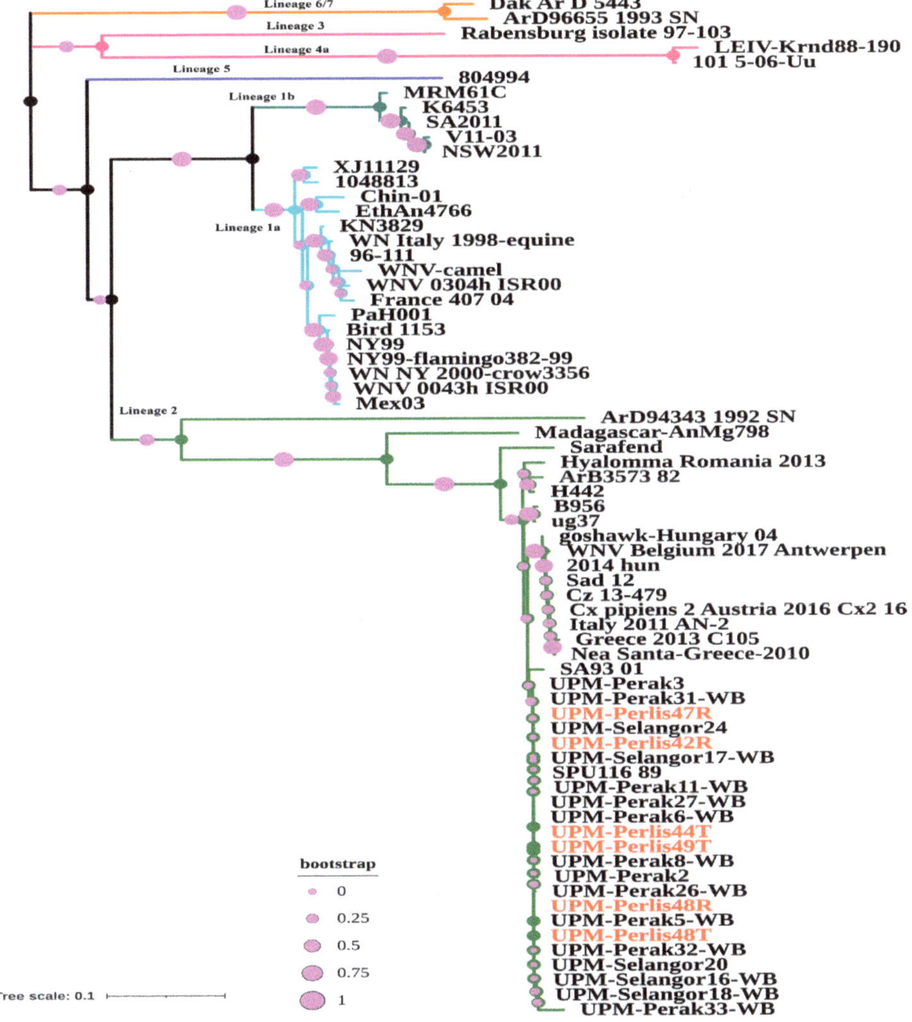

Figure 2. Phylogenetic analysis of partial sequence between capsid (C) and pre-membrane (prM) genes of West Nile Virus detected in bats in Malaysia with other global WNV isolates. The analysis involved 69 nucleotide sequences including six sequences from this study (highlighted in red), which have a 470-bp nucleotide length obtained from the bats. The tree branches were colored by the proposed WNV lineages; orange: Lineage 6/7; hibiscus: Lineage 3 and lineage 4a; blue: Lineage 5; dark cyan: Lineage 1b; sky blue: Lineage 1a; and green: Lineage 2. All the WNV sequences detected from bats are identical to the previous study, which was detected in migratory birds and the SPU 116 98 strain from South Africa. Branches are scaled bootstrap.

4. Discussion

During the mosquito breeding seasons, there is an increase in the reported cases of West Nile Virus outbreaks in the USA, India, and Europe, involving humans and horses [20]. However, in tropical countries such as Malaysia, that have hot, humid, and monsoon seasons, the active mosquito breeding occurs throughout the year [21]. The infection rate is substantial since mosquitoes readily transmit deadly viruses during blood feeding. Although there is no evidence of WNV outbreaks in Malaysia thus far, unprecedented outbreaks of the WNV neuro-invasive disease leading to deaths in humans and animals have occurred elsewhere, due to the lack of early detection and intervention

Apart from mosquito vector control programmes, the WNV surveillance in wildlife is essential for elucidating the status of WNV shedding, with the aim of preventing a virus spillover into the humans and animals population. In the WNV infection, wild birds are a prominent WNV amplifier and reservoir [22]. Nevertheless, the role of wildlife animals, such as Malaysian macaques and bats are unknown. By considering this fact, the bio surveillance of the WNV antibody and RNA were conducted in archived sera and swab samples. The exposure to WNV was determined using a WNV IgG based competitive ELISA. The gold standard to confirm the presence of the WNV antibody was achieved by the neutralization test [23], however, it was not performed in this study due to the unavailability of the high containment biosafety facility required for the test. To rule out the cross reactivity with other Flavivirus species, a specific JEV based ELISA was included in the present study.

This study indicated that 29.63% (24/81) of long-tailed macaques were seropositive towards WNV. Although WNV RNA were not detected in macaques, they have been previously exposed to the virus as evidenced by the presence of the WNV IgG-antibody. This seropositive rate could possibly be due to the natural proximity between *Culex* mosquitoes and the macaque species within the forest habitat, thus causing WNV-specific antibodies to remain longer in macaques [8,24]. Several studies from around the world have reported a varied WNV seroprevalence rate ranging from low to moderate in macaques. In Louisiana, the WNV seroprevalence of 51.4%, 39.4%, 20.3%, 12.5% (2/16), and 6.6% (3/45) were detected in captive baboons (*Papio* spp.), rhesus macaques (*Macaca mulatta*), southern pig-tailed macaques (*Macaca nemestrina*), Japanese macaques (*Macaca fuscata*), and Georgian sooty mangabeys (*Cercocebusatys*), respectively [7,25,26].

Defining exactly the vulnerability to the WNV infection is among the most crucial indicators to be studied for effective WNV control and preventive measures. Therefore, oropharyngeal and rectal swabs obtained from bats were subjected to the one-step RT-PCR and partial sequencing targeting the WNV genes between the capsid and pre-membrane. In the present study, bats from the Pteropodidae, Emballonuridae, and Rhinolophidae families were shown to shed WNV. Among these three families, Pteropodidae and Rhinolophidae were found to be the reservoir for the Nipah Virus (*Paramyxovirus*) and SARS-CoV (Coronaviridae) [27–29]. As evidenced by this study, bats might possibly be infected with WNV either from blood-feeding mosquitoes or directly from mosquito ingestion, as bats are insectivores [30].

Thus far, lineage 1 and 2 have been associated with profound effects during outbreaks in humans [31]. The partial sequencing analysis from the present study demonstrated that the bats shed WNV from lineage 2, particularly from SPU116/89 strains from South Africa. This strain was previously recovered from a liver of human fatal hepatitis in South Africa. The same strain has been discovered in equine in Africa and showed an extreme neuro-invasive effect in experimental mice [32,33]. Interestingly, WNV lineage 2 from the same strain was also found in migratory and water birds found in Malaysia based on a recent study [18]. Further sequencing of the whole virus genome, or at least the hypervariable domain may possibly explain the intercontinental relatedness of WNV isolates from Malaysia and that of South Africa. Ecological links such as migratory factors could also provide a possible translocation of the virus or provide a link between bats in South Africa and Asian regions.

The current study demonstrates a relatively moderate prevalence rate of WNV in macaques and bats despite having a smaller sample size even in the first attempt of the study. This might be due to the natural exposure to Flavivirus in the wildlife mammals [34]. The pronounced exposure of the Malaysian wildlife to the mosquito-borne illness is anticipated, since a single species of mosquitoes exhibits the ability to carry more than one pathogen. For example, *Culex* spp. has been well recognized as the main vector for WNV transmission, as well as Chikungunya and JEV [35–37]. Moreover, staying in an ecosystem that favors mosquito breeding such as a mangrove forest or cave engenders them constantly to be exposed to mosquito bites [38]. Although the exact weather of the actual sampling period is unknown, it is believed that tropical countries such as Malaysia with a rainy, hot, and humid climate contribute to the high population of mosquitoes from the origin of the samples obtained. Collectively, contributing factors such as climate, ecosystem, habitat loss, antigenic properties of the virus, and the availability of vectors and host favor Malaysia to be prevalent to mosquito-borne illness.

5. Conclusions

To summarize, this study substantially highlights the first evidence of WNV infection in bats and long-tailed macaques in Malaysia. Therefore, a constant bio surveillance screening of WNV in the wildlife in Malaysia is a proactive attempt. This study was aligned with the Malaysian government's mission under the Malaysia Strategy for Emerging Diseases and Public Health Emergencies (MYSED) II (2017–2021) under the Ministry of Health priorities in order to enhance the regional capability to rapidly and accurately survey, detect, diagnose, and report outbreaks of pathogens and diseases of security concern.

Supplementary Materials: The following are available online at http://www.mdpi.com/2076-2615/10/12/2367/s1, Table S1: Sequence and location used for RT-PCR to amplify WNV gene, Table S2: List of references strains used in the phylogenetic and pairwise analysis. Figure S1: Pairwise percent identity between nucleotide (black) and amino acid (blue) sequences of selective WNV strains with local detection. The identity percentage range was set from high, medial and low similarity as indicated by light green, yellow and red color, respectively.

Author Contributions: Writing—original draft preparation, methodology, software, M.Y.A.-N.; supervision, writing—original draft preparation, methodology, conceptualization, writing—reviewing and editing, A.R.Y.; conceptualization, supervision A.R.O., S.S.A., J.A., F.B., and H.O.M.; software and editing, K.K., J.A.N. and M.N.M.; visualization, investigation, M.-L.A. and J.J.R.-R. All authors have read and agreed to the published version of the manuscript.

Funding: This research paper is based on the work supported by Grantsmanship Universiti Putra Malaysia (UPM), under the research grant UPM 700-2/1/GP-IPM/2016/9510500 and UPM 700-2/1/GP-IPS/2017/9547800.

Acknowledgments: We would like to express our sincere gratitude to the Department of Veterinary Service, (DVS) Malaysia in smoothing the process of WNV notification and reporting in macaques and bats to OIE. We would also like to thank all the staff of the Department of Wildlife and National Park (DWNP), Peninsular Malaysia for granting the macaques and bats archive samples.

Conflicts of Interest: The authors declare no conflict of interest. The funders had no role in the design of the study; in the collection, analyses, or interpretation of data; in the writing of the manuscript, or in the decision to publish the results.

References

1.   Bunde, J.M.; Heske, E.J.; Mateus-Pinilla, N.E.; Hofmann, J.E.; Novak, R.J. A survey for West Nile virus in bats from Illinois. *J. Wildl. Dis.* **2006**, *42*, 455–458. [CrossRef]

2.   Kulkarni, M.A.; Berrang-Ford, L.; Buck, P.A.; Drebot, M.A.; Lindsay, L.R.; Ogden, N.H. Major emerging vector-borne zoonotic diseases of public health importance in Canada. *Emerg. Microbes Infect.* **2015**, *4*, 1–7. [CrossRef] [PubMed]

3.   Lwande, O.W.; Lutomiah, J.; Obanda, V.; Gakuya, F.; Mutisya, J.; Mulwa, F.; Michuki, G.; Chepkorir, E.; Fischer, A.; Venter, M.; et al. Isolation of tick and mosquito-borne arboviruses from ticks sampled from livestock and wild animal hosts in Ijara District, Kenya. *Vector Borne Zoonotic Dis.* **2013**, *13*, 637–642. [CrossRef] [PubMed]

4.   Londono-Renteria, B.; Colpitts, T.M. A Brief Review of West Nile Virus Biology. *Methods Mol Biol.* **2016**, *1435*, 1–13. [CrossRef] [PubMed]

5.   Chancey, C.; Griney, A.; Volkova, E.; Rios, M. The global ecology and epidemiology of West Nile virus. *BioMed Res. Int.* **2015**, 1–20. [CrossRef] [PubMed]

6.   Root, J.J. West Nile virus associations in wild mammals: A synthesis. *Arch. Virol.* **2013**, *158*, 735–752. [CrossRef] [PubMed]

7.   Ølberg, R.A.; Barker, I.K.; Crawshaw, G.J.; Bertelsen, M.F.; Drebot, M.A.; Andonova, M. West Nile virus encephalitis in a Barbary macaque (*Macaca sylvanus*). *Emerg. Infect Dis.* **2004**, *10*, 712–714. [CrossRef]

8. Ratterree, M.S.; Gutierrez, R.A.; Travassos da Rosa, A.P.; Dille, B.J.; Beasley, D.W.C.; Bohm, R.P.; Desai, S.M.; Didier, P.J.; Bikenmeyer, L.G.; Dawson, G.J.; et al. Experimental infection of rhesus macaques with West Nile virus: Level and duration of viremia and kinetics of the antibody response after infection. *J. Infect. Dis.* **2004**, *189*, 669–676. [CrossRef]

9. Kading, R.C.; Schountz, T. Flavivirus infections of bats: Potential role in Zika virus ecology. *Am. J. Trop. Med. Hyg.* **2016**, *95*, 993–996. [CrossRef]

10. Pilipski, J.D.; Pilipskl, L.M.; Risley, L.S. West nile virus antibodies in bats from New Jersey and New York. *J. Wildl. Dis.* **2004**, *40*, 335–337. [CrossRef] [PubMed]

11. Plowright, R.K.; Parrish, C.R.; McCallum, H.; Hudson, P.J.; Ko, A.I.; Graham, A.L.; Lloyd-Smith, J.O. Pathways to zoonotic spillover. *Nat. Rev. Microbiol.* **2017**, *15*, 502–510. [CrossRef] [PubMed]

12. Low, V.L.; Chen, C.D.; Lee, H.L.; Lim, P.E.; Leong, C.S.; Sofian-Azirun, M. Nationwide distribution of *Culex* mosquitoes and associated habitat characteristics at residential areas in Malaysia. *J. Am. Mosq. Control Assoc.* **2012**, *28*, 160–169. [CrossRef] [PubMed]

13. Marlina, S.; Muhd Radzi, S.F.; Lani, R.; Sieng, K.C.; Rahim, N.F.A.; Hassan, H.; Li-Yen, C.; AbuBakar, S.; Zandi, K. Seroprevalence screening for the West Nile virus in Malaysia's orang asli population. *Parasites Vectors* **2014**, *7*, 1–7. [CrossRef] [PubMed]

14. Rais, M.N.; Omar, A.R.; Jalila, A.; Hussni, O.M. Prevalence of West Nile virus antibody in captive bird populations in selected areas in Selangor, Malaysia. In Proceedings of the 6th Seminar on Veterinary Sciences, Faculty of Veterinary Medicine, Universiti Putra Malaysia, Selangor, Malaysia, 11–14 January 2011; Vulume 127.

15. Ain-Najwa, M.Y.; Yasmin, A.R.; Omar, A.R.; Arshad, S.S.; Abu, J.; Mohammed, H.O.; Kumar, K.; Loong, S.K.; Rovie-Ryan, J.J.; Mohd-Kharip-Shah, A.-K. Evidence of West Nile virus infection in migratory and resident wild birds in West Coast of Peninsular Malaysia. *One Health.* **2020**, *10*, 1–7. [CrossRef]

16. Ching, C.Y.; Casals, J.; Bowen, E.T.; Simpson, D.I.; Platt, G.S.; Way, H.J.; Smith, C.E. Arbovirus infections in Sarawak: The isolation of Kunjin virus from mosquitoes of the *Culex pseudovishnui* group. *Ann. Trop. Med. Parasitol.* **1970**, *64*, 263–268. [CrossRef]

17. Mohammed, M.N.; Yasmin, A.R. West Nile virus: Measures against emergence in Malaysia. *J. Vet. Sci. Res.* **2019**, *4*, 1–6. [CrossRef]

18. Ain-Najwa, M.Y.; Yasmin, A.R.; Omar, A.R.; Arshad, S.S.; Abu, J.; Mohammed, H.O. West Nile Virus Infection in Human and Animals: Potential Risks in Malaysia. *Sains Malays.* **2019**, *48*, 2727–2735. [CrossRef]

19. Tamura, K.; Nei, M.; Kumar, S. Prospects for inferring very large phylogenies by using the neighbor-joining method. *Proc. Natl. Acad. Sci. USA* **2004**, *101*, 30–11030. [CrossRef]

20. Marini, G.; Rosà, R.; Pugliese, A.; Rizzoli, A.; Rizzo, C.; Russo, F.; Montarsi, F.; Capelli, G. West Nile virus transmission and human infection risk in Veneto (Italy): A modelling analysis. *Sci. Rep.* **2018**, *8*, 1–10. [CrossRef]

21. Hii, Y.L.; Zaki, R.A.; Aghamohammadi, N.; Rocklöv, J. Research on climate and dengue in Malaysia: A systematic review. *Curr. Environ. Health Rep.* **2016**, *3*, 81–90. [CrossRef]

22. Di Sabatino, D.; Bruno, R.; Sauro, F.; Danzetta, M.L.; Cito, F.; Ianetti, S.; Narcisi, V.; De Massis, F.; Calistri, P. Epidemiology of West Nile virus disease in Europe and in the Mediterranean Basin from 2019 to 2013. *BioMed Res. Int.* **2014**, *2014*, 1–11. [CrossRef] [PubMed]

23. Murata, R.; Hashiguchi, K.; Yoshii, K.; Kariwa, H.; Nakajima, K.; Ivanov, G.N.; Leonova, G.N.; Takashima, I. Seroprevalence of West Nile virus in wild birds in far eastern Russia using a focus reduction neutralization test. *Am. J. Trop. Med. Hyg.* **2011**, *84*, 461–465. [CrossRef] [PubMed]

24. Mangudo, C.; Aparicio, J.P.; Rossi, G.C.; Gleiser, R.M. Tree hole mosquito species composition and relative abundances differ between urban and adjacent forest habitats in north western Argentina. *Bull. Entomol. Res.* **2018**, *108*, 203–212. [CrossRef] [PubMed]

25. Ratterree, M.S.; Amelia, P.A.; Travassos da Rosa, A.P.A.; Bohm Jr, R.P.; Cogswell, F.B.; Phillippi, K.M.; Caillouet, K.; Schwanberger, S.; Shope, R.E.; Tesh, R.B. West Nile virus infection in nonhuman primate breeding colony, concurrent with the human epidemic, Southern Louisiana. *Emerg. Infect. Dis.* **2003**, *9*, 1388–1394. [CrossRef] [PubMed]

26. Cohen, J.K.; Kilpatrick, A.M.; Stroud, F.C.; Paul, K.; Wolf, F.; Else, J.G. Seroprevalence of West Nile virus in nonhuman primates as related to mosquito abundance at two national primate research centres. *Comp. Med.* **2007**, *57*, 115–119.

27. Gay, N.; Olival, K.J.; Bumrungsri, S.; Siriaroonrat, B.; Bourgarel, M.; Morand, S. Parasite and viral species richness of Southeast Asian bats: Fragmentation of area distribution matters. *Int. J. Parasitol. Parasites Wildl.* **2014**, *3*, 161–170. [CrossRef]

28. Field, H.; Epstein, J.H.; Henipavirus. Investigating the Role of Bats in Emerging Zoonoses, Balancing Ecology, Conservation and Public Health Interests. In *FAO Animal Production and Health Manual*; Newman, S., Field, H.E., De Jong, C.E., Epstein, J.H., Eds.; Univerza v Mariboru: Maribor, Slovenia, 2011; Volume 12, p. 64.

29. Balboni, A.; Battilani, M.; Prosperi, S. The SARS-like coronaviruses: The role of bats and evolutionary relationships with SARS coronavirus. *New Microbiol.* **2012**, *35*, 1–16.

30. Barclay, R.M.R.; Bell, G.P. Marketing and observational techniques. In *Ecological and Behavioural Methods for the Study of Bats*, 2nd ed.; Smithsonian Institution Press: Washington, DC, USA, 1990; pp. 59–76.

31. May, F.J.; Davis, C.T.; Tesh, R.B.; Barrett, A.D.T. Phylogeography of West Nile virus: From the cradle of evolution in Africa to Eurasia, Australia, and the Americas. *J. Virol.* **2011**, *85*, 2964–2974. [CrossRef]

32. Venter, M.; Human, S.; Zaayman, D.; Gerdes, G.H.; Williams, J.; Steyl, J.; Leman, P.A.; Paweska, J.T.; Setzkorn, H.; Rous, G.; et al. Lineage 2 West Nile Virus as Cause of Fatal Neurologic Disease in Horses. *S. Afr. Emerg. Infect. Dis.* **2009**, *15*, 877–884. [CrossRef]

33. Botha, E.M.; Markotter, W.; Wolfaardt, M.; Paweska, J.T.; Swanepoel, R.; Palacios, G.; Nel, L.H.; Venter, M. Genetic determinants of virulence in pathogenic lineage 2 West Nile virus strains. *Emerg. Infect. Dis.* **2008**, *14*, 222–230. [CrossRef]

34. Petruccelli, A.; Zottola, T.; Ferrara, G.; Iovane, V.; Russo, C.D.; Pagnini, U.; Montagnaro, S. West Nile virus and related flavivirus in European wild boar (*Suc scrofa*), Latium region, Italy: A retrospective study. *Animals* **2020**, *10*, 494. [CrossRef] [PubMed]

35. Vogel, C.B.F.; Göertz, G.P.; Pjilman, G.P.; Koenraadt, C.J. Vector competence of European mosquitoes for West Nile virus. *Emerg. Microbes Infect.* **2017**, *6*, 1–13. [CrossRef] [PubMed]

36. Lindahl, J.; Chirico, J.; Boqvist, S.; Thu, H.T.V.; Magnusson, U. Occurrence of Japanese Encephalitis virus mosquito vectors in relation to urban pig holdings. *Am. J. Trop. Med. Hyg.* **2012**, *87*, 1076–1082. [CrossRef] [PubMed]

37. Richards, S.L.; Anderson, S.L.; Smartt, C.T. Vector competence of Florida mosquitoes for chikungunya virus. *J. Vector Ecol.* **2010**, *35*, 439–443. [CrossRef] [PubMed]

38. Marini, G.; Poletti, P.; Giacobini, M.; Pugliese, A.; Merler, S.; Rosà, R. The role of climatic and density-dependent factors in shaping mosquito population dynamics: The case of *Culex pipiens* in Northwestern Italy. *PLoS ONE* **2016**, *11*, e0154018. [CrossRef] [PubMed]

Publisher's Note: MDPI stays neutral with regard to jurisdictional claims in published maps and institutional affiliations.

Article

Ecotyping of *Anaplasma phagocytophilum* from Wild Ungulates and Ticks Shows Circulation of Zoonotic Strains in Northeastern Italy

Laura Grassi, Giovanni Franzo, Marco Martini, Alessandra Mondin, Rudi Cassini, Michele Drigo, Daniela Pasotto, Elena Vidorin and Maria Luisa Menandro *

Department of Animal Medicine, Production and Health (MAPS), University of Padova, 35020 Legnaro, Padova, Italy; laura.grassi.2@phd.unipd.it (L.G.); giovanni.franzo@unipd.it (G.F.); marco.martini@unipd.it (M.M.); alessandra.mondin@unipd.it (A.M.); rudi.cassini@unipd.it (R.C.); michele.drigo@unipd.it (M.D.); daniela.pasotto@unipd.it (D.P.); elena.vidorin@gmail.com (E.V.)
* Correspondence: marialuisa.menandro@unipd.it

Simple Summary: Tick-borne infectious diseases represent a rising threat both for human and animal health, since they are emerging worldwide. Among the bacterial infections, Anaplasma phagocytophilum has been largely neglected in Europe. Despite its diffusion in ticks and animals, the ecoepidemiology of its genetic variants is not well understood. The latest studies identify four ecotypes of Anaplasma phagocytophilum in Europe, and only ecotype I has shown zoonotic potential. The aim of the present study was to investigate the genetic variants of Anaplasma phagocytophilum in wild ungulates, the leading reservoir species, and in feeding ticks, the main vector of infection. The analyzed samples were collected in northeastern Italy, the same area where the first Italian human cases of anaplasmosis in the country were reported. Using biomolecular tools and phylogenetic analysis, ecotypes I and II were detected in both ticks (Ixodes ricinus species) and wild ungulates. Specifically, ecotype II was mainly detected in roe deer and related ticks; and ecotype I, the potentially zoonotic variant, was detected in Ixodes ricinus ticks and also in roe deer, red deer, chamois, mouflon, and wild boar. These findings reveal not only the wide diffusion of Anaplasma phagocytophilum, but also the presence of zoonotic variants.

Abstract: *Anaplasma phagocytophilum* (*A. phagocytophilum*) is a tick-borne pathogen causing disease in both humans and animals. Human granulocytic anaplasmosis (HGA) is an emerging disease, but despite the remarkable prevalence in European ticks and wild animals, human infection appears underdiagnosed. Several genetic variants are circulating in Europe, including the zoonotic ecotype I. This study investigated *A. phagocytophilum* occurrence in wild ungulates and their ectoparasites in an area where HGA has been reported. Blood samples from wild ungulates and ectoparasites were screened by biomolecular methods targeting the *mps2* gene. The *groEL* gene was amplified and sequenced to perform genetic characterization and phylogenetic analysis. A total of 188 blood samples were collected from different wild ungulates species showing an overall prevalence of 63.8% (88.7% in wild ruminants and 3.6% in wild boars). The prevalence of *A. phagocytophilum* DNA in ticks (manly *Ixodes ricinus*), and keds collected from wild ruminants was high, reflecting the high infection rates obtained in their hosts. Among ticks collected from wild boars (*Hyalomma marginatum* and *Dermacentor marginatus*) no DNA was detected. Phylogenetic analysis demonstrated the presence of ecotype I and II. To date, this is the first Italian report of ecotype I in alpine chamois, mouflon, and wild boar species. These findings suggest their role in HGA epidemiology, and the high prevalence detected in this study highlights that this human tick-borne disease deserves further attention.

Keywords: *Anaplasma phagocytophilum*; zoonosis; tick; wild ungulates; phylogenesis; molecular epidemiology

Citation: Grassi, L.; Franzo, G.; Martini, M.; Mondin, A.; Cassini, R.; Drigo, M.; Pasotto, D.; Vidorin, E.; Menandro, M.L. Ecotyping of *Anaplasma phagocytophilum* from Wild Ungulates and Ticks Shows Circulation of Zoonotic Strains in Northeastern Italy. *Animals* **2021**, *11*, 310. https://doi.org/10.3390/ani11020310

Academic Editor: David González-Barrio

Received: 4 December 2020
Accepted: 22 January 2021
Published: 26 January 2021

Publisher's Note: MDPI stays neutral with regard to jurisdictional claims in published maps and institutional affiliations.

1. Introduction

Anaplasma phagocytophilum (*A. phagocytophilum*) is a small, Gram-negative, obligate intracellular bacterium belonging to the Anaplasmataceae family, order Rickettsiales [1]. It infects white blood cells, mainly neutrophilic granulocytes, and is the causal agent of human granulocytic anaplasmosis (HGA), tick-borne fever (TBF), or pasture fever of domestic ruminants and granulocytic anaplasmosis of horses (equine granulocytic anaplasmosis, EGA), dogs (canine granulocytic anaplasmosis, CGA) and cats (feline anaplasmosis, FA) [2,3].

Anaplasma phagocytophilum is a tick-borne pathogen mainly transmitted, in Europe, by the *Ixodes* genus, especially *I. ricinus*. This bacterium has been detected in other hard tick species, but their vectorial efficiency seems negligible [3–5]. Occasionally, HGA can be caused by blood transfusion, transplacental transmission, or direct contact with infected blood of wild ungulates at butchering [6]. However, tick bites represent the main route of transmission, and once infection is established, ticks maintain *A. phagocytophilum* through trans-stadial transmission, while the transovarial route seems not to be efficient [7]. Accordingly, mainly nymphal and adult stages are involved in bacterial transmission [8]. *Ixodes ricinus* is therefore considered an efficient vector but not a reservoir [4], thus vertebrate species, especially wildlife, seem to be pivotal in maintaining *A. phagocytophilum* circulation [9,10].

Prevalence in *I. ricinus* ticks is reported to increase according to the number of blood meals. Adult ticks therefore show higher infection rates than nymphs and larvae [4], reaching even 100% in adult stages collected from wild ungulates [11]. Wild ungulates harbor relevant amounts of all tick stages [11] and play a critical role in the transmission cycle of anaplasma, as they act as both tick hosts and reservoir species of *A. phagocytophilum* [12]. In Europe, roe deer (*Capreolus capreolus*) and red deer (*Cervus elaphus*) in particular show very high *A. phagocytophilum* prevalence [4]. Although other wild ruminants may display a non-negligible infection frequency, their role has not yet been established [8]. In wild boar, *A. phagocytophilum* infection is less common, but this ungulate has been suggested as a reservoir of zoonotic strains [11,13]. Indeed, the genetic diversity of *A. phagocytophilum* strains has been demonstrated, and the genetic variants seem to display a host preference, showing different epidemiological and clinical pictures [8,10]. The genes commonly used to genotype and distinguish these variants are 16S *rRNA*, *groEL*, *ankA*, and *msp4*. Among these, the *groEL* gene, belonging to *groESL* heat shock operon, seems to be a good marker of *A. phagocytophilum* genetic diversity. It is mostly studied in Europe, where the genetic variability is greater than that observed in other geographical areas, such as North America [3,10].

Based on the *groEL* classification, four ecotypes, subdivided into eight clusters, have been identified in different tick and vertebrate species [10,14]. Ecotype I has the widest host range, and is found in livestock, dogs, cats, horses, and wild species, including carnivores, small mammals, wild boars, and other ungulates. Also, human infections are driven by ecotype I, which is of primary importance in the epidemiology of HGA [3,14]. Other ecotypes seem to be preferentially associated with certain species. Ecotype II is detected mainly in roe deer and only sporadically in other species, ecotype III is linked to small mammals, especially rodents, and ecotype IV is found mainly in birds. This distribution seems to also be related to the host preference of different species of ticks. Ecotypes I and II are mainly detected in *I. ricinus* [10]. Both ecotypes have also been identified in other ectoparasites such as deer keds (*Lipoptena cervi*, *L. cervi*) [14]. The role of this species in the transmission of anaplasmosis has been suggested but not demonstrated [15]. Ecotypes III and IV have been detected in *Ixodes trianguliceps* and *Ixodes frontalis*, which usually feed on rodents and birds, respectively [14]. The distribution of European ecotypes thus seems to be affected by vertebrate host and vector species, their interaction and, consequently, geographic origin [3,10]. Although ecotypes seem more represented in some preferential animal species, many studies show their overlap within the same mammal species, suggesting that *A. phagocytophilum* variants may not be host-specific at all [3,16].

Human granulocytic anaplasmosis has been reported worldwide, but is mainly detected in the northern hemisphere [17]. The first human cases in the USA and Europe were identified in 1990 and 1996, respectively [18,19]. In Italy, the first HGA cases were reported in the Friuli–Venezia–Giulia region in 2000 [20], followed by other cases reported in Sicily and Sardinia [21,22]. Despite high infection rates of *A. phagocytophilum* in ticks and in different wild and domestic species, HGA seems to be rare at present. In Italy, 16 different tick species have been reported to bite humans, in Northeastern Italy, almost all cases are attributed to *I. ricinus* [23–25]. A possible reason for the low HGA prevalence in Italy is that it is not a notifiable disease, which could lead to underestimation of disease incidence [24]. In addition, the subclinical or poorly specific nature of *A. phagocytophilum*-induced symptoms further decreases the recognition and reporting infections. However, if not treated, it could evolve into more severe and even fatal syndromes. Immunocompromised patients are at higher risk of high morbidity and mortality [26]. Based on these findings, many authors agree that HGA is neglected and underdiagnosed, and thus underestimated [3,9,26].

Although the studied area, northeastern Italy, is the same place where the first Italian cases of HGA were reported [20,27], to date, there are few epidemiological and no phylogenetic studies on this pathogen in wild ungulates, which may act as an important reservoir of *A. phagocytophilum* variants related to HGA. Thus, this study aims to fill this gap, evaluating the presence and distribution of zoonotic *A. phagocytophilum* ecotypes in wildlife and in their associated ectoparasites.

2. Materials and Methods

2.1. Sampling Sites and Specimens Collection

The area under investigation comprises several different sites in northeastern Italy. The first comprises pre-alpine and alpine localities in the province of Udine in the Friuli–Venezia–Giulia region. The other two, both located in the Veneto region, included pre-alpine and alpine areas in Belluno province, and hilly areas in Euganean Hills Regional Park. The highest mountain peak of alpine areas reaches an elevation of 2500 m above sea level (a.s.l.), while Euganean hills are characterized by a maximum elevation of 600 m a.s.l. The hunting reserves are all characterized by the presence of wild ungulate populations, that increased remarkably in the last decades as a consequence of animal protection and reintroduction policies or illegal release of gaming animals [28]. In particular, free-ranging wild ruminants are particularly present in pre-alpine and alpine areas, where the most abundant are roe deer and red deer, followed by chamois and mouflon; these areas are also colonized by free-ranging wild boar populations [29,30]. In contrast, only free-ranging wild boars and fallow deer are present in the hilly areas of Euganean Hills Regional Park [31].

Blood and ectoparasites were collected from culled wild ungulates by hunters during the hunting seasons of 2017–2018 and 2018–2019, lasting from spring to winter. Hunters were instructed and trained on sampling procedures and data collection, reporting the features of killed animals (species, sex, and estimated age), and the place and the date of hunting. To avoid cross-contamination linked to field conditions, a selection of samples was carried out, i.e., collected when only one animal was killed per day for each hunter. Whole-blood samples were retrieved by cutting the jugular vein or main vessels of the chest cavity and collected in 9 mL Vacumed® with K3EDTA tubes (FL Medical s.r.l., Torreglia, Padova, Italy), refrigerated as soon as possible, and brought to the veterinary infectious disease laboratory (Department of Animal Medicine, Production and Health, Padua University, Italy) within 4 days. The blood samples were divided into 200 µL aliquots and stored at −80 °C until analysis.

When present, the matching ectoparasites were also collected before animal exsanguination and brought to the laboratory in sterile plastic tubes. The identification of tick species was performed with stereomicroscope and microscope, following the identification keys described by Cringoli et al., then the parasites were stored at −80 °C until analysis [32].

A total of 188 blood samples were collected from wild ungulates: 17 from Euganean Hills Regional Park, 62 from an alpine area of the Veneto region, and 109 from an alpine area of the Friuli– Venezia–Giulia region. Six species were sampled: Alpine chamois (*Rupicapra rupicapra*; n = 9), roe deer (*Capreolus capreolus*; n = 74), red deer (*Cervus elaphus*; n = 39), mouflon (*Ovis musimon*; n = 8), fallow deer (*Dama dama*; n = 3), and wild boar (*Sus scrofa*; n = 55). A subsample of ticks (up to a maximum of 10) was collected from all infested ungulates. With the exception of alpine chamois, wild ruminant species had a higher tick burden compared to wild boar. In total, 277 ticks were collected and analyzed. The following tick species were identified: *I. ricinus* (n = 258), *Dermacentor marginatus* (*D. marginatus*) (n = 18) and *Hyalomma marginatum* (*H. marginatum*) (n = 1). In addition, also 15 deer keds (*L. cervi*) were found. When more than one tick was collected from the same host, ticks were analyzed in pools composed of a maximum of 2 specimens characterized by the same species, gender, and stage. Thus, a total of 216 samples were tested for *A. phagocytophilum*: 201 tick samples (pooled ticks = 76; individual ticks = 125) and 15 deer keds.

2.2. Biomolecular Analysis

DNA was extracted from 200 μL of whole blood and from tick homogenates using the DNeasy Blood & Tissue Kit (QIAGEN, Hilden, Germany), according to the manufacturer's instructions. A canine blood sample, certified as negative to *A. phagocytophilum* infection by the Italian authority and research organization for animal health and food safety (Istituto Zooprofilattico delle Venezie, IZSVe), was included in each extraction run as an extraction negative control and thereafter tested together with the diagnostic samples. Prior to extraction, an exogenous DNA internal control (supplied by Quanti-Nova Pathogen + IC kit, QIAGEN) was spiked in all blood and ectoparasites specimens. To exclude false negative results due to the presence of polymerase inhibitors, all DNA samples were checked in duplicate for the presence of the internal control DNA, according to the manufacturer's instructions. DNA samples showing no inhibition were then screened for the presence of *A. phagocytophilum* using real-time PCR targeting a 77 bp portion of the *msp2* gene (Table 1) [33].

Table 1. List of primers and probes used for *Anaplasma phagocytophilum* detection and sequencing.

Biomolecular Method	Target Gene	Primer	Nucleotide Sequence 5′–3′	Reference
Real-time PCR	*msp2*	ApMSP2f	ATGGAAGGTAGTGTTGGTTATGGTATT	[33]
		ApMSP2r	TTGGTCTTGAAGCGCTCGTA	-
		ApMSP2p	HEX-TGGTGCCAGGGTTGAGCTTGAGATTG-BHQ1	-
PCR	*groEL*	groEL643f	ACTGATGGTATGCARTTTGAYCG	[34]
		groEL1236r	TCTTTRCGTTCYTTMACYTCAACTTC	-

All real-time PCR reactions were performed on a LightCycler96 instrument (Roche, Basel, Switzerland) using QuantiNova Pathogen + IC Kit reagents (QIAGEN, Hilden, Germany) in a final reaction volume of 7 μL, consisting of 0.8 μM of each primer, 0.25 μM of probe and 2 μL of DNA sample. The following thermal protocol was used: Preincubation at 95 °C for 120 s, followed by 50 cycles of denaturation at 95 °C for 5 s, annealing at 60 °C for 5 s, and extension at 72 °C for 30 s. Internal, positive, and negative controls were included in each run. Samples were considered positive if both replicates showed an amplification curve with a mean Cq value lower to 40.00 or equal.

To genetically characterize *A. phagocytophilum* strains, positive samples were tested by conventional PCR using primers targeting a 600 bp portion of *groEL* gene (Table 1) [34]. As a target, 5 μL of DNA was used in a 25 μL reaction mixture containing 0.3 μM of each primer and 1× Phire Hot Start II PCR Master Mix (Thermo Fischer Scientific, Milano, Italy). Amplifications were carried out in a TGradient thermal cycler (Biometra, Analytic Jena GmbH, Jena, Germany) in the following conditions: Initial denaturation at 98 °C for 60 s, followed by 50 cycles of denaturation at 98 °C for 5 s, annealing at 60 °C for 5 s, and extension at 72 °C for

7 s. Negative and positive controls were included in each run. PCR products were visualized by 2% agarose gel electrophoresis containing SybrSafe DNA (Thermo Fischer Scientific, Milano, Italy) gel stain in Tris-borate-EDTA (TBE) buffer (Merk KGaA, Darmstadt, Germany) using GelDoc EZ Imager (Bio–Rad Laboratories, Milano, Italy). Amplicons were enzymatically purified with ExoSap-IT Express PCR Product Cleanup (Thermo Fischer Scientific, Milano, Italy) and delivered to an external laboratory (StarSEQ®, Mainz, Germany) to be sequenced in both directions using the same PCR primers.

2.3. Statistical Analysis

Logistic regression models were fitted to evaluate the association between *A. phagocytophilum* infection and host features. The following categorical variables were considered in univariable and multivariable models: Area (Euganean Hills Regional Park, Friuli–Venezia–Giulia Alps, Veneto Alps), season (spring, summer, autumn, winter), gender (female, male), age (<1 year, ≥1 year), and species (chamois, red deer, roe deer, wild boar). Due to the small number of tested individuals, fallow deer and mouflon were not included among the considered species. The analysis was performed also considering the variable species as dichotomous (ruminant, wild boar), also including in that case fallow deer and mouflons. A model was preferred over a simpler one when a significant improvement on the overall fit was demonstrated by likelihood ratio test. All analyses were performed in SPSS v23.0 (IBM Corporation, Armonk, NY, USA): The significance level was set to $p < 0.05$.

2.4. Sequence Analysis

Consensus sequences were generated using ChromasPro Software v.2.1.8 (Technelysium Pty Ltd., South Brisbane, QLD, Australia) and compared with representative sequences available in the National Center for Biotechnology GenBank database with the BLASTn tool [35]. The genetic analysis was performed according to the classification proposed by Jahafari et al. and implemented by Jaarsma et al., based on *groEL* gene sequencing [10,14].

DNA sequences corresponding to the region between nucleotide positions 533 and 1033 of the CP015376 *groEL* open reading frame obtained in the present study, were aligned to 1998 sequences representative of the 4 ecotypes and 8 clusters previously described [10] using MAFFT v7.450 [36]. Additionally, the Italian sequences reported by Di Domenico et al., were also included in the study [37]. To reduce the tree complexity and computational burden, only one sequence representative of all identical ones was identified using CD-HIT and included in the phylogenetic tree (Table S1) [38]. The sequence suitability for phylogenetic analysis was assessed by likelihood mapping analysis, performed using IQ-TREE, and a phylogenetic tree was reconstructed with the same software, selecting as a substitution model the one with the lowest Akaike information criterion (AIC), value calculated using JmodelTest [39,40]. The reliability of inferred clades was investigated by performing 10,000 ultrafast bootstrap replicates.

3. Results

3.1. Wild Ungulates

The overall prevalence of *A. phagocytophilum* in wild ungulates was 63.8% (95% CI: 56.7–70.3%; 120 out of 188). However, the prevalence was above 75% in all ruminant species and significantly higher (overall ruminants prevalence = 88.7% (95% CI: 83.4–94.1; 118 out of 133) than that observed in wild boar (3.6%; 95% CI: 0–8.6; 2 out of 55). Wild ruminants thus have 208 times higher odds of being infected (95% CI: 57.1–504.1; $p < 0.001$) (Table 2). No significant differences were identified among wild ruminant species ($p = 0.454$).

Table 2. Prevalence of *Anaplasma phagocytophilum* detected by real-time PCR in ungulate species grouped according to different variables. EHR, Euganean Hills Regional Park; FVG, Friuli–Venezia–Giulia.

Sample Features		Roe Deer Pos/N Prevalence	Red Deer Pos/N Prevalence	Mouflon Pos/N Prevalence	Alpine Chamois Pos/N Prevalence	Fallow Deer Pos/N Prevalence	Wild Boar Pos/N Prevalence
Ungulate		68/74 (91.9%)	34/39 (87.2%)	6/8 (75.0%)	7/9 (77.8%)	3/3 (100%)	2/55 (3.6%)
Gender	Female	22/24 (91.7%)	14/15 (93.3%)	3/3 (100%)	1/2 (50.0%)	1/1 (100%)	1/20 (5%)
	Male	46/50 (92%)	20/24 (83.3%)	3/5 (60.0%)	6/7 (85.71%)	2/2 (100%)	1/35 (2.9%)
Age *	<1 year	25/26 (96.2%)	10/11 (90.9%)	2/2 (100%)	2/2 (100%)	1/1 (100%)	1/25 (4.0%)
	>1 year	41/46 (89.1%)	23/27 (85.2%)	4/6 (66.7%)	5/7 (71.4%)	2/2 (100%)	0/28 (0%)
Season	Spring	6/7 (85.71%)	1/1 (100%)	-	-	-	1/18 (5.6%)
	Summer	38/42 (90.5%)	8/9 (88.9%)	1/1 (100%)	2/3 (66.7%)	1/1 (100%)	0/18 (0%)
	Autumn	20/21 (95.2%)	22/26 (84.6%)	5/7 (71.4%)	4/5 (80.0%)	2/2 (100%)	0/13 (0%)
	Winter	4/4 (100%)	3/3 (100%)	-	1/1 (100%)	-	1/6 (16.7%)
Area	EHR Park	-	-	-	-	3/3 (100%)	0/14 (0%)
	Veneto Alps	23/24 (95.8%)	19/22 (86.4%)	3/4 (75.0%)	-	-	2/12 (16.7%)
	FVG Alps	45/50 (90%)	1/17 (5.9%)	3/4 (75.0%)	7/9 (77.8%)	-	0/29 (0%)

* Ages of four positive animals (2 roe deer, 1 red deer, and 1 wild boar) and one negative wild boar were not available.

Other statistically significant differences in *A. phagocytophilum* infection prevalence were observed in univariable analysis among the three different areas. Particularly, the odds ratios were 8.4 (95% CI: 2.5–37.9; $p = 0.001$) and 14.6 (95% CI: 4.1–69.9; $p < 0.001$) higher in Friuli–Venezia–Giulia and the Veneto Alps than in the Euganean Hills Regional Park, respectively. These differences should be evaluated considering the species distribution, which can act as confounder. In particular, wild boar samples were collected especially from the Euganean Hills Regional Park during spring. In fact, taking into account host distribution in the multivariable analysis, the differences among geographic areas became non-significant. A significant difference was observed among seasons, with summer, autumn, and winter showing odds of 4.7 (95% CI: 1.8–12.8; $p = 0.001$), 5.6 (95% CI: 2.2–15.7; $p < 0.001$) and 4.0 (95% CI: 1.1–17.2; $p = 0.046$) higher than spring. Also in this case, seasonal patterns of wild species culling can be considered a confounder, and no significant difference in seasonality was observed when the host species was included in the multivariable model. No significant differences were observed according to gender ($p = 0.871$) or age ($p = 0.639$).

3.2. Ectoparasites

Out of 201 tick samples, 122 were *I. ricinus* adult ticks, mainly female (142/182); adult male (38/182) and nymphal stages (2/182) were less frequent. All *D. marginatus* (10/18 females and 8/18 males) and *H. marginatum* (1 male) were collected from wild boars at Euganean Hills Regional Park. *L. cervi* keds were found on roe deer (9/15), red deer (3/15), and wild boars (3/15).

Real-time PCR positive results were obtained in 63.7% of ticks (95% CI: 56.5–70.2; 128 out of 201) of ticks and in 40.0% of deer keds (95% CI: 17.4–67.1; 6 out of 15). Bacterial DNA was not detected in *D. marginatus* and *H. marginatum* that were collected from negative wild boars. One out of two nymphal pools tested positive.

Results of real-time PCR analysis of tick *A. phagocytophilum* were grouped according to several variables (Table 3). Twelve host–parasite pairs displayed at least one missing feature (e.g., sex, engorgement, etc.). Therefore, a total of 189 records were included in this final analysis. Tick positivity to *Anaplasma* infection was differently distributed according to host species, host infectious status, and tick sex (Table 3).

Table 3. Real-time PCR positivity of *Anaplasma phagocytophilum* in ticks according to different host and vector variables. Only records with complete data for both hosts and ectoparasites are included.

Sample Features		Tick Infection: Neg. (N = 69)	Tick Infection: Pos. (N = 120)
Host Species	Roe deer	35/69 (50.7%)	75/120 (62.5%)
	Red deer	5/69 (7.2%)	37/120 (30.8%)
	Mouflon	4/69 (5.8%)	6/120 (5.0%)
	Alpine chamois	1/69 (1.4%)	0/120 (0.0%)
	Fallow deer	2/69 (2.9%)	2/120 (1.7%)
	Wild boar	22/69 (31.9%)	0/120 (0.0%)
Host Infection	Neg.	28/69 (40.6%)	12/120 (10.0%)
	Pos.	41/69 (59.4%)	108/120 (90.0%)
Tick Sex	Female	43/69 (62.3%)	102/120 (85.0%)
	Male	26/69 (37.7%)	18/120 (15.0%)

3.3. Ecotypes and Clusters

Only samples positive with real-time PCR screening (120 blood samples, 128 tick samples, and 6 keds) were further analyzed by conventional PCR, targeting a portion of the *groEL* gene to ecotype them. Fifty-seven good quality sequences were obtained when mean Cq values between the two replicates in real-time PCR were below 33.00. Of these 57 sequences, 48 were from blood samples and 9 from *I. ricinus* feeding ticks (GenBank accession numbers from MT473440 to MT 473496) and were used for bioinformatics analysis (Figure 1 and Figure S1 and Table S2).

Two *A. phagocytophilum* from ticks and 28 from animals were classified as ecotype I, cluster 1 (n = 30). These were derived from wild boar (n = 1), alpine chamois (n = 1), mouflon (n = 3), roe deer (n = 6), and red deer (n = 17). The ticks infected with ecotype I were hosted by one roe deer and one red deer. All other *A. phagocytophilum* sequences (n = 27) were classified as ecotype 2, and all were grouped in cluster 3, sampled from roe deer (n = 20) and related ticks (n = 7). Of note, a tick infected with *A. phagocytophilum* belonging to ecotype I was feeding on an ecotype II positive roe deer. Further details are shown in Supplementary Files (Figure S1 and Table S2).

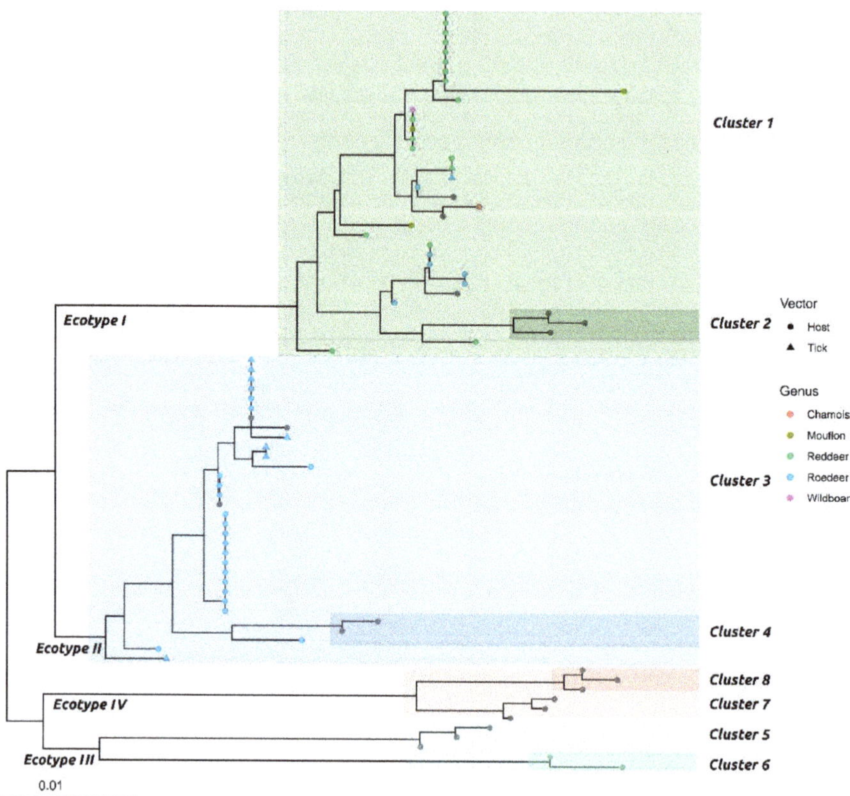

Figure 1. Maximum likelihood phylogenetic tree based on *groEL* gene of Italian strains collected from wild ungulates (dots) and their *Ixodes ricinus* ticks (triangles). Species of origin are color-coded. Ecotypes and respective clusters are highlighted with different colors. For graphical reasons, only a subset of reference sequences [10] is included.

4. Discussion

In European countries *A. phagocytophilum* infection has been reported in a wide range of wild animals, including several species of rodents, hares, wild carnivores, birds, reptiles, and especially wild ungulates species [4,8]. This research investigated the epidemiology and phylogenesis of *A. phagocytophilum* infection in wild ungulates and related ectoparasites. We report widespread circulation of *A. phagocytophilum* in sylvatic cycles in northeastern Italy.

The overall *A. phagocytophilum* prevalence detected in wild ungulates was 63.8%. A high prevalence was reported in roe deer (91.9%) and red deer (87.2%), irrespective of the age (Table 2). This evidence supports their reservoir role [16,41]. Previous Italian studies reported both similar [37,42,43] and much lower prevalence data in wild ungulates [44–48]. *A. phagocytophilum* was also frequently detected in the other wild ruminant species, i.e., mouflon (75%), chamois (77.8%), and fallow deer (100%). Although only a limited number of samples of these species were analyzed in the present study, which were not representative of the target population, our data are in line with other studies [12,37,49], suggesting that more attention should be given to their impact on the anaplasmosis infectious cycle. An opposite role emerged for wild boars, with a much lower prevalence (3.6%) observed. This finding seems related to the lower quantity of ticks harbored by wild boars [11,50], thus decreasing the probability of wild boar infection.

No statistical association has been observed regarding the host prevalence and sex, age and, interestingly, seasonality. Actually, ticks have been collected from wild ruminants throughout the year, at similar frequency. This finding could be due to an increased tick survival and activity even during the winter months because of global warming [51,52].

Our research confirms the pivotal role of *I. ricinus* as the main vector of *A. phagocytophilum*, compared to other hard tick species, as previously reported [8,53]. The high prevalence (63.7%) of *A. phagocytophilum* detected in ticks collected from wild ungulates in northeastern is related to the reservoir role of these animals and the engorgement status of ticks [11]. Previous studies on *I. ricinus* removed from wild ungulates in other Italian regions, however, reported a lower prevalence, ranging from 5.1 to 31.2% [37,44,54–56], even when the same analytical methods was used. This evidence highlights that the prevalence of this bacterium depends not only on the presence of *I. ricinus* and wild ungulates, but on complex interactions involving the entire ecosystem, resulting in different enzootic cycles depending on the studied area, even within the same country [14,57]. Twelve positive ticks were collected from negative animals (Table 3) and three were non-engorged males. Therefore, infection may have been acquired in a previous stages and maintained through trans-stadial transmission, as previously reported by other authors [3,14]. However, the higher percentage of *A. phagocytophilum* infected ticks found on positive wild ungulates confirms the efficient transmission between these reservoir species and the vector. In this study mainly adult ticks were collected from ungulates, but larvae and nymphs can also parasitize these animals and some of them could maintain the infection through trans-stadial transmission [11]. Previous studies conducted on questing ticks in the same study area showed a percentage of *A. phagocytophilum* infection up to 9% in *I. ricinus* ticks from Belluno province [58,59] and up to 9.9% in ticks from a neighboring area [60]. A lower prevalence (1.5%) was found when a more extensive territory, including sites near the plain, was investigated [61]. On the other hand, the lack of *A. phagocytophilum* detection in *Dermacentor* and *Hyalomma* ticks supports a secondary or negligible role in anaplasmosis transmission [3,8]. Alternatively, they could be involved in the anaplasma infectious cycle only in areas where *I. ricinus* populations are particularly rare or absent, which is not the case of the considered region [5,57]. Moreover, *Dermacentor* and *Hyalomma* ticks have been found only on wild boars in the Euganean Hills Regional Park, an area characterized by a maximum altitude of 600 m a.s.l. where wild boars are more represented than ruminants, consisting mainly of fallow deer. Ecological conditions thus may be poorly comparable with other wilder areas where several wild ruminant populations share the same habitat and where mainly *I. ricinus* is detected. Therefore, a potential biasing effect of the host species and ecological niche cannot be excluded.

Of interest is the 40% *A. phagocytophilum* positivity in *L. cervi*. These ectoparasites were collected from positive red and roe deer. Although their role in the anaplasmosis infectious cycle is still under debate, other authors report deer keds as potential carriers of *A. phagocytophilum*, raising questions about their possible vectorial capability [15,62].

The genetic analysis of the 57 sequences performed, following the classification proposed by Jahafari et al. and Jaarsma et al., showed different distributions of ecotypes and clusters [10,14]. Of the 26 sequences from roe deer, 20 belonged to ecotype II/cluster 3. Ecotype II thus seems strictly related to this species [14,16]. On the other hand, ecotype I/cluster 1 showed the widest host range, having been found in roe deer (6/26), red deer (17/17), chamois (1/1), mouflon (3/3), and wild boar (1/1) species. Interestingly, the same clade includes genetic sequences of *A. phagocytophilum* strains from human cases in Europe; sequences MT473492, MT473484, MT473481, MT473458, MT473457, MT473453, MT473452, and MT473443 from red deer and sequence MT473485 from roe deer obtained in the present study were genetically identical to strains previously sampled from humans in Belgium (BelgiumPatient2014) and the Netherlands (LT06052), respectively (Table S1). Our finding suggests that these species could be a reservoir for zoonotic *A. phagocytophilum* variants. Even if an intense debate is still going on among authors on the actual role of red deer in ecotype I ecology and zoonotic risk [16], the high prevalence and association with this ecotype observed in our study favors their relevance and should thus be considered when evaluating the epidemiological risk for HGA [9,11,14]. The detection of ecotype I in wild boar highlights that it might be involved in anaplasmosis epidemiology, with the ability to harbor zoonotic variants [13,63]. However, the role of this species in anaplasmosis epidemiology needs further investigation, as it quickly eliminates *A. phagocytophylum* infection and has a lower tick burden [8].

Although chamois and mouflon were not considered to be pivotal in anaplasmosis diffusion, the remarkable prevalence herein reported and the demonstration that they can harbor ecotype I deserve more attention. To the best of the authors' knowledge, this is the first Italian report of ecotype I in wild boar, alpine chamois, and mouflon.

Most of the sequences obtained from *I. ricinus* ticks belonged to ecotype II/cluster 3 (7/9) and were collected from roe deer, while the remaining (2/9) were associated with ecotype I/cluster 1. In detail, one of these two ecotype I positive ticks was collected from a red deer and, interestingly, the other one was collected from a roe deer that tested positive to ecotype II, suggesting the potential simultaneous presence of different *A. phagocytophilum* variants.

In line with these results, some authors reported that *I. ricinus* ticks do not play a role in the host specificity of *A. phagocytophilum* variants, as it has been demonstrated that they may harbor all *A. phagocytophilum* ecotypes [10]. Furthermore, the generalist and teletropic feeding behavior of *I. ricinus* could facilitate the continuous exchange of ecotypes, and this behavior could explain the wide diffusion of ecotype I in almost all tested ungulates [5]. At the same time, our findings support that host specificity appears to be driven mainly by the variants themselves rather than tick feeding habits and host infection susceptibility. The presence of an overlap in ecotype host distribution further complicates the estimation of circulating ecotypes. In the present study, this emerged clearly in roe deer in which both ecotypes have been detected, accordingly to the findings of Remesar et al. [64]. Since the simultaneous presence of more than one variant can occur [65], it cannot be under-emphasized that the routinely applied Sanger sequencing did not allow identification of multiple strains, just the most abundant one, which may have resulted in an underreporting of such co-infections.

The findings of this study highlight the concrete risk of humans acquiring *A. phagocytophilum* infection from tick bites in the investigated area. Moreover, other transmission routes should be also considered. Contact with animal infected blood has been suggested as a potential infection source, particularly for people manipulating carcasses of ungulates or their meat [6]. Because of the subclinical infection in immunocompetent people, HGA should also be taken in account in blood transfusions. Unfortunately, currently HGA occurrence has rarely been investigated, especially through adequate genetic characterization, and further studies will be necessary to estimate its actual prevalence and relevance.

5. Conclusions

A. phagocytophilum was detected in all tested wild ungulates, with a higher prevalence in wild ruminants. Wild ungulates are involved in the *A. phagocytophilum* infectious cycle as the pivotal hosts for *I. ricinus* development and reproduction, harboring different *A. phagocytophilum* genetic variants. Of the detected variants, ecotype II/cluster 3 showed a host specificity for roe deer, while a broader tropism and the presence of overlapping niches emerged for ecotype I. Some sequences of ecotype I/cluster 1 from red deer and roe deer were identical to sequences from European human cases, highlighting their zoonotic potential. HGA itself may not cause severe disease, but it has been detected in co-infection cases with other tick-borne pathogens, leading to more severe symptomatology or atypical clinical presentations [27,66]. Due to the presence of zoonotic variants in the investigated area, the risk of infection should be communicated to categories of people at risk, such as hunters, veterinarians, and forest rangers, among others. Moreover, the often neglected

HGA deserves more attention and should be considered in the differential diagnosis of human tick-borne diseases.

Supplementary Materials: The following are available online at https://www.mdpi.com/2076-2615/11/2/310/s1, Figure S1: Phylogenetic tree of *Anaplasma phagocytophilum* strains. Table S1: List of representative sequence used for phylogenetic tree. Table S2: Accession numbers of new Italian *Anaplasma phagocytophilum* sequences deposited into GeneBank.

Author Contributions: Conceptualization, M.L.M.; methodology, A.M. and G.F.; formal analysis, M.M. and G.F.; investigation, L.G., E.V., and R.C.; resources, M.L.M., L.G., and E.V.; data curation, M.L.M. and G.F.; writing—draft preparation, review, and editing L.G., M.L.M., and G.F.; visualization, D.P. and M.D.; supervision, M.L.M.; funding acquisition, M.L.M. All authors have read and agreed to the published version of the manuscript.

Funding: This research was funded by the Integrated Budget for Interdepartmental Research BIRD177004/17, "Emerging Infectious Diseases of Wildlife in Northeastern Italy", and by grant DOR1949799/19 from the Department of Animal Medicine, Productions and Health, University of Padova.

Institutional Review Board Statement: Game animals were hunted during licensed hunting seasons and not culled specifically for the purpose of this study. Therefore, approval of the ethics committee was not required.

Informed Consent Statement: Not applicable.

Data Availability Statement: The data presented in this study are available in the text and in Tables 2 and 3 and Figure 1 of this article and in the Figure S1, Tables S1 and S2.

Acknowledgments: The authors are grateful to the hunting reserve directors, hunters, and the practitioner of Euganean Hills Regional Park for their kind collaboration and efforts in collecting the samples. Particularly, we thank Valerio Pituelli, Claudio Sabbadini, and Franco Micoli, presidents of hunting districts Tarvisiano, Valli del Natisone, and Colline Moreniche, respectively; and we thank the director of Euganean Hills Regional Park, Michele Gallo, whose coordination efforts made this work possible, Renato Sinigaglia, and Roberto Trevisan.

Conflicts of Interest: The authors declare no conflict of interest.

References

1. Dumler, J.S.; Barbet, A.F.; Bekker, C.P.J.; Dasch, G.A.; Palmer, G.H.; Ray, S.C.; Rikihisa, Y.; Rurangirwa, F.R. Reorganization of genera in the families *Rickettsiaceae* and *Anaplasmataceae* in the order *Rickettsiales*: Unification of some species of *Ehrlichia* with *Anaplasma*, *Cowdria* with *Ehrlichia* and *Ehrlichia* with *Neorickettsia*, descriptions of six new species combinations and designation of *Ehrlichia equi* and 'HGE agent' as subjective synonyms of *Ehrlichia phagocytophila*. *Int. J. Syst. Evol. Microbiol.* **2001**, *51*, 2145–2165. [CrossRef]
2. Schäfer, I.; Kohn, B. *Anaplasma phagocytophilum* infection in cats—A literature review to raise clinical awareness. *J. Feline Med. Surg.* **2020**, *22*, 428–441. [CrossRef] [PubMed]
3. Matei, I.A.; Estrada-Peña, A.; Cutler, S.J.; Vayssier-Taussat, M.; Varela-Castro, L.; Potkonjak, A.; Zeller, H.; Mihalca, A.D. A review on the eco-epidemiology and clinical management of human granulocytic anaplasmosis and its agent in Europe. *Parasit. Vectors* **2019**, *12*, 1–19. [CrossRef] [PubMed]
4. Stuen, S.; Granquist, E.G.; Silaghi, C. *Anaplasma phagocytophilum*—A widespread multi-host pathogen with highly adaptive strategies. *Front. Cell. Infect. Microbiol.* **2013**, *3*, 1–33. [CrossRef] [PubMed]
5. Dugat, T.; Leblond, A.; Keck, N.; Lagrée, A.-C.; Desjardins, I.; Joulié, A.; Pradier, S.; Durand, B.; Boulouis, H.-J.; Haddad, N. One particular *Anaplasma phagocytophilum* ecotype infects cattle in the Camargue, France. *Parasit. Vectors* **2017**, *10*, 1–6. [CrossRef] [PubMed]
6. Bakken, J.S.; Dumler, J.S. Human Granulocytic Anaplasmosis. *Infect. Dis Clin. N. Am.* **2015**, *29*, 341–355. [CrossRef] [PubMed]
7. Hauck, D.; Jordan, D.; Springer, A.; Schunack, B.; Pachnicke, S.; Fingerle, V.; Strube, C. Transovarial transmission of *Borrelia* spp., *Rickettsia* spp. and *Anaplasma phagocytophilum* in *Ixodes ricinus* under field conditions extrapolated from DNA detection in questing larvae. *Parasit. Vectors* **2020**, *13*. [CrossRef]
8. Dugat, T.; Lagrée, A.-C.; Maillard, R.; Boulouis, H.-J.; Haddad, N. Opening the black box of *Anaplasma phagocytophilum* diversity: Current situation and future perspectives. *Front. Cell. Infect. Microbiol.* **2015**, *5*, 1–18. [CrossRef]
9. Tomassone, L.; Berriatua, E.; De Sousa, R.; Duscher, G.G.; Mihalca, A.D.; Silaghi, C.; Sprong, H.; Zintl, A. Neglected vector-borne zoonoses in Europe: Into the wild. *Vet. Parasitol.* **2018**, *251*, 17–26. [CrossRef]

10. Jaarsma, R.I.; Sprong, H.; Takumi, K.; Kazimirova, M.; Silaghi, C.; Mysterud, A.; Rudolf, I.; Beck, R.; Földvári, G.; Tomassone, L.; et al. *Anaplasma phagocytophilum* evolves in geographical and biotic niches of vertebrates and ticks. *Parasit. Vectors* **2019**, *12*, 1–17. [CrossRef]

11. Kazimírová, M.; Hamšíková, Z.; Špitalská, E.; Minichová, L.; Mahríková, L.; Caban, R.; Sprong, H.; Fonville, M.; Schnittger, L.; Kocianová, E. Diverse tick-borne microorganisms identified in free-living ungulates in Slovakia. *Parasit. Vectors* **2018**, *11*. [CrossRef] [PubMed]

12. Kauffmann, M.; Rehbein, S.; Hamel, D.; Lutz, W.; Heddergott, M.; Pfister, K.; Silaghi, C. *Anaplasma phagocytophilum* and *Babesia* spp. in roe deer (*Capreolus capreolus*), fallow deer (*Dama dama*) and mouflon (*Ovis musimon*) in Germany. *Mol. Cell. Probes* **2017**, *31*, 46–54. [CrossRef] [PubMed]

13. Michalik, J.; Stańczak, J.; Cieniuch, S.; Racewicz, M.; Sikora, B.; Dabert, M. Wild boars as osts of buman- pathogenic *Anaplasma phagocytophilum* variants. *Emerg. Infect. Dis.* **2012**, *18*, 998–1001. [CrossRef] [PubMed]

14. Jahfari, S.; Coipan, E.C.; Fonville, M.; Van Leeuwen, A.D.; Hengeveld, P.; Heylen, D.; Heyman, P.; Van Maanen, C.; Butler, C.M.; Földvári, G.; et al. Circulation of four *Anaplasma phagocytophilum* ecotypes in Europe. *Parasit. Vectors* **2014**, *7*, 1–11. [CrossRef]

15. Bezerra-Santos, M.A.; Otranto, D. Keds, the enigmatic flies and their role as vectors of pathogens. *Acta Trop.* **2020**, *209*. [CrossRef] [PubMed]

16. Stigum, V.M.; Jaarsma, R.I.; Sprong, H.; Rolandsen, C.M.; Mysterud, A. Infection prevalence and ecotypes of *Anaplasma phagocytophilum* in moose *Alces alces*, red deer *Cervus elaphus*, roe deer *Capreolus capreolus* and *Ixodes ricinus* ticks from Norway. *Parasit. Vectors* **2019**, *12*, 1–8. [CrossRef]

17. Boulanger, N.; Boyer, P.; Talagrand-Reboul, E.; Hansmann, Y. Ticks and tick-borne diseases. *Med. Mal. Infect.* **2019**, *49*, 87–97. [CrossRef]

18. Dumler, J.S.; Choi, K.S.; Garcia-Garcia, J.C.; Barat, N.S.; Scorpio, D.G.; Garyu, J.W.; Grab, D.J.; Bakken, J.S. Human granulocytic anaplasmosis and *Anaplasma phagocytophilum*. *Emerg. Infect. Dis.* **2005**, *11*, 1828–1834. [CrossRef]

19. Petrovec, M.; Furlan, S.L.; Zupanc, T.A.; Strle, F.; Brouqui, P.; Roux, V.; Dumler, J.S. Human disease in Europe caused by a granulocytic *Ehrlichia* species. *J. Clin. Microbiol.* **1997**, *35*, 1556–1559. [CrossRef]

20. Ruscio, M.; Cinco, M. Human Granulocytic Ehrlichiosis in Italy. *Ann. N. Y. Acad. Sci.* **2003**, *990*, 350–352. [CrossRef]

21. De la Fuente, J.; Torina, A.; Naranjo, V.; Caracappa, S.; Di Marco, V.; Alongi, A.; Russo, M.; Maggio, A.; Kocan, K. Infection with *Anaplasma phagocytophilum* in a seronegative patient in Sicily, Italy: Case Report. *Ann. Clin. Microbiol. Antimicrob.* **2005**, *4*. [CrossRef] [PubMed]

22. Mastrandrea, S.; Mura, M.S.; Tola, S.; Patta, C.; Tanda, A.; Porcu, R.; Masala, G. Two cases of human granulocytic ehrlichiosis in Sardinia, Italy confirmed by PCR. *Ann. N. Y. Acad. Sci.* **2006**, *1078*, 548–551. [CrossRef] [PubMed]

23. Otranto, D.; Dantas-Torres, F.; Giannelli, A.; Latrofa, M.S.; Cascio, A.; Cazzin, S.; Ravagnan, S.; Montarsi, F.; Zanzani, S.A.; Manfredi, M.T.; et al. Ticks infesting humans in Italy and associated pathogens. *Parasit. Vectors* **2014**, *7*. [CrossRef] [PubMed]

24. Beltrame, A.; Laroche, M.; Degani, M.; Perandin, F.; Bisoffi, Z.; Raoult, D.; Parola, P. Tick-borne pathogens in removed ticks Veneto, northeastern Italy: A cross-sectional investigation. *Travel Med. Infect. Dis.* **2018**, *26*, 58–61. [CrossRef] [PubMed]

25. Battisti, E.; Zanet, S.; Boraso, F.; Minniti, D.; Giacometti, M.; Duscher, G.G.; Ferroglio, E. Survey on tick-borne pathogens in ticks removed from humans in Northwestern Italy. *Vet. Parasitol. Reg. Stud. Rep.* **2019**, *18*. [CrossRef] [PubMed]

26. Madison-Antenucci, S.; Kramer, L.D.; Gebhardt, L.L.; Kauffman, E. Emerging Tick-Borne Diseases. *Clin. Microbiol. Rev.* **2020**, *33*. [CrossRef] [PubMed]

27. Beltrame, A.; Ruscio, M.; Arzese, A.; Rorato, G.; Negri, C.; Londero, A.; Crapis, M.; Scudeller, L.; Viale, P. Human granulocytic anaplasmosis in Northeastern Italy. *Ann. N. Y. Acad. Sci.* **2006**, *1078*, 106–109. [CrossRef] [PubMed]

28. Franzo, G.; Grassi, L.; Tucciarone, C.M.; Drigo, M.; Martini, M.; Pasotto, D.; Mondin, A.; Menandro, M.L. A wild circulation: High presence of Porcine circovirus 3 in different mammalian wild hosts and ticks. *Transbound Emerg. Dis.* **2019**, *66*, 1548–1557. [CrossRef]

29. Provincia di Belluno—Piano Faunistico Venatorio Provinciale. Available online: https://www.distrettivenatoribelluno.it/normative.html (accessed on 25 January 2021).

30. Regione Autonoma Friuli Venezia Giulia—Dati sui Censimenti e Piani di Abbattimento. Available online: http://www.regione.fvg.it/rafvg/cms/RAFVG/ambiente-territorio/tutela-ambiente-gestione-risorse-naturali/gestione-venatoria/FOGLIA9/ (accessed on 28 December 2020).

31. Mammiferi—Parco Regionale dei Colli Euganei. Available online: http://www.parcocollieuganei.com/pagina.php?id=30 (accessed on 25 January 2021).

32. Cringoli, G.; Iori, A.; Rinaldi, L.; Veneziano, V.; Genchi, C. *Zecche. Mappe parassitologiche*; Rolando Editore: Napoli, Italy, 2005; ISBN 3904144987. Available online: https://www.parassitologia.unina.it/ricerca/mappe.xhtml?page=5 (accessed on 25 January 2021).

33. Courtney, J.W.; Kostelnik, L.M.; Zeidner, N.S.; Massung, R.F. Multiplex real-time PCR for detection of *Anaplasma phagocytophilum* and *Borrelia burgdorferi*. *J. Clin. Microbiol.* **2004**, *42*, 3164–3168. [CrossRef]

34. Guillemi, E.C.; Tomassone, L.; Farber, M.D. Tick-borne Rickettsiales: Molecular tools for the study of an emergent group of pathogens. *J. Microbiol. Methods* **2015**, *119*, 87–97. [CrossRef]

35. Altschul, S.F.; Gish, W.; Miller, W.; Myers, E.W.; Lipman, D.J. Basic Local Alignment Search Tool. *J. Mol. Biol.* **1990**, *215*, 403–410. [CrossRef]

36. Katoh, K.; Standley, D.M. MAFFT Multiple Sequence Alignment Software Version 7: Improvements in performance and usability. *Mol. Biol. Evol.* **2013**, *30*, 772–780. [CrossRef] [PubMed]

37. Di Domenico, M.; Pascucci, I.; Curini, V.; Cocco, A.; Dall'Acqua, F.; Pompilii, C.; Cammà, C. Detection of *Anaplasma phagocytophilum* genotypes that are potentially virulent for human in wild ruminants and Ixodes ricinus in Central Italy. *Ticks Tick. Borne. Dis.* **2016**, *7*, 782–787. [CrossRef] [PubMed]

38. Weizhong, L.; Godzik, A. Cd-hit: A fast program for clustering and comparing large sets of protein or nucleotide sequences. *Bioinform. Appl. Note* **2006**, *22*, 1658–1659. [CrossRef]

39. Nguyen, L.-T.; Schmidt, H.A.; von Haeseler, A.; Minh, B.Q. IQ-TREE: A Fast and Effective Stochastic Algorithm for Estimating Maximum-Likelihood Phylogenies. *Mol. Biol. Evol.* **2015**, *32*, 268–274. [CrossRef] [PubMed]

40. Darriba, D.; Taboada, G.L.; Doallo, R.; Posada, D. jModelTest 2: More models, new heuristics and high-performance computing. *Nat. Methods* **2012**, *9*, 772. [CrossRef] [PubMed]

41. Razanske, I.; Rosef, O.; Radzijevskaja, J.; Bratchikov, M.; Griciuviene, L.; Paulauskas, A. Prevalence and co-infection with tick-borne *Anaplasma phagocytophilum* and *Babesia* spp. in red deer (*Cervus elaphus*) and roe deer (*Capreolus capreolus*) in Southern Norway. *IJP Parasites Wildl.* **2019**, *8*, 127–134. [CrossRef] [PubMed]

42. Ebani, V.V.; Cerri, D.; Fratini, F.; Ampola, M.; Andreani, E. *Anaplasma phagocytophilum* infection in a fallow deer (*Dama dama*) population in a preserve of central Italy. *New Microbiol.* **2007**, *30*, 161–165.

43. Ebani, V.V.; Bertelloni, F.; Cecconi, G.; Sgorbini, M.; Cerri, D. Zoonotic tick-borne bacteria among wild boars (*Sus scrofa*) in Central Italy. *Asian Pacific J. Trop. Dis.* **2017**, *7*, 141–143. [CrossRef]

44. Veronesi, F.; Galuppi, R.; Tampieri, M.P.; Bonoli, C.; Mammoli, R.; Fioretti, D.P. Prevalence of *Anaplasma phagocytophilum* in fallow deer (*Dama dama*) and feeding ticks from an Italy preserve. *Res. Vet. Sci.* **2011**, *90*, 40–43. [CrossRef]

45. Torina, A.; Alongi, A.; Naranjo, V.; Scimeca, S.; Nicosia, S.; Di Marco, V.; Caracappa, S.; Kocan, K.M.; De La Fuente, J. Characterization of *Anaplasma* infections in Sicily, Italy. *Ann. N. Y. Acad. Sci.* **2008**, *1149*, 90–93. [CrossRef] [PubMed]

46. Beninati, T.; Piccolo, G.; Rizzoli, A.; Genchi, C.; Bandi, C. *Anaplasmataceae* in wild rodents and roe deer from Trento Province (northern Italy). *Eur. J. Clin. Microbiol. Infect. Dis.* **2006**, *25*, 677–678. [CrossRef] [PubMed]

47. Ebani, V.V.; Rocchigiani, G.; Bertelloni, F.; Nardoni, S.; Leoni, A.; Nicoloso, S.; Mancianti, F. Molecular survey on the presence of zoonotic arthropod-borne pathogens in wild red deer (*Cervus elaphus*). *Comp. Immunol. Microbiol. Infect. Dis.* **2016**, *47*, 77–80. [CrossRef] [PubMed]

48. Menandro, M.L.; Martini, M.; Dotto, G.; Mondin, A.; Ziron, G.; Pasotto, D. Tick-borne zoonotic bacteria in fallow deer (*Dama dama*) in Euganean Hills Regional Park of Italy. *Int. J. Infect. Dis.* **2016**, *53*, 61. [CrossRef]

49. Hornok, S.; Sugár, L.; Fernández de Mera, I.G.; de la Fuente, J.; Horváth, G.; Kovács, T.; Micsutka, A.; Gönczi, E.; Flaisz, B.; Takács, N.; et al. Tick- and fly-borne bacteria in ungulates: The prevalence of *Anaplasma phagocytophilum*, haemoplasmas and rickettsiae in water buffalo and deer species in Central Europe, Hungary. *BMC Vet. Res.* **2018**, *14*. [CrossRef] [PubMed]

50. Takumi, K.; Sprong, H.; Hofmeester, T.R. Impact of vertebrate communities on *Ixodes ricinus*-borne disease risk in forest areas. *Parasit. Vectors* **2019**, *12*. [CrossRef] [PubMed]

51. Dantas-Torres, F. Climate change, biodiversity, ticks and tick-borne diseases: The butterfly effect. *Int. J. Parasitol. Parasites Wildl.* **2015**, *4*, 452–461. [CrossRef]

52. Medlock, J.M.; Hansford, K.M.; Bormane, A.; Derdakova, M.; Estrada-Peña, A.; George, J.-C.; Golovljova, I.; Jaenson, T.G.T.; Jensen, J.-K.; Jensen, P.M.; et al. Driving forces for changes in geographical distribution of *Ixodes ricinus* ticks in Europe. *Parasit. Vectors* **2013**, *6*. [CrossRef]

53. Grech-Angelini, S.; Stachurski, F.; Vayssier-Taussat, M.; Devillers, E.; Casabianca, F.; Lancelot, R.; Moutailler, S. Tick-borne pathogens in ticks (Acari: *Ixodidae*) collected from various domestic and wild hosts in Corsica (France), a Mediterranean island environment. *Transbound. Emerg. Dis.* **2020**, *67*, 745–757. [CrossRef]

54. Ebani, V.V.; Bertelloni, F.; Turchi, B.; Filogari, D.; Cerri, D. Molecular survey of tick-borne pathogens in *Ixodid* ticks collected from hunted wild animals in Tuscany, Italy. *Asian Pac. J. Trop. Med.* **2015**, *8*, 714–717. [CrossRef]

55. Baráková, I.; Derdáková, M.; Selyemová, D.; Chvostáč, M.; Špitalská, E.; Rosso, F.; Collini, M.; Rosà, R.; Tagliapietra, V.; Girardi, M.; et al. Tick-borne pathogens and their reservoir hosts in northern Italy. *Ticks Tick. Borne. Dis.* **2018**, *9*, 164–170. [CrossRef] [PubMed]

56. Carpi, G.; Bertolotti, L.; Pecchioli, E.; Cagnacci, F.; Rizzoli, A. *Anaplasma phagocytophilum groEL* gene heterogeneity in *Ixodes ricinus* larvae feeding on roe deer in Northeastern Italy. *Vector-borne Zoonotic Dis.* **2009**, *9*, 179–184. [CrossRef] [PubMed]

57. Chisu, V.; Zobba, R.; Lecis, R.; Sotgiu, F.; Masala, G.; Foxi, C.; Pisu, D.; Alberti, A. *GroEL* typing and phylogeny of *Anaplasma* species in ticks from domestic and wild vertebrates. *Ticks Tick. Borne. Dis.* **2018**, *9*, 31–36. [CrossRef] [PubMed]

58. Piccolin, G.; Benedetti, G.; Doglioni, C.; Lorenzato, C.; Mancuso, S.; Papa, N.; Pitton, L.; Ramon, M.C.; Zasio, C.; Bertiato, G. A Study of the Presence of *B. burgdorferi*, *Anaplasma* (Previously *Ehrlichia*) *phagocytophilum*, *Rickettsia*, and *Babesia* in *Ixodes ricinus* collected within the territory of Belluno, Italy. *Vector-borne Zoonotic Dis.* **2006**, *6*, 24–31. [CrossRef]

59. Favia, G.; Cancrini, G.; Carfi, A.; Grazioli, D.; Lillini, E.; Iori, A. Molecular identification of *Borrelia valaisiana* and HGE-like *Ehrlichia* in *Ixodes ricinus* ticks sampled in north-eastern Italy: First report in Veneto region. *Parassitologia* **2001**, *43*, 143–146. [CrossRef]

60. Mantelli, B.; Pecchioli, E.; Hauffe, H.C.; Rosà, R.; Rizzoli, A. Prevalence of Borrelia burgdorferi s.l. and Anaplasma phagocytophilum in the wood tick Ixodes ricinus in the Province of Trento, Italy. *Eur. J. Clin. Microbiol. Infect. Dis.* **2006**, *25*, 737–739. [CrossRef]
61. Capelli, G.; Ravagnan, S.; Montarsi, F.; Ciocchetta, S.; Cazzin, S.; Porcellato, E.; Babiker, A.M.; Cassini, R.; Salviato, A.; Cattoli, G.; et al. Occurrence and identification of risk areas of *Ixodes ricinus*-borne pathogens: A cost-effectiveness analysis in north-eastern Italy. *Parasit. Vectors* **2012**, *5*. [CrossRef]
62. Víchová, B.; Majláthová, V.; Nováková, M.; Majláth, I.; Čurlík, J.; Bona, M.; Komjáti-Nagyová, M.; Peťko, B. PCR detection of re-emerging tick-borne pathogen, *Anaplasma phagocytophilum*, in deer ked (Lipoptena cervi) a blood-sucking ectoparasite of cervids. *Biologia* **2011**, *66*, 1082–1086. [CrossRef]
63. Hrazdilová, K.; Lesiczka, P.M.; Bardoň, J.; Vyroubalová, Š.; Šimek, B.; Zurek, L.; Modrý, D. Wild boar as a potential reservoir of zoonotic tick-borne pathogens. *Ticks Tick. Borne. Dis.* **2021**, *12*. [CrossRef]
64. Remesar, S.; Díaz, P.; Prieto, A.; García-Dios, D.; Fernández, G.; López, C.M.; Panadero, R.; Díez-Baños, P.; Morrondo, P. Prevalence and molecular characterization of *Anaplasma phagocytophilum* in roe deer (*Capreolus capreolus*) from Spain. *Ticks Tick. Borne. Dis.* **2020**, *11*. [CrossRef]
65. Jouglin, M.; Chagneau, S.; Faille, F.; Verheyden, H.; Bastian, S.; Malandrin, L. Detecting and characterizing mixed infections with genetic variants of *Anaplasma phagocytophilum* in roe deer (*Capreolus capreolus*) by developing an *ankA* cluster-specific nested PCR. *Parasit. Vectors* **2017**, *10*. [CrossRef] [PubMed]
66. Hoepler, W.; Markowicz, M.; Schoetta, A.-M.; Zoufaly, A.; Stanek, G.; Wenisch, C. Molecular diagnosis of autochthonous human anaplasmosis in Austria—An infectious diseases case report. *BMC Infect. Dis.* **2020**, *20*. [CrossRef] [PubMed]

Article

Interaction Patterns between Wildlife and Cattle Reveal Opportunities for Mycobacteria Transmission in Farms from North-Eastern Atlantic Iberian Peninsula

Lucía Varela-Castro [1,*], Iker A. Sevilla [1], Ariane Payne [2], Emmanuelle Gilot-Fromont [3] and Marta Barral [1]

1   Animal Health Department, NEIKER-Basque Institute for Agricultural Research and Development, Basque Research and Technology Alliance (BRTA), Parque Científico y Tecnológico de Bizkaia, P812, E-48160 Derio, Spain; isevilla@neiker.eus (I.A.S.); mbarral@neiker.eus (M.B.)

2   Unité Sanitaire de la Faune, OFB/DGDPCE/DRAS, 45100 Orléans, France; ariane.payne@ofb.gouv.fr

3   UMR 5558 LBBE, VetAgro Sup, Université de Lyon, 1 Avenue Bourgelat, F-69280 Marcy-l'Etoile, France; emmanuelle.gilotfromont@vetagro-sup.fr

*   Correspondence: lvarela@neiker.eus

Citation: Varela-Castro, L.; Sevilla, I.A.; Payne, A.; Gilot-Fromont, E.; Barral, M. Interaction Patterns between Wildlife and Cattle Reveal Opportunities for Mycobacteria Transmission in Farms from North-Eastern Atlantic Iberian Peninsula. *Animals* **2021**, *11*, 2364. https://doi.org/10.3390/ani11082364

Academic Editors: Laila Darwich Soliva and Julie Arsenault

Received: 28 June 2021
Accepted: 5 August 2021
Published: 10 August 2021

Publisher's Note: MDPI stays neutral with regard to jurisdictional claims in published maps and institutional affiliations.

Simple Summary: Mycobacteria can cause medically and socio-economically significant diseases, including several non-tuberculous infections and tuberculosis, and are considered a One Health challenge for their impact on public and animal health. These microorganisms are maintained and shared between the environment, domestic and wild animals, and humans. The aim of this research was to characterize the interactions that take place between several wild mammals and cattle through camera-trapping in order to provide insights into the dynamics of mycobacteria transmission opportunities in the environment of cattle farms located in Atlantic habitats from northern Iberian Peninsula. Camera traps were set during a one-year period in three cattle farms and visits of six wild species were modelled. We demonstrated that cross-species mycobacteria transmission, if occurring, would be mainly maintained through indirect interactions and most likely occur in pastures. In contrast to previous studies, wildlife visits were abundant but brief, and food and water resources did not attract wild animals. We suggest that badger latrines might act as aggregation points and sources of exposure to mycobacteria for badgers, wild boars, foxes, and cattle. This knowledge can contribute to designing and implementing effective measures aimed at controlling the spread of mycobacterioses in the environment–wild–domestic–human interface.

Abstract: Interactions taking place between sympatric wildlife and livestock may contribute to interspecies transmission of the *Mycobacterium tuberculosis* complex or non-tuberculous mycobacteria, leading to the spread of relevant mycobacterioses or to interferences with the diagnosis of tuberculosis. The aim of this study was to characterize the spatiotemporal patterns of interactions between wildlife and cattle in a low bovine tuberculosis prevalence Atlantic region. Camera traps were set during a one-year period in cattle farms with a history of tuberculosis and/or non-tuberculous mycobacterioses. The frequency and duration of wildlife visits, and the number of individuals per visit, were analysed through generalized linear mixed models. The seasons, type of place, type of point, and period of the day were the explanatory variables. A total of 1293 visits were recorded during 2741 days of camera observation. Only 23 visits showed direct contacts with cattle, suggesting that mycobacteria transmission at the wildlife–livestock interface would occur mainly through indirect interactions. Cattle pastures represented the most appropriate habitat for interspecies transmission of mycobacteria, and badgers' latrines appear to be a potential hotspot for mycobacteria circulation between badgers, wild boars, foxes, and cattle. According to both previous epidemiological information and the interaction patterns observed, wild boars, badgers, foxes, and small rodents are the species or group most often in contact with livestock, and thus may be the most involved in the epidemiology of mycobacterioses in the wildlife–livestock interface in this area.

Keywords: camera-traps; interactions; wildlife-livestock interface; tuberculosis; non-tuberculous mycobacteria

1. Introduction

Multi-host pathogens are often of wide concern because of the complexity that entails their control [1]. This control may become harder to manage when wild species are involved in their maintenance and transmission and even more difficult when poor or lacking farm biosecurity measures enable the occurrence of interactions between livestock and wildlife. In order to improve biosecurity, it is necessary to identify the places, moments, and circumstances that entail highest risk. Generally speaking, the rate of interactions between species tends to increase when scarce water or food sources are shared by domestic and wild species, such as in Mediterranean ecosystems, due to a high spatial and/or temporal overlap between them [2,3] and so does the probability of pathogen spread and transmission. Indirect transmission, more likely than direct transmission, tends to be involved across a community of host species [4]. However, when several host species are involved in the transmission of the same pathogen, it is crucial to identify the most important epidemiological connections between species and where/when these connections occur. Understanding interactions that can potentially lead to pathogen transmission at the wildlife-livestock interface is therefore a key for the implementation of appropriate disease control strategies in a multi-host system. However, this is often difficult to assess.

Animal tuberculosis (TB) is a worldwide zoonotic disease caused mainly by *Mycobacterium bovis* and other mycobacteria belonging to the *Mycobacterium tuberculosis* complex (MTC). Although cattle are considered its main and most well-studied host, *M. bovis* represents the perfect example of a multi-host pathogen with a complex and diverse spectrum of both domestic and wild hosts. In fact, a recent study has demonstrated that TB systems in some regions of Europe are dominated by non-bovine domestic and wild species [5]. *M. bovis* survival in the environment is highly variable according to environmental conditions but may last for several months [6], enhancing the likelihood of interspecies transmission within shared habitats (mainly through indirect contacts) [2,3]. On the other hand, the emerging prevalence of non-tuberculous mycobacteria (NTM) has become a matter of concern [7], even in countries reporting a low TB incidence [8]. Some of these NTM are associated with opportunistic or major mycobacterioses affecting humans and several domestic and wild species, as well as with interferences in the diagnosis of bovine TB [9]. NTM are widely distributed in a broad variety of aquatic and terrestrial environments [10]. Some species of veterinary relevance, such as *Mycobacterium avium* subsp. *paratuberculosis*, are able to persist in the environment for long periods [11].

In the Iberian Peninsula, multiple domestic and wild hosts are implicated in the epidemiology of animal TB. Among domestic species, cattle is still considered the main reservoir [12], despite the fact that other livestock can also play this epidemiological role (e.g., goats [13], sheep [14], and pigs) [15]. Although Spain is far from being considered officially TB free, the herd-level prevalence in cattle has been greatly reduced since the introduction of the National Eradication Program in 1987. In Atlantic regions in particular, this prevalence has been kept below one per cent [16] for the last twelve years. However, eradication has not been accomplished yet. In spite of the absence of a mandatory NTM surveillance program, NTM-infected cattle have also been detected in these regions during the national TB eradication campaigns among cattle showing false positive reactions to the tuberculin skin test [17,18]. The interactions between cattle and competent cohabiting wild hosts could contribute to this epidemiological picture of Atlantic Iberian Peninsula. There, the European badger (*Meles meles*) has been described as a potential wild reservoir of TB [19,20]. Furthermore, occasional TB cases have been detected in red deer (*Cervus elaphus*), and wild boar (*Sus scrofa*) seems to be implicated in the epidemiology of the disease (its role still being under debate) [17,21,22]. Besides, several species of NTM have been detected in these three wild species as well as in roe deer (*Capreolus capreolus*), wood mice (*Apodemus sylvaticus*), fox (*Vulpes vulpes*), and other carnivores such as the stone marten (*Martes foina*) and the mink (Varela-Castro, unpublished data) [17,18,23]. Before designing and implementing strategies aimed at reducing pathogens transmission between wild and domestic animals, deepening our current understanding of wild-domestic interaction dynamics is necessary. Among the current tools available for this purpose, camera trapping is a non-invasive technique useful for the assessment of a broad variety of ecological phenomena [24–27], which can be helpful to delve into disease transmission mechanisms.

The aims of the present research were (1) to study through camera trapping the nature of interactions (direct or indirect, frequency, duration and number of animals per wildlife visit, and observed behaviours) between cattle and wild mammal species from the Basque Country, a low bovine TB prevalence Atlantic region, and (2) to investigate whether these interactions may vary according to season, period of the day, places and points sampled. The results will provide useful information for assessing the risk of transmission of mycobacteria that could help in designing potential control strategies adapted to this specific scenario.

2. Materials and Methods

2.1. Study Area

This study was carried out in the Basque Country, northern Iberian Peninsula, where the annual prevalence of TB among cattle herds has been less than one per cent for the last 17 years and kept below 0.1% since 2017 [16]. According to official censuses from 2018 [28], there are 134,611 cattle in 4703 farms. These animals graze in the pastures regardless of the management system. Even when managed under an intensive production system, enclosures are open to the field and no biosafety measures such as fencing are always implemented. Therefore, cattle may share the pastures with cohabiting wildlife. Traditional husbandry practices are still maintained by some farmers in the Basque Country, being communal pastures shared by cattle and other domestic species such as horses and sheep, during the summer. MTC infection among wild mammals from this region has been detected in wild boar (1.12%) and red deer (2.40%) [22]. Several species of NTM able to infect cattle and interfere with the diagnosis of bovine tuberculosis have been also detected in wood mice [18] and other wild species from the study area (Varela-Castro, unpublished data: wild boar, red deer, roe deer, badgers, foxes and stone martens).

Three cattle farms located in the municipalities of Kexaa, Kortezubi and Deba (named A, B and C, respectively (see Figure 1)) were selected to represent farms where TB and other mycobacterioses, mainly provoked by *M. avium* subsp. *paratuberculosis* and/or *Mycobacterium avium* subsp. *avium*, have been recently diagnosed. These farms hosted tuberculin skin test-reactor cattle that were subsequently confirmed as *M. bovis*-infected or as false positive cases. Almost half of the MTC-positive cattle of the last ten years in the Basque Country were detected in farms A and C and NTM were also detected in these farms, while in farm B only *M. avium* subsp. *avium* was detected [18]. Farms A and B are dairy farms and farm C is a fighting bull farm. All three farms follow a free-range system. Cattle from farm A can either graze in the pastures or stay indoors, since facilities are open all year long. Facilities from farm B are completely closed and cattle are kept indoors during autumn and winter. Bulls of farm C are always kept outdoors. There are no other domestic species in these farms that could be important in terms of MTC transmission.

2.2. Camera Trap Survey

During a one-year period (January to November 2017), a total of twenty-three infrared motion-triggered camera traps (CTs) (Trophy Cam HD Aggressor, Bushnell, Overland Park, KS, USA) were used for the detection of wild mammal visits in different places of the farms while cattle were present (either in the field or inside facilities connected with outdoors). The field design comprised a two-week sampling period per farm and season except for farm B, where cattle are kept in closed facilities with no contact with outdoors during autumn and winter; thus, no sampling periods were recorded for those seasons. Overall, 10 sampling periods were recorded. CTs owned movement detection up to 25 m and a response time of 0.2 s. They were programmed to work day and night, recording 10 s videos each time a movement was detected with a triggered interval of 5 s. Date and time were displayed for each video. CTs were tied on trees, spikes, fences, or walls at ≈50 cm above the ground or up to 150–200 cm with a downward inclination, depending on the sampling point. When needed, branches that fell in the field of vision were removed. There was no overlap between the CTs' field of view.

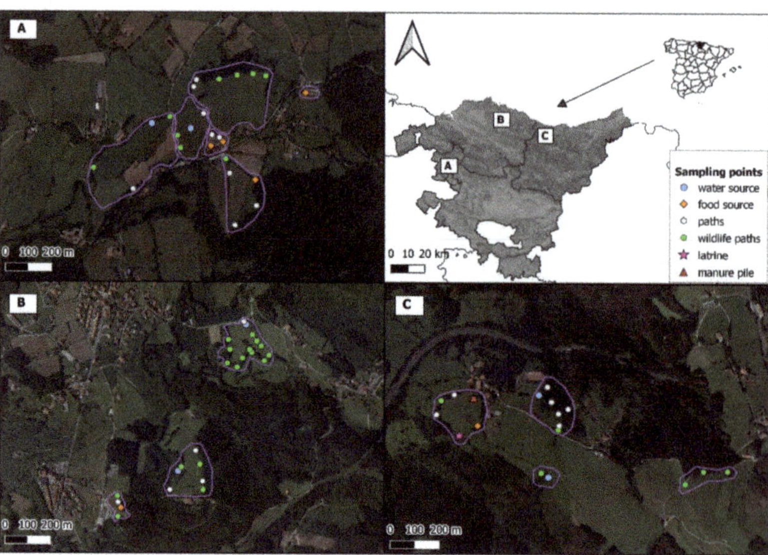

Figure 1. Field sampling design. The map shows the location of the studied farms (**A–C**) within the Basque Country (northern Iberian Peninsula). The spatial distribution of the sampling points recorded within farms (**A–C**) is displayed on the satellite photographs (**A–C**), respectively (see legend for sampling point description). Purple lines surround the sites (1 to 13) included in the models as a random factor.

The sampled places were cattle pastures, bushy edges between pastures, farm buildings, and a pine forest. In each place CTs were set in one to several points that could be *a priori* attractive for some wild species, such as water or food sources, a badger latrine and a manure pile, as well as in points that could potentially indicate the presence of wildlife, such as wildlife paths or paths that could be used by both cattle and wild species (see Figure 1). Water sources were all located outdoors and could be either a stream, a pond or cattle troughs settled with a certain height but surrounded by a flooded ground. Food sources were located indoors and outdoors. Those located indoors could be cattle feeders settled on the ground or feed-storages with some spillage of grain, while those located outdoors were piles of straw or hay delivered on the ground or a hazelnut trees plantation. Because of husbandry practices such as the rotation of herds among different pastures, the number of cameras varied between sampling periods and some points were not recorded during all the sampling periods of each farm. Overall, we sampled 67 points located in 17 places (Table 1). Due to the unbalanced availability among sampling places and points in the study area, pastures and wildlife paths were the type of place and point of the survey with longer surveillance time (Table 1). All videos were checked for species identification. If a wild mammal was detected, the number of animals, their behaviour and the duration of the visit was also registered.

Table 1. Number of visits, surveillance recording hours and sessions grouped by type of place and point.

Type of Place	Number Surveyed	Type of Point	Number Surveyed	Number of Surveillance Hours	Number of Sessions
Pasture (1070)	11	Water source (114)	6	4514.83	17
		Food source (10)	3	939.60	4
		Manure (12)	1	1343.82	4
		Latrine (54)	1	511.27	2
		Wildlife path (703)	30	20,115.23	77
		Path (177)	13	8590.40	30
Farm building (22)	2	Food source (8)	3	2564.35	11
		Path (16)	2	1144.52	7
Forest (61)	1	Wildlife path (12)	1	360.23	1
		Path (49)	2	2756.07	8
Edge (140)	3	Wildlife path (17)	1	1455.08	4
		Path (109)	3	3689.23	12
		Water source (14)	1	1027.97	3
Total	17		67	49,012.60	180

Numbers in brackets indicate the number of wild mammal visits.

2.3. Variables Definition

Since some CTs were located relatively close to each other, their observations could be non-independent. In each farm, we thus defined three to six sites, a "site" being a spatial unit corresponding to either a farm building or a pasture, including its bushy edges and the forest when present (see Figure 1). Each site (n = 13) was thus considered an independent area from other sites from the same farm, while the non-independence of observations within a site was accounted for in the analysis (see below). Distances between sampling points situated within a site varied among sites (range = 16 to 393 m).

We defined a "session" as a continuous period of monitoring on the same sampling point with the same camera. Although sessions were planned to last two weeks, some of them terminated earlier due to cattle moving CTs, thefts and unexpected battery depletion. For this reason, the duration (in hours) of each session was taken into account for subsequent analyses. We defined independent visits as (1) consecutive videos of individuals of different species; (2) consecutive videos of individuals of the same species more than 30 min apart; or (3) non-consecutive videos of a different or same species [29]. The number of visits per wild species, their duration (interval between the time displayed at the beginning of the first video and at the end of the last video included in the same visit, in minutes), and the number of animals per visit were the dependent variables. Animals could not always be individually identified, so the maximum number of individuals seen simultaneously in any of the videos of each visit was recorded. A direct interaction was defined as the simultaneous presence of cattle and at least one wild mammal on the same video. All other visits were considered as indirect interactions with cattle, since all points included areas used by cattle. Explanatory variables were the season (spring: April–June, summer: July–September, autumn: October–November, and winter: January–March), the period of the day (dawn, day, night and sunset, being the time slots determined according to the season where visits were observed), the place (pasture, farm building, forest, and edge) and the sampling point (water source, food source, manure, latrine, path, and wildlife path).

2.4. Statistical Analysis

The observed behaviours were classified focusing on those that could represent a risk of mycobacteria acquisition or excretion (Table 2). When none of these behaviours was observed, animals were considered as "moving through". If more than one behaviour were detected during a visit, either by one or more individuals, they were all recorded, except for "moving through" [30]. The percentage of occurrence of the different behaviours was calculated for each species. Then, the frequencies of wildlife visits were described, for each species, in terms of means and standard errors (SEs) by computing the number of visits per month, based on the observations obtained for each session. The means and SEs were also computed for the duration of the visits and the number of individuals per visit. Afterwards, a description of the direct interactions between each species and cattle was performed.

Table 2. Description of behaviours observed among wild species.

Behavior	Description
Grazing (for roe deer)	Feeding from grass, plants or fruits from a surface
Foraging (for badger and wild boar)	Searching for food by digging the ground with the snout
Sniffing (for all species)	Smelling the ground to search for food or to explore a surface/object
Excreting (for all species)/Scent marking (for all carnivores)	Urinating or defecating. For carnivores, lifting the tail and approaching the pelvis to the ground
Grooming/Scratching/Wallowing (for all species)	Applying tongue or paws to parts of the body in repeated motions, shaking the body, scraping against a surface, rolling in a water point
Drinking (for all species)	Drinking from water sources
Moving through (for all species)	Passing through a sampling point without performing any of the aforementioned behaviours

Then, generalized linear mixed models (GLMMs) were used to analyse how the number of visits per each species, their duration and the number of individuals varied among seasons, periods of the day, type of place and type of point, using the farms (A, B and C) and the sites (1–13) within each farm as random effects in order to take into account the likely dependence of wildlife visits within farms and each site of every farm. For the number of visits, a model adjusted to a Poisson distribution was used and, in order to consider the sampling effort of the sessions, the logarithm of the number of surveillance hours per session was included in the model as an offset. For the number of animals and the duration of visits, models adjusted to Poisson and Gamma distributions were used respectively. A total of 18 models were initially fit, one per response variable and species. For each one, a maximal model including all the variables was first created. Hereafter, the dredge function of R software was used to generate a selection table of models with combinations of the fixed variables originally included in the maximal model. The selection of the best combination was made following the parsimony principle [31]: among models that had similar AIC values (delta < 2), the one with fewest parameters was selected.

Finally, we used the overdisp.glmer function of R in order to check whether overdispersion was still present in the residuals of the selected Poisson models [32]. Nakagawa and Schielzeth R-squared were used to determine the variability explained by the fixed and random parts of the selected models (using the r.squaredGLMM function of R software). All of the statistical analyses were performed using the R 4.0.0 software [33]. The data sets employed for the statistical analyses are submitted as Supplementary Material: Tables S1 and S2.

3. Results

3.1. Data Collected from the Field Samplings

Data were recorded during 2741 camera days (i.e., data obtained from a given camera over a given day) distributed into 180 sessions (mean duration ± standard error: 271.76 h ± 7.73). A total of 127,091 videos were recorded. Among them, 48,976 involved only cattle, 1329 other domestic species (cats, dogs, and horses), 4942 birds, 2320 wild mammals, and 4 reptiles. In 71 videos, it was not possible to identify the species. Wild mammal videos involved wild boar, roe deer, badger, fox, other carnivores (hereafter OC group, which includes genet (*Genetta genetta*), stone martens and pine martens (*Martes martes*)), small rodents (mouse-like), hedgehogs (*Erinaceus europaeus*), squirrels (*Sciurus vulgaris*), and bats. After excluding those species without previous epidemiological data on mycobacterial infection in the study area (hedgehogs, squirrels and bats), 2182 videos of wild mammals were retained for the analyses.

A total of 1293 visits by wild species of interest were registered, each visit being recorded by 1 to 33 videos. All species visited the farms during all seasons. Pastures and wildlife paths received the highest number of visits (Table 1). Since the observed species were mainly nocturnal, most of the visits (85%), including direct contacts with cattle, took place at night. Visits occurred in 64 out of the 67 sampling points. The three points that did not receive any visits were food sources located inside a farm building (2 points) and in a pasture (1 point). Wild boar, fox and small rodents were the only visitors of farm buildings (Table 3).

Table 3. Description of wild mammal visits. Mean ± SE and range are shown for frequency of visits per month, visit duration, and number of individuals per visit.

	Badger (*n* = 315)	Wild Boar (*n* = 304)	Roe Deer (*n* = 175)	Fox (*n* = 376)	Other Carnivores (*n* = 38)	Small Rodents (*n* = 85)
Frequency of visits (all visits, number per month)	4.73 ± 0.61 0–46.67	4.41 ± 0.58 0–46.33	2.76 ± 0.64 0–83.72	5.69 ± 0.66 0–59.20	0.54 ± 0.16 0–19.66	1.85 ± 0.74 0–117.2
Frequency of visits in buildings only (number per month)	0	1.07 ± 0.76 0–12	0	1.30 ± 0.76 0–12.12	0	2.04 ± 2.04 0–36.76
Visit duration (min)	0.89 ± 0.19 0.17–31	1.64 ± 0.26 0.17–38	2.33 ± 0.60 0.17–56	1.31 ± 0.31 0.17–86	0.47 ± 0.13 0.17–4	5.12 ± 1.68 0.17–95
Number of individuals per visit	1.05 ± 0.01 1–4	2.57 ± 0.10 1–10	1.16 ± 0.03 1–3	1.04 ± 0.01 1–3	1 ± 0 1–1	1.08 ± 0.04 1–3

Numbers in brackets indicate the total number of visits.

3.2. Frequency and Characterization of Visits Per Species

Figure 2 shows the proportion of occurrence of the behaviours exhibited per species. The most frequent behaviour was moving through (60% of the visits), followed by sniffing (31%), being both behaviours displayed by all species. Table 3 describes wild mammal visits in terms of frequency, number of individuals per visit, and duration of visits. The frequency of visits was highest for foxes, followed by badgers, wild boar, roe deer, small rodents, and the OC group. Small rodents were the group that showed longest visits on average (5.12 min ± 1.68), while the OC group showed the shortest on average (0.47 min ± 0.13). The species that showed up in more numerous groups was the wild boar (2.57 individuals ± 0.10, up to 10 individuals), while the rest of the species showed mainly solitary incursions (82% of visits performed by a single individual) or appeared, punctually, in small groups (up to four badgers, up to three roe deer, foxes, and small rodents). Even though visits longer than half an hour occurred sporadically (1% of the visits) except for the OC group, short visits (less than 5 min) were predominant (93% of the visits). Twenty-three direct contacts with cattle were recorded (see Table 4). Thus, the other 1270 visits were considered as indirect interactions. The fox was the species which showed most direct contacts with cattle (eight), followed by small rodents (six) and wild boar (six), badger (two) and roe deer (one). No direct interaction was recorded for the OC group. More than half of these direct contacts took place during autumn (13/23) and within pastures (17/23), being more frequent in wildlife paths (9/23). Those that took place in farm buildings were mostly between small rodents and cattle (4/5). Even though the most common behaviour recorded during the visits was moving through, when a direct interaction took place, wild animals showed other behaviours such as sniffing, scent marking or foraging, except for small rodents (Table 4).

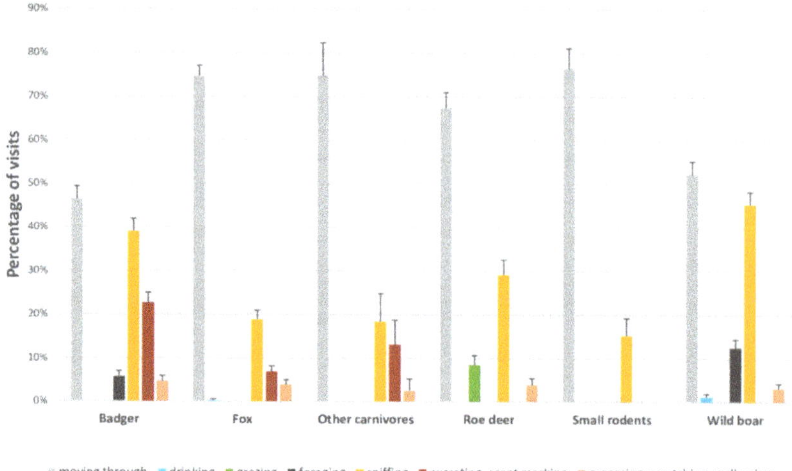

Figure 2. Percentage of the occurrence of each behaviour exhibited per species.

Table 4. Description of direct contacts between wild mammals and cattle.

	Badger	Wild Boar	Roe Deer	Fox	Small Rodents
Number of direct interactions	2	6	1	8	6
Most frequent season	Autumn (2)	Autumn (5)	Summer	Summer (5)	Autumn (4)
Most frequent place	Pasture (2)	Pasture (5)	Pasture	Pasture (7)	Farm building (4)
Most frequent point	Latrine (2)	Manure (2)/Wildlife path (2)	Wildlife path	Wildlife path (6)	Path (4)
Behaviors observed	Sniffing/ scent marking	Sniffing/foraging/moving through	Sniffing	Moving through/sniffing	Moving through

Numbers in brackets indicate the number of direct contacts.

Table 5 shows the outputs of the models selected to explain the frequency of visits, the number of individuals per visit, and the duration of visits for each wild mammal. Models related to the number of individuals could only be fit for wild boar data, due to the quasi-absence of variability for other species. The overdispersion of residuals was limited for all Poisson models, ranging from 0.6 to 3.03. The random effects (farms and sites within farms) accounted for 0 to 91.5% of the variations.

Table 5. Models selected per wild species and response variable. For each model, the table gives the percentage of variation explained by fixed and random parts of the model, and the OR, estimate and *p*-value of Wald test for each contrast between the reference level and the given level. The number of individuals was analyzed for wild boar only due to the quasi-absence of variability for other species.

Species	Response Variable	V.E by Fixed Part	V.E by Random Part	Fixed Effect	Level	OR (95% CI)	Estimate	*p*-Value
Badger	Frequency of visits	28.35%	40.94%	Season	Autumn	0.87 (0.59–1.26)	−0.14	0.454
				(ref: summer)	Winter	2.49 (1.75–3.54)	0.91	<0.001 ***
					Spring	1.16 (0.86–1.58)	0.15	0.325
				Place	Edge	0.56 (0.34–0.91)	−0.59	0.020 *
				(ref: pasture)	Forest	1.06 (0.53–2.11)	0.06	0.875
				Point	Latrine	3.81 (1.97–7.35)	1.34	<0.001 ***
				(ref: wildlife path)	Manure	0.04 (0.01–0.31)	−3.17	0.002 **
					Path	0.46 (0.31–0.68)	−0.78	<0.001 ***
					Water source	0.81 (0.54–1.20)	−0.21	0.287
	Duration of visits	1.03%	9.09%	Place	Edge	0.43 (0.20–0.89)	−0.85	0.024 *
				(ref: pasture)	Forest	0.35 (0.15–0.80)	−1.06	0.013 *
				Point	Latrine	0.43 (0.18–1.03)	−0.84	0.059
				(ref: wildlife path)	Manure	0.27 (0.02–3.04)	−1.29	0.292
					Path	0.41 (0.25–0.70)	−0.88	<0.001 ***
					Water source	1.60 (0.95–2.70)	0.47	0.079
Wild boar	Frequency of visits	29.08%	19.64%	Season	Autumn	1.57 (1.13–2.18)	0.45	0.007 **
				(ref: summer)	Winter	0.39 (0.23–0.67)	−0.94	<0.001 ***
					Spring	1.38 (1.02–1.88)	0.32	0.038 *
				Point	Food source	0.34 (0.13–0.94)	−1.07	0.037 *
				(ref: wildlife path)	Latrine	9.07 (4.47–18.42)	2.20	<0.001 ***
					Manure	1.30 (0.49–3.41)	0.26	0.599
					Path	0.77 (0.55–1.08)	−0.26	0.128
					Water source	1.23 (0.86–1.76)	0.21	0.260
	Number of animals	10.85%	0.60%	Season	Autumn	1.52 (1.20–1.94)	0.42	<0.001 ***
				(ref: summer)	Winter	0.70 (0.46–1.07)	−0.35	0.103
					Spring	1.04 (0.83–1.30)	0.04	0.746
				Period of the day	Dawn	0.70 (0.34–1.43)	−0.35	0.332
				(ref: night)	Sunset	0.59 (0.38–0.92)	−0.53	0.020 *
	Duration of visits	0.32%	10.88%	Season	Autumn	1.12 (0.60–2.07)	0.11	0.730
				(ref: summer)	Winter	3.41 (1.54–7.57)	1.23	0.003 **
					Spring	1.21 (0.71–2.06)	0.19	0.480

Table 5. *Cont.*

Species	Response Variable	V.E by Fixed Part	V.E by Random Part	Fixed Effect	Level	OR (95% CI)	Estimate	*p*-Value
Roe deer	Frequency of visits	4.87%	91.50%	Season	Autumn	1.57 (0.83–2.97)	0.45	0.162
				(ref: summer)	Winter	0.54 (0.25–1.18)	−0.61	0.123
					Spring	0.30 (0.21–0.44)	−1.19	<0.001 ***
	Duration of visits	6.21%	20.20%	Season	Autumn	0.28 (0.10–0.81)	−1.26	0.020 *
				(ref: summer)	Winter	0.95 (0.33–2.72)	−0.05	0.925
					Spring	0.10 (0.06–0.16)	−2.32	<0.001 ***
				Point	Food source	22.15 (4.46–110.05)	3.10	<0.001 ***
				(ref: wildlife path)	Path	1.43 (0.59–3.46)	0.35	0.434
					Water source	1.85 (0.42–8.11)	0.62	0.413
Fox	Frequency of visits	28.42%	6.43%	Point	Food source	0.15 (0.05–0.42)	−1.89	<0.001 ***
				(ref: wildlife path)	Latrine	2.79 (1.42–5.49)	1.03	0.003 **
					Manure	0.48 (0.19–1.23)	−0.73	0.127
					Path	1.02 (0.79–1.32)	0.02	0.863
					Water source	0.62 (0.42–0.93)	−0.47	0.020 *
	Duration of visits	3.88%	3.89%	Season	Autumn	0.44 (0.20–0.97)	−0.83	0.042 *
				(ref: summer)	Winter	0.86 (0.43–1.73)	−0.15	0.670
					Spring	2.02 (1.10–3.68)	0.70	0.022 *
				Period of the day	Dawn	0.89 (0.25–3.19)	−0.12	0.859
				(ref: night)	Day	3.76 (1.61–8.78)	1.33	0.002 **
					Sunset	0.60 (0.25–1.44)	−0.51	0.254
				Place	Edge	0.38 (0.15–0.97)	−0.97	0.042 *
				(ref: pasture)	Farm building	1.48 (0.13–16.54)	0.39	0.750
					Forest	0.45 (0.16–1.25)	−0.80	0.125
				Point	Food source	2.22 (0.15–33.41)	0.80	0.563
				(ref: wildlife path)	Latrine	0.23 (0.04–1.34)	−1.47	0.103
					Manure	0.12 (0.01–1.03)	−2.13	0.053.
					Path	0.68 (0.33–1.41)	−0.39	0.301
					Water source	0.32 (0.13–0.83)	−1.13	0.019 *

Table 5. *Cont.*

Species	Response Variable	V.E by Fixed Part	V.E by Random Part	Fixed Effect	Level	OR (95% CI)	Estimate	*p*-Value
Other carnivores	Frequency of visits	2.68%	3.58%	Place (ref: edge)	Pasture	0.19 (0.06–0.60)	−1.68	0.005 **
				Point	Water source	0.20 (0.06–0.67)	−1.61	0.009 **
				(ref: path)	Wildlife path	0.34 (0.11–1.02)	−1.08	0.054.
	Duration of visits	6.9%	0.00%	Place (ref: edge)	Pasture	0.36 (0.20–0.66)	−1.02	<0.001 ***
Small rodents	Frequency of visits	23.92%	18.23%	Season	Summer	0.03 (0.01–0.09)	−3.64	<0.001 ***
				(ref: autumn)	Winter	0.11 (0.05–0.23)	−2.19	<0.001 ***
					Spring	0.15 (0.08–0.29)	−1.91	<0.001 ***
				Place (ref: pasture)	Edge	0.33 (0.15–0.72)	−1.11	0.005 **
					Farm building	1.86 (0.16–21.88)	0.62	0.623
				Point	Food source	0.20 (0.03–1.13)	−1.63	0.068
				(ref: wildlife path)	Path	0.53 (0.31–0.92)	−0.63	0.023 *
					Water source	0.42 (0.17–1.04)	−0.86	0.060
	Duration of visits	11.08%	0.00%	Season	Summer	0.02 (0.00–0.14)	−3.74	<0.001 ***
				(ref: autumn)	Winter	0.04 (0.01–0.11)	−3.31	<0.001 ***
					Spring	0.02 (0.01–0.06)	−3.74	<0.001 ***

V.E = Variation explained. ref = reference level of the fixed effect. *: *p*-value ≤ 0.05. **: *p*-value ≤ 0.01. ***: *p*-value ≤ 0.001.

3.2.1. Badger

No badger visit was recorded during the day, inside farm buildings, or in food sources. The two direct interactions with cattle were both observed in the latrine (Table 4). Pastures were the place where badger visits were most frequent and longest. The frequency of visits was also significantly higher in winter than in summer (OR = 2.49, 95% CI: 1.75–3.54). Moreover, visits were significantly more frequent in the badger latrine than in wildlife paths (OR = 3.81, 95% CI: 1.97–7.35), but significantly less frequent in paths (OR = 0.46, 95% CI: 0.31–0.68) or in the manure pile (OR = 0.04, 95% CI: 0.01–0.31) than in wildlife paths. The duration of visits was also significantly shorter in paths (OR = 0.41, 95% CI: 0.25–0.70) compared to wildlife paths (Table 5).

3.2.2. Wild Boar

No diurnal visit was recorded. Apart from one case that was recorded in a farm building, direct interactions with cattle took place in pastures (5/6) (Table 4), including one in the latrine. The frequency of visits was significantly higher in autumn (OR = 1.57, 95% CI: 1.13–2.18) and spring (OR = 1.38, 95% CI: 1.02–1.88) compared to summer, while it was significantly lower during winter (OR = 0.39, 95% CI: 0.23–0.67). The same tendency was also observed for the number of animals per visit, being significantly higher in autumn compared to summer (OR = 1.52, 95% CI: 1.20–1.94). However, the visits were significantly longer in winter (OR = 3.41, 95% CI: 1.54–7.57). Wild boar visits were significantly more frequent in the badger's latrine (OR = 9.07, 95% CI: 4.47–18.42) and significantly less frequent in food sources (OR = 0.34, 95% CI: 0.13–0.94). Lastly, wild boars were significantly less numerous during the dawn than at night (OR = 0.59, 95% CI: 0.38–0.92) (Table 5).

3.2.3. Roe Deer

No visits were recorded in farm buildings, in the manure pile or in the latrine. The only direct interaction with cattle recorded took place in summer, during the day and in a wildlife path located in pastures (Table 4). Visits were significantly less frequent during spring (OR = 0.30, 95% CI: 0.21–0.44) and shorter during spring (OR = 0.10, 95% CI: 0.06–0.16) and autumn (OR = 0.28, 95% CI: 0.10–0.81), if compared with summer. Visits were significantly longer in food sources than in wildlife paths (OR = 22.15, 95% CI: 4.46–110.05) (Table 5).

3.2.4. Fox

Foxes were seen at all periods of the day, type of places and type of points. All direct interactions with cattle but two that were recorded in the latrine and in a path took place in wildlife paths (6/8) (Table 4). Compared to wildlife paths, visits were significantly more frequent in badger's latrine (OR = 2.79, 95% CI: 1.42–5.49) and less frequent in food (OR = 0.15, 95% CI: 0.05–0.42) or water sources (OR = 0.62, 95% CI: 0.42–0.93). Water sources received shorter visits (OR = 0.32, 95% CI: 0.13–0.83) than wildlife paths did. In comparison to summer, these were significantly longer in spring (OR = 2.02, 95% CI: 1.10–3.68) and shorter in autumn (OR = 0.44, 95% CI: 0.20–0.97). Moreover, they were significantly longer during the day than during the night (OR = 3.76, 95% CI: 1.61–8.78), and shorter in edges than in pastures (OR = 0.38, 95% CI: 0.15–0.97) (Table 5).

3.2.5. Other Carnivores

Most of these visits were performed by genets (33/38), even though stone martens (3/38) and martens (2/38) could be sporadically observed. All visits by these species were recorded in the pastures and their edges. No visit was recorded in food sources, in the latrine, or in the manure pile. Visits were significantly less frequent (OR = 0.19, 95% CI: 0.06–0.60) and shorter (OR = 0.36, 95% CI: 0.20–0.66) in pastures than in edges, and less frequent in water sources than in paths (OR = 0.20, 95% CI: 0.06–0.67) (Table 5).

3.2.6. Small Rodents

No visit was recorded during sunset, in the forest, in the manure pile, or in the latrine. Unlike the rest of the species, direct interactions between cattle and small rodents were more frequent inside farm buildings (4/6) (Table 4). The selected models showed that frequency and duration of visits were significantly lower in summer (OR = 0.03, 95% CI: 0.01–0.09), winter (OR = 0.11, 95% CI: 0.05–0.23) and spring (OR = 0.15, 95% CI: 0.08–0.29) compared to autumn. Their frequency was also lower in edges compared to pastures (OR = 0.33, 95% CI: 0.15–0.72) and in paths compared to wildlife paths (OR = 0.53, 95% CI: 0.31–0.92) (Table 5).

4. Discussion

4.1. Methodology

Camera trapping has proved to be a useful tool for studying interactions with a minimal disturbance to animals. However, failures in the detection due to intrinsic characteristics of the CTs (distance detection, response time), the position angle when hanging the devices, camera malfunctioning, and loss of battery power or adverse weather conditions may have led to an underestimation of the number and duration of visits and the number of individuals. Hence, our observations correspond to a minimum of what is actually occurring in these farms. On the other hand, due to husbandry practices and organization issues, there was an imbalance in the field sampling, since all points were not sampled at all seasons. This could have limited our ability to detect seasonal variations. Besides, the small sample size of some types of places and points (e.g., forest, latrine; see Table 1) has narrowed the information obtained from them and may be underrepresented.

4.2. Spatiotemporal Patterns of Wildlife-Cattle Interactions

Our observations confirmed that as well as in other regions [2,3,30,34], indirect interactions between wildlife and cattle can be considered more frequent than direct interactions. Different spatial and temporal patterns were observed depending on the species surveyed.

Winter was the most favourable season for badger visits to occur. This season is a period of food scarcity for this species, whose diet in the Basque Country is mainly based on earthworms and garden fruits [35]. However, badgers did not approach farm buildings or food sources, probably due to the fact that cattle resources (e.g., silage or hay) lack attractiveness for them. These findings are consistent with those previously reported in a medium density population area, where badgers clearly avoided farmyards [36] but differ with some British [37,38] and French studies [30], where high rates of building use were described. Resource availability and badger population density might account for these differences. In agreement with previous reports [39], pastures were the preferred place for badgers in our study area. Earthworm intake might explain the attractiveness of pastures, since their soft ground is suitable for foraging. The most attractive point for this mustelid was the badger latrine. Moreover, the few direct interactions recorded between cattle and badgers always took place in the latrine. For these reasons, this point might be considered as a potential hotspot for both indirect or direct interactions between these two species. However, in the absence of other latrines, it is not clear whether this high frequency of visits was due to the latrine itself or to another specific feature of this particular point.

The seasonal differences observed in the frequency and duration of wild boar visits, as well as in the number of individuals per visit, may be due to different factors. In winter, wild boar density is at its lowest and the duration of incursions may be longer when searching for food. However, farm food sources were the less attractive point for this species. This could be related to a higher availability of natural resources in the study area, at least compared to areas from southern Spain, where baited points turned out to be very attractive to wild boar [2]. As well as with badgers, wild boar visits were more frequent in the badger latrine, possibly due to an attraction effect of its characteristic scent. Thus, the latrine could also represent a significant point for wild boar to interact with other species.

Roe deer visits were most frequent and longest in summer, coinciding with the mating period of this species. Some of the activities during this period, such as the defence of the territory, the avoidance of dangerous fights and the chasing of females by bucks [40] might make them more visible. Actually, the only direct contact recorded between roe deer and cattle took place in pastures during summer and by day. Longer visits were observed in food sources than in other sampling points. Roe deer is mainly a browser, not a grazer [41], so they are not expected to use the same food resources as cattle. Indeed, all roe deer visits to food sources were recorded in the hazelnut trees plantation, which represents the only resource not interesting for cattle.

As for badgers and wild boar, the badger latrine was the most visited point by foxes and also the scene of a direct interaction with cattle, which supports considering the badger latrine a potential hotspot for intra and interspecies interactions in our study area. Despite visiting all types of points, foxes showed less interest for food and water sources. This may reflect their interest in other food supplies, such as small mammals to prey on. Conversely to other species, fox visits were longer during the day because they spent a long time resting on the pastures. Lastly, visits were shorter in autumn and longer in spring compared to the summer. These findings could be also related to their rest times, which were longer during the warmest seasons.

OC group species turned out to be less often seen and always alone. Visits were most frequent and longest in the edges, which is consistent with their search for protection from predators and unfavourable weather conditions [42]. The majority of visits were performed by genets, and almost half of them were recorded in one specific path within the edge of one pasture. Genets spend most of their time resting in the same place [42]. Thus, the potential existence of a resting site close to this path might explain the output of the models.

Although small rodent species could not be determined, in a previous study conducted in the same farms the wood mouse was the species most frequently captured [18]. Small rodents visits occurred more often and were longer in autumn than during other seasons, probably since this is a period when most rodent populations, such as the widely distributed wood mouse population, are at their maximal abundance [43]. These wild rodents typically move along field margins of farmlands and are known to be common in hedgerows [44], which might explain their preferences for pastures and wildlife paths from the study area.

4.3. Opportunities of Mycobacteria Transmission

Since indirect interactions were much more common than direct interactions, mycobacteria transmission at the wildlife-livestock interface, if occurring, would be mainly held through indirect interactions. In general, pastures represent the most appropriate place for interspecies transmission of mycobacteria in the study area. Our results suggest that badger latrines can be suitable places for both indirect and direct contacts at least between badgers, wild boar, foxes and cattle. During the visits to the latrine, individuals of these three wild species showed behaviours related to possible excretion of or exposure to pathogens such as sniffing or scent marking, where cattle was also seen sniffing or grazing. Consequently, these points could be considered potential hotspots for mycobacteria circulation in this habitat. However, a single latrine was found and recorded, and therefore, further studies are needed to confirm this hypothesis.

Some remarkable differences have been identified between this study and previous reports. In our study, whatever the wild species considered, the average duration of the visits was shorter (<5 min) than in studies from France [29,30]. For instance, the average duration of wild boar visits was significantly shorter in our study area (1.64 min) than in a bovine TB-infected area in France (14.5 min) [30]. The most common behaviour in our study area was "moving through". Wild species mainly move around shared habitats with cattle, but resources such as water or food supplies do not act as aggregation points, conversely to the results of previous studies [2,29,30,45]. A higher availability of natural resources throughout the whole year may account for these differences. These findings, together with the absence of MTC-infected individuals among wildlife, except for wild boar, and the low TB prevalence reported for this wild ungulate and cattle from the Basque Country [22], suggest that the risk of MTC transmission between wild animals and cattle would be, overall, low. On the contrary, we suspect that a risk of indirect NTM transmission could be more feasible in the study area, since this group of mycobacteria have been detected in all species and the prevalence observed in some of them was significant.

Since interaction patterns and infection figures differed among wild species, some of them might be more involved than others in the epidemiology of mycobacterioses in the Basque Country. Thus, depending on the species and the situation, different control strategies could be implemented to maximize effectiveness. Foxes, badgers and wild boars were the species observed most frequently. Even though not considered as a TB reservoir in the Atlantic Iberian Peninsula, wild boar is the only species observed in this study and found to be infected with MTC in the study area, since red deer distribution is limited to a few settings that do not encompass these farms. Wild boar has shown an unexpectedly high MTC seroprevalence of 17% in this region [21] and, despite the low prevalence detected by culture (<2%) and the absence of animals with disseminated lesions/infection, a potential geographical link was found between spoligotypes identified in cattle and wild boar [22]. Furthermore, culture methods revealed a 9% prevalence of NTM in this species. Considering this information, the high frequency of visits and the high proportion of individuals per visit, wild boar could contribute to the dynamics of mycobacteria transmission in the Basque Country. Conversely, the badger is already considered a potential reservoir of TB in neighbouring Atlantic regions [19,20] but no infected individual was found in our study area [22]. However, a high prevalence of NTM infection (17%) was detected in this mustelid, as well as in other regions of northern Iberian Peninsula [23]. Since the ability of badger to transmit MTC is already confirmed, either as a TB maintenance host or as a bridge between other species through its latrines [46], the potential role of this carnivore in the epidemiology of mycobacterioses in the Basque Country should not be ruled out. MTC-infected foxes have been sporadically found in Spain [47] but not in the Basque Country, where 46 individuals were analysed throughout a 10-year survey [22]. The fox is currently considered a spill-over host of TB in Europe [48] (i.e., populations cannot maintain infection on the long-term, but may transmit it to other species), even though the prevalence reported in foxes ranged from 9% in four TB endemic areas of France [49] to 26% in Portugal [50,51]. In addition, the prevalence of NTM in foxes from the Basque Country (4.3%) was lower compared to badgers and wild boar. Nevertheless, the fox was the species most often observed and for which most direct contacts with cattle were recorded, so its behaviour could counteract its apparent irrelevance in the epidemiology of mycobacterioses. A study carried out earlier in the same three farms proved that small rodents such as *A. sylvaticus* can carry potentially pathogenic NTM with the ability to cross-react with TB diagnosis in cattle, reporting an overall prevalence of 6.5% [18]. However, no species belonging to the MTC were detected. The scarce literature available on the epidemiology of natural *M. bovis* infection in small rodents suggests that these animals could be dead-end hosts (i.e., not able to transmit infection to other species [52–54]). However, the field vole (*Microtus agrestis*) is considered as a natural maintenance host for *M. microti*, a role that other small rodents like the wood mouse might play, maintaining the infection and spreading the bacteria through wounds inflicted to their predators or by indirect transmission through sputum, saliva or skin crusts [55,56]. These routes should not be ruled out for NTM transmission. Although most of the visits in our study were recorded in pastures and wildlife paths, small rodents were also observed inside the farm buildings and, conversely to the rest of the studied species, most of their direct contacts with cattle took place inside the enclosures. Roe deer visits were on average more frequent than those of small rodents and longer compared to badgers, wild boars and foxes. In the Basque Country, no cases of TB were detected [22] and the prevalence of NTM in roe deer was 4.70%. Like the fox, the roe deer has been considered a spill-over host, particularly in endemic areas [57]. However, TB cases in roe deer are reported even more sporadically [58]. The behaviour of this species during this study (mostly solitary, not observed close to farm buildings, preference for food sources disregarded by other species and almost no direct contacts with cattle) suggests that roe deer is unlikely to play any role in the epidemiology of TB in this low prevalence area. Accordingly, its relevance in the epidemiology of other mycobacterial infections seems to be limited. Finally, OC group could be considered the least threatening in terms of mycobacteria transmission risk in the study area. If we focus on the species observed in this study, *M. avium* was detected in one stone marten out of 18 specimens analysed in the Basque Country. Besides, to the best of our knowledge the only cases of MTC infection reported in Europe belonged to one stone marten and two genets from Portugal [51]. These epidemiological features, as well as the behaviour observed (lowest frequency and duration of visits that are always performed by one individual, no direct interaction with cattle and preference of edges over pastures) support our statement.

The findings of this study together with previous results in wildlife from the Basque Country and the low TB infection prevalence observed in cattle do not show a strong justification for intervention to reduce the risk of mycobacteria transmission at the wildlife–livestock interface. However, monitoring has proved to be an essential tool for defining the most appropriate measures if the situation changes. To reduce wild visits to farms, combined strategies rather than a single one would be more effective [59]. This study suggests that biosafety should particularly concern pastures, for example by activating electric fencing at night and at different heights, taking into account the anatomy of the species of interest. Furthermore, according to our results the presence of badger latrines inside pastures should be at least identified and hereafter reduced or cattle access to them should be avoided. These measures may reduce indirectly the presence of wild boar and foxes in these particular points as well. Because measures directed at minimizing contacts between cattle and small rodents are difficult to implement, rodent population control strategies may indirectly help to reduce these interactions. Notwithstanding, MTC and *M. leprae* put aside and in spite of a few recognized pathogens such as *M. avium* subsp. *paratuberculosis* and *M. avium* subsp. *avium* [60,61], the vast majority of NTM are generally regarded as environmental and ubiquitous or opportunistic pathogens at the most [62,63]. This fact makes NTM control even more challenging. Strategies to avoid exposure to these mycobacteria may rely more strongly on hygiene than on preventing contact between animals. Possible measures include avoidance of animal-driven farm environmental contamination, water contamination, biofilm formation, and surface spreading [63].

5. Conclusions

The results of the present study combined with the information derived from our previous epidemiological surveys suggest that four wild species or groups might be most involved in the epidemiology of mycobacterioses in the Basque Country: wild boar, badgers, foxes, and small rodents. Cattle pastures were the most frequently visited habitat and indirect interactions represent the most likely route for the potential transmission of mycobacteria between cattle and wild species. Conversely to previous studies, food and water sources did not attract wild species in this region, while badger latrines could have acted as aggregation points and as a source of mycobacteria exposure for badgers, wild boars, foxes and cattle. Further studies are first needed to confirm that interactions between wild species and cattle in pastures occur preferentially on the latrines. If so, their identification and management would be a key to avoid inter-species transmission on pastures in this area. Moreover, analysing interactions among these three wild species around latrines and in other interfaces (elsewhere than in farm environment) will be needed to completely understand mycobacteria transmission dynamics. In agreement with the low TB prevalence of the study area, the risk of MTC transmission at the wildlife-livestock interface is expected to be low. However, the risk of NTM interspecies transmission in the Basque Country is more likely than that of MTC, which could result in the maintenance and spread of potentially pathogenic mycobacteria that could also affect the tuberculin test specificity in cattle. The current quantification and qualification of connections among several hosts provides a valuable insight into the dynamics of transmission within the wildlife-livestock interface.

Supplementary Materials: The following are available online at https://www.mdpi.com/article/10.3390/ani11082364/s1, Table S1: Sessions database; Table S2: Visits database.

Author Contributions: Conceptualization, M.B.; methodology, M.B., L.V.-C., E.G.-F. and A.P.; formal analysis, L.V.-C.; investigation, M.B., I.A.S. and L.V.-C.; resources, M.B., E.G.-F. and A.P.; data curation, L.V.-C., E.G.-F. and A.P.; writing—original draft preparation, L.V.-C.; writing—review and editing, M.B., L.V.-C., I.A.S., E.G.-F. and A.P.; visualization, L.V.-C.; supervision, M.B., I.A.S. and E.G.-F.; project administration, M.B. and I.A.S.; funding acquisition, M.B. and I.A.S. All authors have read and agreed to the published version of the manuscript.

Funding: This research was funded by the National Institute for Agricultural and Food Research and Technology (INIA) (Research Project RTA2014-00002-C02-02), the "Departamento de Desarrollo Económico, Sostenibilidad y Medio Ambiente" of the Basque Government and the Interreg-POCTEFA EFA357-19-INNOTUB Project (FEDER co-funded). LV holds a pre-doctoral fellowship from the National Institute for Agricultural and Food Research and Technology (INIA), grant number CPD2016-0006.

Institutional Review Board Statement: Ethical review and approval were waived for this study, since camera trapping allows for collection of information on wild animals without capturing or handling them. Permissions for the field work were obtained from the competent authorities (corresponding approval numbers and dates: 6387/2917 in December 2017, 1907 in March 2017 and 183 in February 2017).

Data Availability Statement: The data presented in this study are contained within the Supplementary Materials Tables S1 and S2.

Acknowledgments: We want to thank the Provincial Councils of Araba, Bizkaia, and Gipuzkoa for giving us the permissions needed for camera trapping activities. A sincere appreciation goes to Diana Paniagua, Andrés Illana and Jorge Echegaray, members of GADEN, for their implication in the field work design and for the substantial work they performed on the identification of the species appearing in all the videos. Special thanks go to Vega Álvarez, Xeider Gerrikagoitia, Olalla Torrontegi and Miriam Martinez de Egidua for their collaboration with the field work, as well as to José Luis Lavín for helping to build the initial database in a record time. We also want to thank the laboratory UMR 5558 Laboratoire de Biométrie et Biologie Évolutive for receiving Lucía Varela-Castro for a training on the statistical methods used in this work. We are grateful to the farmers because this study would not have been possible without their collaboration.

Conflicts of Interest: The authors declare no conflict of interest.

References

1. Roche, B.; Eric Benbow, M.; Merritt, R.; Kimbirauskas, R.; McIntosh, M.; Small, P.L.C.; Williamson, H.; Guégan, J.F. Identifying the Achilles heel of multi-host pathogens: The concept of keystone "host" species illustrated by *Mycobacterium ulcerans* transmission. *Environ. Res. Lett.* **2013**, *8*. [CrossRef]
2. Kukielka, E.; Barasona, J.A.; Cowie, C.E.; Drewe, J.A.; Gortazar, C.; Cotarelo, I.; Vicente, J. Spatial and temporal interactions between livestock and wildlife in South Central Spain assessed by camera traps. *Prev. Vet. Med.* **2013**, *112*, 213–221. [CrossRef]
3. Carrasco-Garcia, R.; Barasona, J.A.; Gortazar, C.; Montoro, V.; Sanchez-Vizcaino, J.M.; Vicente, J. Wildlife and livestock use of extensive farm resources in South Central Spain: Implications for disease transmission. *Eur. J. Wildl. Res.* **2015**, *62*, 65–78. [CrossRef]
4. Triguero-Ocaña, R.; Barasona, J.A.; Carro, F.; Soriguer, R.C.; Vicente, J.; Acevedo, P. Spatio-temporal trends in the frequency of interspecific interactions between domestic and wild ungulates from Mediterranean Spain. *PLoS ONE* **2019**, *14*, e0211216. [CrossRef]
5. Santos, N.; Richomme, C.; Nunes, T.; Vicente, J.; Alves, P.C.; de la Fuente, J.; Correia-neves, M.; Boschiroli, M.L.; Delahay, R.; Gortázar, C. Quantification of the animal tuberculosis multi-host community offers insights for control. *Pathogens* **2020**, *9*, 421. [CrossRef] [PubMed]
6. Rodríguez-Hernández, E.; Pizano-Martínez, O.E.; Canto-Alarcón, G.; Flores-Villalva, S.; Quintas-Granados, L.I.; Milián-Suazo, F. Persistence of *Mycobacterium bovis* under environmental conditions: Is it a real biological risk for cattle? *Rev. Med. Microbiol.* **2016**, *27*, 20–24. [CrossRef]

7. Baldwin, S.L.; Larsenid, S.E.; Ordway, D.; Cassell, G.; Coler, R.N. The complexities and challenges of preventing and treating nontuberculous mycobacterial diseases. *PLoS Negl. Trop. Dis.* **2019**, *13*, e0007083. [CrossRef] [PubMed]
8. Hoefsloot, W.; Van Ingen, J.; Andrejak, C.; Ängeby, K.; Bauriaud, R.; Bemer, P.; Beylis, N.; Boeree, M.J.; Cacho, J.; Chihota, V.; et al. The geographic diversity of nontuberculous mycobacteria isolated from pulmonary samples: An NTM-NET collaborative study. *Eur. Respir. J.* **2013**, *42*, 1604–1613. [CrossRef]
9. Biet, F.; Boschiroli, M.L. Non-tuberculous mycobacterial infections of veterinary relevance. *Res. Vet. Sci.* **2014**, *97*, S69–S77. [CrossRef] [PubMed]
10. Hruska, K.; Kaevska, M. Mycobacteria in water, soil, plants and air: A review. *Vet. Med.* **2012**, *57*, 623–679. [CrossRef]
11. Eppleston, J.; Begg, D.J.; Dhand, N.K.; Watt, B.; Whittington, R.J. Environmental survival of *Mycobacterium avium* subsp. *paratuberculosis in different climatic zones of eastern Australia. Appl. Environ. Microbiol.* **2014**, *80*, 2337–2342. [CrossRef]
12. Gortazar, C.; Boadella, M. Animal tuberculosis in Spain: A multihost system. In *Zoonotic Tuberculosis: Mycobacterium Bovis and Other Pathogenic Mycobacteria*, 3rd ed.; John Wiley & Sons, Inc.: New York, NY, USA, 2014; pp. 349–356. ISBN 9781118474310.
13. Napp, S.; Allepuz, A.; Mercader, I.; Nofrarias, M.; Lopez-Soria, S.; Domingo, M.; Romero, B.; Bezos, J.; De Val Perez, B. Evidence of goats acting as domestic reservoirs of bovine tuberculosis. *Vet. Rec.* **2013**, *172*, 663. [CrossRef] [PubMed]
14. Muñoz-Mendoza, M.; Romero, B.; del Cerro, A.; Gortázar, C.; García-Marín, J.F.; Menéndez, S.; Mourelo, J.; de Juan, L.; Sáez, J.L.; Delahay, R.J.; et al. Sheep as a Potential Source of Bovine TB: Epidemiology, Pathology and Evaluation of Diagnostic Techniques. *Transbound. Emerg. Dis.* **2015**, *63*, 635–646. [CrossRef] [PubMed]
15. Parra, A.; Fernández-Llario, P.; Tato, A.; Larrasa, J.; García, A.; Alonso, J.M.; Hermoso De Mendoza, M.; Hermoso De Mendoza, J. Epidemiology of *Mycobacterium bovis* infections of pigs and wild boars using a molecular approach. *Vet. Microbiol.* **2003**, *97*, 123–133. [CrossRef]
16. Ministerio de Agricultura Pesca y Alimentación Programa Nacional de Erradicación de Tuberculosis Bovina Presentado Por España Para el Año 2020. Available online: https://www.mapa.gob.es/es/ganaderia/temas/sanidad-animal-higiene-ganadera/pnetb_2020final_tcm30-523317.PDF (accessed on 7 October 2020).
17. Muñoz-Mendoza, M.; Marreros, N.; Boadella, M.; Gortázar, C.; Menéndez, S.; de Juan, L.; Bezos, J.; Romero, B.; Copano, M.F.; Amado, J.; et al. Wild boar tuberculosis in Iberian Atlantic Spain: A different picture from Mediterranean habitats. *BMC Vet. Res.* **2013**, *9*, 176. [CrossRef] [PubMed]
18. Varela-Castro, L.; Torrontegi, O.; Sevilla, I.A.; Barral, M. Detection of wood mice (*Apodemus sylvaticus*) carrying non-tuberculous mycobacteria able to infect cattle and interfere with the diagnosis of bovine tuberculosis. *Microorganisms* **2020**, *8*, 374. [CrossRef]
19. Acevedo, P.; Prieto, M.; Quirós, P.; Merediz, I.; de Juan, L.; Infantes-Lorenzo, J.A.; Triguero-Ocaña, R.; Balseiro, A. Tuberculosis Epidemiology and Badger (*Meles meles*) Spatial Ecology in a Hot-Spot Area in Atlantic Spain. *Pathogens* **2019**, *8*, 292. [CrossRef]
20. Blanco Vázquez, C.B.; Barral, T.D.; Romero, B.; Queipo, M.; Merediz, I.; Quirós, P.; Armenteros, J.Á.; Juste, R.; Domínguez, L.; Domínguez, M.; et al. Spatial and Temporal Distribution of *Mycobacterium tuberculosis* Complex Infection in Eurasian Badger (*Meles meles*) and Cattle in Asturias, Spain. *Animals* **2021**, *11*, 1294. [CrossRef]
21. Varela-Castro, L.; Alvarez, V.; Sevilla, I.A.; Barral, M. Risk factors associated to a high *Mycobacterium tuberculosis* complex seroprevalence in wild boar (*Sus scrofa*) from a low bovine tuberculosis prevalence area. *PLoS ONE* **2020**, *15*, e0231559. [CrossRef] [PubMed]
22. Varela-Castro, L.; Gerrikagoitia, X.; Alvarez, V.; Geijo, M.V.; Barral, M.; Sevilla, I.A. A long-term survey on *Mycobacterium tuberculosis* complex in wild mammals from a bovine tuberculosis low prevalence area. *Eur. J. Wildl. Res.* **2021**, *67*, 1–8. [CrossRef]
23. Balseiro, A.; Merediz, I.; Sevilla, I.A.; García-Castro, C.; Gortázar, C.; Prieto, J.M.; Delahay, R.J. Infection of Eurasian badgers (*Meles meles*) with *Mycobacterium avium* complex (MAC) bacteria. *Vet. J.* **2011**, *188*, 231–233. [CrossRef]
24. Rowcliffe, J.M.; Field, J.; Turvey, S.T.; Carbone, C. Estimating animal density using camera traps without the need for individual recognition. *J. Appl. Ecol.* **2008**, *45*, 1228–1236. [CrossRef]
25. Niedballa, J.; Wilting, A.; Sollmann, R.; Hofer, H.; Courtiol, A. Assessing analytical methods for detecting spatiotemporal interactions between species from camera trapping data. *Remote Sens. Ecol. Conserv.* **2019**. [CrossRef]
26. Niedballa, J.; Sollmann, R.; Courtiol, A.; Wilting, A. camtrapR: An R package for efficient camera trap data management. *Methods Ecol. Evol.* **2016**, *7*, 1457–1462. [CrossRef]
27. Triguero-Ocaña, R.; Vicente, J.; Palencia, P.; Laguna, E.; Acevedo, P. Quantifying wildlife-livestock interactions and their spatio-temporal patterns: Is regular grid camera trapping a suitable approach? *Ecol. Indic.* **2020**, *117*, 106565. [CrossRef]
28. Gobierno Vasco euskadi.eus. Available online: https://www.euskadi.eus/inicio/ (accessed on 5 May 2020).
29. Payne, A.; Philipon, S.; Hars, J.; Dufour, B.; Gilot-Fromont, E. Wildlife Interactions on Baited Places and Waterholes in a French Area Infected by Bovine Tuberculosis. *Front. Vet. Sci.* **2017**, *3*, 122. [CrossRef]
30. Payne, A.; Chappa, S.; Hars, J.; Dufour, B.; Gilot-Fromont, E. Wildlife visits to farm facilities assessed by camera traps in a bovine tuberculosis-infected area in France. *Eur. J. Wildl. Res.* **2016**, *62*, 33–42. [CrossRef]
31. Burnham, K.P.; Anderson, D.R. *Model Selection and Multimodel Inference: A Practical Information-Theoretic Approach*, 2nd ed.; Springer: Berlin/Heidelberg, Germany, 2002.
32. Zuur, A.F.; Ieno, E.N.; Walker, N.; Saveliev, A.A.; Smith, G.M. *Mixed Effects Models and Extensions in Ecology with R*; Springer: New York, NY, USA, 2009.
33. *R Core Team 2020: R: A Language and Environment for Statistical Computing*; R Foundation for Statistical Computing: Vienna, Austria, 2021; Available online: https://www.R-project.org/ (accessed on 31 July 2021).

34. Drewe, J.A.; O'Connor, H.M.; Weber, N.; McDonald, R.A.; Delahay, R.J. Patterns of direct and indirect contact between cattle and badgers naturally infected with tuberculosis. *Epidemiol. Infect.* **2013**, *141*, 1467–1475. [CrossRef] [PubMed]

35. Zabala, J.; Garin, I.; Zuberogoitia, I.; Aihartza, J. Habitat selection and diet of badgers (*Meles meles*) in Biscay (northern Iberian Peninsula). *Ital. J. Zool.* **2002**, *69*, 233–238. [CrossRef]

36. Mullen, E.M.; MacWhite, T.; Maher, P.K.; Kelly, D.J.; Marples, N.M.; Good, M. The avoidance of farmyards by European badgers *Meles meles* in a medium density population. *Appl. Anim. Behav. Sci.* **2015**, *171*, 170–176. [CrossRef]

37. Tolhurst, B.A.; Delahay, R.J.; Walker, N.J.; Ward, A.I.; Roper, T.J. Behaviour of badgers (*Meles meles*) in farm buildings: Opportunities for the transmission of *Mycobacterium bovis* to cattle? *Appl. Anim. Behav. Sci.* **2009**, *117*, 103–113. [CrossRef]

38. Judge, J.; Mcdonald, R.A.; Walker, N.; Delahay, R.J. Effectiveness of Biosecurity Measures in Preventing Badger Visits to Farm Buildings. *PLoS ONE* **2011**, *6*, e28941. [CrossRef] [PubMed]

39. Woodroffe, R.; Donnelly, C.A.; Ham, C.; Jackson, S.Y.B.; Moyes, K.; Chapman, K.; Stratton, N.G.; Cartwright, S.J. Badgers prefer cattle pasture but avoid cattle: Implications for bovine tuberculosis control. *Ecol. Lett.* **2016**, *19*, 1201–1208. [CrossRef] [PubMed]

40. Hoem, S.A.; Melis, C.; Linnell, J.D.C.; Andersen, R. Fighting behaviour in territorial male roe deer *Capreolus capreolus*: The effects of antler size and residence. *Eur. J. Wildl. Res.* **2007**, *53*, 1–8. [CrossRef]

41. Duncan, P.; Tixier, H.; Hofmann, R.; Lechner-Doll, M. Feeding strategies and the physiology of digestion in roe deer. In *The European Roe Deer: The Biology of Success*; Scandinavian University Press: Oslo, Norway, 1998; pp. 91–116.

42. Palomares, F.; Delibes, M. Spatio-temporal Ecology and Behaviour of European Genets in Southwestern Spain. *J. Mammal.* **1994**, *75*, 714–724. [CrossRef]

43. Torre, I.; Arrizabalaga, A.; Díaz, M. Ratón de campo *Apodemus sylvaticus* (Linnaeus, 1758). *Galemys* **2002**, *14*, 1–26.

44. Montgomery, W.I.; Dowie, M. The Distribution and Population Regulation of the Wood Mouse *Apodemus sylvaticus* on Field Boundaries of Pastoral Farmland. *J. Appl. Ecol.* **1993**, *30*, 783. [CrossRef]

45. Barasona, J.A.; Latham, M.C.; Acevedo, P.; Armenteros, J.A.; Latham, A.D.M.; Gortazar, C.; Carro, F.; Soriguer, R.C.; Vicente, J. Spatiotemporal interactions between wild boar and cattle: Implications for cross-species disease transmission. *Vet. Res.* **2014**, *45*, 122. [CrossRef]

46. Caron, A.; Cappelle, J.; Cumming, G.S.; De Garine-Wichatitsky, M.; Gaidet, N. Bridge hosts, a missing link for disease ecology in multi-host systems. *Vet. Res.* **2015**, *46*, 1–11. [CrossRef]

47. Millán, J.; Jiménez, M.Á.; Viota, M.; Candela, M.G.; Peña, L.; León-Vizcaíno, L. Disseminated Bovine Tuberculosis in a Wild Red Fox (*Vulpes vulpes*) in Southern Spain. *J. Wildl. Dis.* **2008**, *44*, 701–706. [CrossRef]

48. Michelet, L.; De Cruz, K.; Hénault, S.; Tambosco, J.; Richomme, C.; Réveillaud, É.; Gares, H.; Moyen, J.L.; Boschiroli, M.L. *Mycobacterium bovis* infection of red fox, France. *Emerg. Infect. Dis.* **2018**, *24*, 1151–1153. [CrossRef]

49. Richomme, C.; Réveillaud, E.; Moyen, J.L.; Sabatier, P.; de Cruz, K.; Michelet, L.; Boschiroli, M.L. *Mycobacterium bovis* infection in red foxes in four animal tuberculosis endemic areas in France. *Microorganisms* **2020**, *8*, 1070. [CrossRef]

50. Matos, A.C.; Figueira, L.; Martins, M.H.; Matos, M.; Morais, M.; Dias, A.P.; Pinto, M.L.; Coelho, A.C. Disseminated *Mycobacterium bovis* Infection in Red Foxes (*Vulpes vulpes*) with Cerebral Involvement Found in Portugal. *Vector-Borne Zoonotic Dis.* **2014**, *14*, 531–533. [CrossRef]

51. Matos, A.C.; Figueira, L.; Martins, M.H.; Pinto, M.L.; Matos, M.; Coelho, A.C. New insights into *Mycobacterium bovis* prevalence in wild mammals in Portugal. *Transbound. Emerg. Dis.* **2016**, *63*, e313–e322. [CrossRef]

52. Mathews, F.; Macdonald, D.W.; Taylor, G.M.; Gelling, M.; Norman, R.A.; Honess, P.E.; Foster, R.; Gower, C.M.; Varley, S.; Harris, A.; et al. Bovine tuberculosis (*Mycobacterium bovis*) in British farmland wildlife: The importance to agriculture. *Proc. R. Soc. B Biol. Sci.* **2006**, *273*, 357–365. [CrossRef] [PubMed]

53. Delahay, R.J.; De Leeuw, A.N.S.; Barlow, A.M.; Clifton-Hadley, R.S.; Cheeseman, C.L. The status of *Mycobacterium bovis* infection in UK wild mammals: A review. *Vet. J.* **2002**, *164*, 90–105. [CrossRef]

54. Delahay, R.J.; Smith, G.C.; Barlow, A.M.; Walker, N.; Harris, A.; Clifton-Hadley, R.S.; Cheeseman, C.L. Bovine tuberculosis infection in wild mammals in the South-West region of England: A survey of prevalence and a semi-quantitative assessment of the relative risks to cattle. *Vet. J.* **2007**, *173*, 287–301. [CrossRef]

55. Smith, N.H.; Crawshaw, T.; Parry, J.; Birtles, R.J. *Mycobacterium microti*: More diverse than previously thought. *J. Clin. Microbiol.* **2009**, *47*, 2551–2559. [CrossRef]

56. Kipar, A.; Burthe, S.J.; Hetzel, U.; Abo Rokia, M.; Telfer, S.; Lambin, X.; Birtles, R.J.; Begon, M.; Bennett, M. *Mycobacterium microti* Tuberculosis in Its Maintenance Host, the Field Vole (*Microtus agrestis*): Characterization of the Disease and Possible Routes of Transmission. *Vet. Pathol.* **2014**, *51*, 903–914. [CrossRef] [PubMed]

57. Lambert, S.; Hars, J.; Réveillaud, E.; Moyen, J.-L.; Gares, H.; Rambaud, T.; Gueneau, E.; Faure, E.; Boschiroli, M.-L.; Richomme, C. Host status of wild roe deer in bovine tuberculosis endemic areas. *Eur. J. Wildl. Res.* **2017**, *63*, 15. [CrossRef]

58. Balseiro, A.; Orusa, A.O.R.; Robetto, S.; Zoppi, Z.; Dondo, A.; Goria, M.; Gortázar, C.; Marín, J.F.G.; Domenis, L. Tuberculosis in roe deer from Spain and Italy. *Vet. Rec.* **2009**, *164*, 468–470. [CrossRef] [PubMed]

59. Gortázar, C.; Diez-Delgado, I.; Barasona, J.A.; Vicente, J.; De La Fuente, J.; Boadella, M. The Wild Side of Disease Control at the Wildlife-Livestock-Human Interface: A Review. *Front. Vet. Sci.* **2015**, *1*, 27. [CrossRef] [PubMed]

60. Turenne, C.Y.; Collins, D.M.; Alexander, D.C.; Behr, M.A. *Mycobacterium avium* subsp. *paratuberculosis* and *M. avium* subsp. *avium* are independently evolved pathogenic clones of a much broader group of *M. avium* organisms. *J. Bacteriol.* **2008**, *190*, 2479–2487. [CrossRef]

61. Turenne, C.Y.; Wallace, R.; Behr, M.A. *Mycobacterium avium* in the Postgenomic Era. *Clin. Microbiol. Rev.* **2007**, *20*, 205–229. [CrossRef]
62. Kasperbauer, S.; Huitt, G. Management of Extrapulmonary Nontuberculous Mycobacterial Infections. *Semin. Respir. Crit. Care Med.* **2013**, *34*, 143–150. [CrossRef]
63. Claeys, T.A.; Robinson, R.T. The Many Lives of Nontuberculous Mycobacteria. *J. Bacteriol.* **2018**, *200*, e00739-17. [CrossRef] [PubMed]

Article

Spatial and Temporal Distribution of *Mycobacterium tuberculosis* Complex Infection in Eurasian Badger (*Meles meles*) and Cattle in Asturias, Spain

Cristina Blanco Vázquez [1], Thiago Doria Barral [2], Beatriz Romero [3], Manuel Queipo [4], Isabel Merediz [5], Pablo Quirós [6], José Ángel Armenteros [6], Ramón Juste [7], Lucas Domínguez [3,8], Mercedes Domínguez [9], Rosa Casais [1] and Ana Balseiro [10,11,*]

[1] Servicio Regional de Investigación y Desarrollo Agroalimentario del Principado de Asturias (SERIDA), 33300 Villaviciosa, Spain; cristina.blancovazquez@serida.org (C.B.V.); rosacg@serida.org (R.C.)
[2] Laboratório de Imunologia e Biologia Molecular, Instituto de Ciências da Saúde, Universidade Federal da Bahia, 40.110-100 Salvador, Bahia, Brazil; tadbarral@gmail.com
[3] Centro de Vigilancia Sanitaria Veterinaria VISAVET, Universidad Complutense, 28040 Madrid, Spain; bromerom@visavet.ucm.es (B.R.); lucasdo@visavet.ucm.es (L.D.)
[4] Servicio de Sanidad y Producción Animal del Principado de Asturias, 33007 Oviedo, Asturias, Spain; manuelantonio.queiporodriguez@asturias.org
[5] Laboratorio Regional de Sanidad Animal del Principado de Asturias, 33201 Gijón, Asturias, Spain; isabel.meredizgutierrez@asturias.org
[6] Dirección General del Medio Natural y Planificación Rural del Principado de Asturias, 33007 Oviedo, Asturias, Spain; pablo.quirosmenendezdeluarca@asturias.org (P.Q.); joseangel.armenterossantos@asturias.org (J.Á.A.)
[7] Animal Health Department, NEIKER-Instituto Vasco de Investigación y Desarrollo Agrario, 48160 Derio, Bizkaia, Spain; rjuste@neiker.eus
[8] Departamento de Sanidad Animal, Facultad de Veterinaria, Universidad Complutense, 28040 Madrid, Spain
[9] Unidad de Inmunología Microbiana, Centro Nacional de Microbiología, Instituto de Salud Carlos III, 28029 Madrid, Spain; mdominguez@isciii.es
[10] Departamento de Sanidad Animal, Facultad de Veterinaria, Universidad de León, 24071 León, Spain
[11] Departamento de Sanidad Animal, Instituto de Ganadería de Montaña (CSIC-Universidad de León), Finca Marzanas, Grulleros, 24346 León, Spain
* Correspondence: abalm@unileon.es

Citation: Blanco Vázquez, C.; Barral, T.D.; Romero, B.; Queipo, M.; Merediz, I.; Quirós, P.; Armenteros, J.Á.; Juste, R.; Domínguez, L.; Domínguez, M.; et al. Spatial and Temporal Distribution of *Mycobacterium tuberculosis* Complex Infection in Eurasian Badger (*Meles meles*) and Cattle in Asturias, Spain. *Animals* **2021**, *11*, 1294. https://doi.org/10.3390/ani11051294

Academic Editor: Nicole Gottdenker

Received: 1 March 2021
Accepted: 28 April 2021
Published: 30 April 2021

Publisher's Note: MDPI stays neutral with regard to jurisdictional claims in published maps and institutional affiliations.

Simple Summary: The aim of the present work was to investigate the prevalence, spatial distribution, and temporal distribution of tuberculosis in 673 free-ranging Eurasian badgers (*Meles meles*) and cattle from Asturias (Atlantic Spain) during a 13-year follow-up. The study objective was to assess the role of badgers as a reservoir of tuberculosis for cattle and other sympatric wild species in the region. During the follow-up, 27/639 badgers (4.23%) were positive for the *Mycobacterium tuberculosis* complex based on bacterial isolation, while 160/673 (23.77%) were positive based on P22 ELISA. Badger infection was spatially and temporally associated with cattle herd infection.

Abstract: The present work investigated the prevalence, spatial distribution, and temporal distribution of tuberculosis (TB) in free-ranging Eurasian badgers (*Meles meles*) and cattle in Asturias (Atlantic Spain) during a 13-year follow-up. The study objective was to assess the role of badgers as a TB reservoir for cattle and other sympatric wild species in the region. Between 2008 and 2020, 673 badgers (98 trapped and 575 killed in road traffic accidents) in Asturias were necropsied, and their tissue samples were cultured for the *Mycobacterium tuberculosis* complex (MTC) isolation. Serum samples were tested in an in-house indirect P22 ELISA to detect antibodies against the MTC. In parallel, data on MTC isolation and single intradermal tuberculin test results were extracted for cattle that were tested and culled as part of the Spanish National Program for the Eradication of Bovine TB. A total of 27/639 badgers (4.23%) were positive for MTC based on bacterial isolation, while 160/673 badgers (23.77%) were found to be positive with the P22 ELISA. The rate of seropositivity was higher among adult badgers than subadults. Badger TB status was spatially and temporally associated with cattle TB status. Our results cannot determine the direction of possible interspecies transmission, but they are consistent with the idea that the two hosts may exert infection pressure on each other. This study highlights the importance of the wildlife monitoring of infection and disease during epidemiological interventions in order to optimize outcomes.

Keywords: *Meles meles*; badger; tuberculosis; *Mycobacterium tuberculosis* complex; P22 ELISA; isolation; serology; cattle; Atlantic Spain

1. Introduction

Animal tuberculosis (TB) is caused by infection with members of the *Mycobacterium tuberculosis* complex (MTC), mainly *M. bovis* and, to a lesser extent, *M. caprae*. TB is a zoonotic disease that subjects livestock worldwide and can cause substantial economic losses [1]. Its main domestic reservoir is cattle, and eradication campaigns spanning more than four decades in developed countries including Spain have been quite but not totally successful [2]. Wildlife hosts are also susceptible to *M. bovis* and can act as reservoirs of the infection for livestock. Badgers (*Meles meles*) and wild boars (*Sus scrofa*) are major wildlife reservoirs of *M. bovis* for bovine TB in several European countries. *M. bovis*-infected badgers have been found in Ireland [3], the United Kingdom (UK) [4], France [5,6], and Spain [7,8]; in Ireland and the UK, badgers are recognized as major reservoirs with the potential to transmit infection to local cattle herds [9,10]. Infection in wild boars has been described in France and Italy, as well as the southern and central Iberian Peninsula [5,11].

In Spain, the southern and central regions, which have a Mediterranean climate, show a TB cattle herd-prevalence as high as 12.3%, while the northwestern region with an Atlantic climate shows an overall prevalence of 0.02–0.08% but with hotspots where prevalence can be as high as 5% [2,12]. This variability may reflect several factors. One is climate: drier, hotter areas in the south favor animal aggregation and therefore disease spread [13]. Another factor is type of cattle: dairy herds, in which TB prevalence is lower, concentrate in the north, while beef and bullfighting cattle are more abundant in the central and southern regions [13]. A third factor is cattle management: beef and bullfighting cattle herds are extensively managed with a lower biosecurity and are therefore at higher risk of infection due to contact with animals in other herds, as well as with other potentially infected domestic (goat/sheep) or wildlife species [14–16].

Recent studies have suggested that badgers may be a reservoir of *M. bovis* infection in hotspots in Atlantic Spain, as reported in Ireland and the UK [12,17]. In addition, wild boar, red deer (*Cervus elaphus*), and, to a lesser extent, fallow deer (*Dama dama*) are major wildlife reservoirs in continental Mediterranean habitats, where wild boars can show a TB prevalence of >50% [14]. In contrast, wild boars in Atlantic habitats show a TB prevalence of around only 5% [16]. Mid- and long-term studies of wildlife reservoirs such as badgers and wild boar can shed light on the spatial and temporal dynamics of TB, as well as on risk to local livestock [14,18,19].

Along these lines, the present work investigated the prevalence, spatial distribution, and temporal distribution of TB in free-ranging Eurasian badgers and cattle from Asturias (Atlantic Spain) during a 13-year follow-up. The study objective was to assess the potential role of badgers as a TB reservoir for cattle and other sympatric wild species such as wild boar in the region.

2. Materials and Methods

2.1. Ethics Statement

All methods were carried out in accordance with relevant guidelines and regulations. All experimental procedures with trapped badgers were approved by the Government of the Principality of Asturias (010/07-01-2011, PROAE 20/2015, PROAE 47/2018).

2.2. Study Area

The study was carried out in the region of Asturias in northwestern Spain. This region has an Atlantic climate, and the temperature ranges from −4 to 8 °C in the coldest months; precipitation is abundant throughout the year, annually reaching 1400–2100 mm [20]. More than 30% of the territory is forest and mainly comprises oaks, beech, and birch woods. Asturias is bordered to the north by the Cantabrian Sea and to the south by the Cantabrian Range. Nowadays, the cattle population is 391,797 animals and 15,856 herds, and the well-studied badger population shows intermediate density of 3.81 adults/km^2 [21].

2.3. Description of TB Detection in Badgers and Cattle in Asturias, 2008–2020

2.3.1. Badgers

From 2008 to 2020, 673 badgers from Asturias were necropsied (see Supplementary Material 1: Raw Data); these animals came from western (n = 77), central (n = 220), and eastern (n = 376) parts of Asturias. Of the 673 animals, 98 were trapped badgers that had been captured from setts located close to TB cattle herds and subsequently euthanized; these 98 animals came nearly equally from the three areas. Badgers were captured in steel mesh box traps baited with peanuts and anesthetized with ketamine hydrochloride (0.1 mL kg^{-1}), medetomidine (Domtor$^®$, Ecuphar Veterinaria, Barcelona, Spain; 0.05 mL kg^{-1}), and butorphanol (Torbugesic$^®$, Zoetis, Madrid, Spain; 0.1 mL kg^{-1}) administered by means of intramuscular injection. Serum samples (2 mL) were taken from anesthetized trapped badgers by jugular venipuncture and collected in serum separation vacutainer tubes. Afterwards, badgers were euthanized with an overdose of sodium pentobarbital for post-mortem examination. The other 575 were killed in road traffic accidents and found by gamekeepers in the region. Serum samples (2 mL) from badgers killed in road traffic accidents were taken from the heart or thoracic cavity and collected in vacutainer tubes.

The location, sex, age class (subadult/adult), and weight were recorded for trapped and road-killed badgers. The entire group contained 301 males and 322 females, comprising 163 subadults and 460 adults based on body size and teeth. Sex and age class were undetermined for 50 individuals killed in road traffic accidents because of insufficient tissue availability. The numbers of animals analyzed per year were as follows: 2008: 21; 2009: 37; 2010: 68; 2011: 45; 2012: 20; 2013: 23; 2014: 43; 2015: 22; 2016: 92; 2017: 57; 2018: 63; 2019: 102; and 2020: 80. Trapping was not carried out in 2012–2015, 2017, or 2020, so all badgers studied in those years were killed in road traffic accidents. Badgers were preserved at 4 °C, and a complete post-mortem examination of each carcass was conducted at the laboratory in less than 24 h.

2.3.1.1. Pathology

Serial sections (0.2 cm) were taken from the lungs and lymph nodes (LNs) of the 673 badgers submitted for necropsy for further macroscopic observation. Gross visible lesions found during necropsy were recorded. Tissue samples from the lungs and retropharyngeal, submandibular, tracheobronchial, mediastinal, hepatic, and mesenteric LNs were further analyzed by histopathology from the nine animals that showed gross lesions. Samples for histopathology were fixed in 10% neutral buffered formalin and processed using standard methods. Sections were cut to a thickness of 4 μm and stained with hematoxylin and eosin and by the Ziehl–Neelsen method for the detection of acid-fast bacteria.

2.3.1.2. Bacteriology and Typing

Bacteriological and molecular studies were performed in 639 out of the 673 badgers (the remaining 34 animals were excluded due to insufficient tissue availability to make the pools). Pools (2 g) of lungs and mandibular, retropharyngeal, tracheobronchial, mediastinal, hepatic, and mesenteric LNs were frozen at −20 °C for no longer than two weeks and subsequently used to isolate potential bacteria as described previously [12]. Members of the MTC or the *Mycobacterium avium* complex (MAC) were isolated using the Mycobacteria Growth Indicator Tube (MGIT) liquid medium system, the Löwenstein–Jensen solid medium with sodium pyruvate, and Coletsos solid media. The samples were decontaminated using the BBL MycoPrep Becton Dickinson kit (BD Diagnostic Systems, Franklin Lakes, NJ, USA) and then incubated at 37 °C in an MGIT liquid medium for at least 6 weeks using the automated BACTEC MGIT 960 (BD Diagnostic Systems). Solid cultures were incubated at 37 °C for at least 10 weeks.

Quantitative PCR to identify MTC species was performed on culture isolates using the MTC forward primer 5′-TAGTGCATGCACCGAATTAGAACGT-3′, the MTC reverse primer 5′-CGAGTAGGTCATGGCTCCTCC-3′, and the TaqMan probe YY/BHQ 5′-AATCGCGTCGCCGGGAGC-3′, which amplifies a 184-bp fragment [22–24]. MTC isolates were characterized using DVR-spoligotyping after hybridizing biotin-labeled PCR products onto a home-made spoligotyping membrane (VISAVET, Madrid, Spain). Results were recorded in SB code followed by a field of four digits according to the *M. bovis* Spoligotype Database [23,24]. In order to confirm similarity between the isolates from cattle and badgers in the same area, MIRU-VNTR typing was performed as described previously [23] using the following nine VNTR markers: ETR-A, ETR-B, ETR-D, ETR-E, MIRU26, QUB11a, QUB11b, QUB26, and QUB3232. A badger was classified as positive if MTC was isolated. MAC isolates were identified as described previously [25].

2.3.1.3. Serology (P22 ELISA Assay)

Serum samples from the 673 badgers were frozen at −20 °C before processing and tested using a previous validated in-house indirect P22 ELISA to detect antibodies against the MTC in badgers [26].

Briefly, plates were coated overnight with P22 in phosphate-buffered saline (PBS) at 4 °C and then blocked with PBS containing 5% powdered skim milk. After three washes with PBS containing 0.05% Tween-20, sera (diluted 1:100 in PBS-powdered skim milk) were added to duplicate wells and incubated for 60 min at 37 °C. Horseradish peroxidase-conjugated anti-badger CF2/HRPo IgG (100 µL) [26] was diluted to 1.5 µg/mL in PBS and added to the plates. Then plates were incubated at room temperature for 15 min in the dark with 3,3′,5,5′-tetramethylbenzidine substrate (Perbio Science, Helsingborg, Sweden). The reaction was stopped by adding 100 µL of 2 M H_2SO_4. Optical density (OD) was measured at 450 nm using an ELISA reader. As a negative control, serum from a TB-free Spanish badger was included in quadruplicate in every plate. As a positive control, serum from a Spanish badger experimentally infected with *M. bovis* was included in every plate. OD readings were expressed as an ELISA percentage (E%): E% = (mean OD/(2 × mean $OD_{negative\ control}$)) × 100%. The cut-off was specifically defined for this study, since the previous P22 ELISA study in badgers suggested that the optimal cut-off (initially established between 100% and 150%) may depend on the epidemiological situation [26]. Indeed, when we applied the three cut-offs tested in that study (100%, 120%, and 150%), we found that sensitivity and specificity were too close to a random (50%) diagnostic value. Instead, we found that a cut-off of 260% gave 55.6% sensitivity and 80.4% specificity, corresponding to a moderate diagnostic value of 68% (Figure 1). In order to define the 260% cut-off, data sera from positive and negative animals (trapped and road-killed badgers) to bacteriological culture (MTC or MAC) were included.

2.3.2. Cattle

Single intradermal tuberculin test data and bacterial isolation were available for cattle that were tested and culled as part of the Spanish National Program for the Eradication of Bovine TB until 2019 [2]. A total of 17,474 herds were studied, of which 4762 belonged to the western area, 8281 belonged to the center area, and 4431 belonged to the eastern area.

2.4. Statistical Analysis

An overall descriptive analysis of absolute frequencies was performed based of tables and chi squared tests with the SAS (Cary, NC, USA) Freq procedure. Differences in MTC isolation, P22 ELISA E% value, and positive frequency in badgers stratified by area, sex, age class, origin, or time were tested using a generalized linear model (GLM) in SAS with Tukey or Tukey–Kramer's t (T or TK t) test for main effects level means comparisons. Differences with a $p < 0.05$ were considered significant. Potential associations between P22 ELISA values and TB prevalence in cattle during follow-up were tested using a Pearson correlation analysis with the SAS CORR procedure. Another GLM model was built to assess a potential association between the ability to isolate MTC or MAC from samples and the sample's P22 ELISA E%.

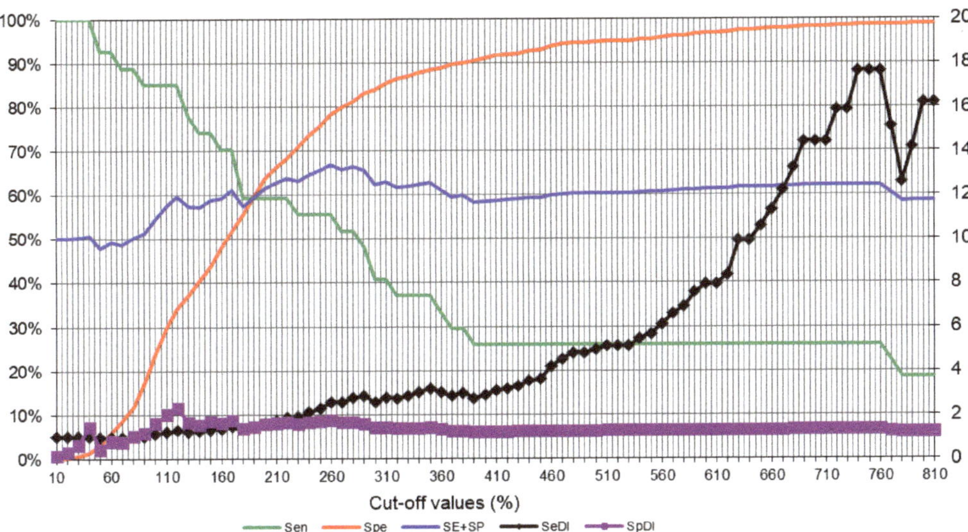

Figure 1. Definition of the P22 ELISA cut-off against a *Mycobacterium tuberculosis* complex (MTC) isolation reference. The graph includes data sera from positive and negative animals to bacteriological culture, as well as both trapped and road-killed animals. Sen: sensitivity; Spe: specificity; SE + SP: semi-sum or diagnostic value of sensitivity and specificity; SeDI: sensitivity discriminatory index (ratio of sensitivity to 1-specificity); SpDI: specificity discriminatory index (ratio of specificity to 1-sensitivity).

3. Results

3.1. Badgers

3.1.1. Gross and Microscopic Lesions

Gross lesions were observed in nine badgers from the eastern area, eight of which had been trapped and one of which had been killed in a road traffic accident. Lesions consisted of granulomatous areas of caseous necrosis and mineralization with diameters from 1 mm to 1 cm in submandibular, retropharyngeal, bronchial, or mediastinal LNs and lungs. Such lesions were also found in mesenteric LNs in two animals and in hepatic LNs in three animals. TB-like lesions were confirmed using histology, which revealed scarce, small granulomas with areas of central necrosis or mineralization. Ziehl–Neelsen staining revealed sparse acid-fast bacteria.

3.1.2. Bacteriological Isolation and Typing

Of the 639 badgers, 27 (4.23%) were positive for MTC, and the distribution of positives across the years was as follows (Figure 2): 2008: 2; 2009: 3; 2010: 4; 2011: 5; 2012: 1; 2016: 1; 2018: 10; and 2019: 1. Isolates were identified as spoligotype SB0828 with VNTR profile 5-5-3-4-5-9-3-3-6 (*n* = 7), spoligotype SB0121 with VNTR profile 5-4-3-3-5-F-2-5-8 (*n* = 3), spoligotype SB0120 (*n* = 2), spoligotype SB01019 (*n* = 2), spoligotype SB1312 (*n* = 1), and spoligotype SB0329 (*n* = 1). Spoligotypes SB0828 and SB0121 identified in badgers were also isolated from cattle in the same areas (Figure 2) [24]. Spoligotype SB0828 with VNTR profile 5-5-3-4-5-9-3-3-6 was also identified in cattle, but spoligotype SB0121 in badgers (VNTR profile 5-4-3-3-5-F-2-5-8) was different, albeit related to that in cattle (VNTR profile 4-4-3-3-5-10-2-5-8). SB1019 was identified in badgers and cattle but in populations from different geographical areas (Figure 3). Neither spoligotype nor VNTR profiles could be determined from other badgers because of insufficient DNA.

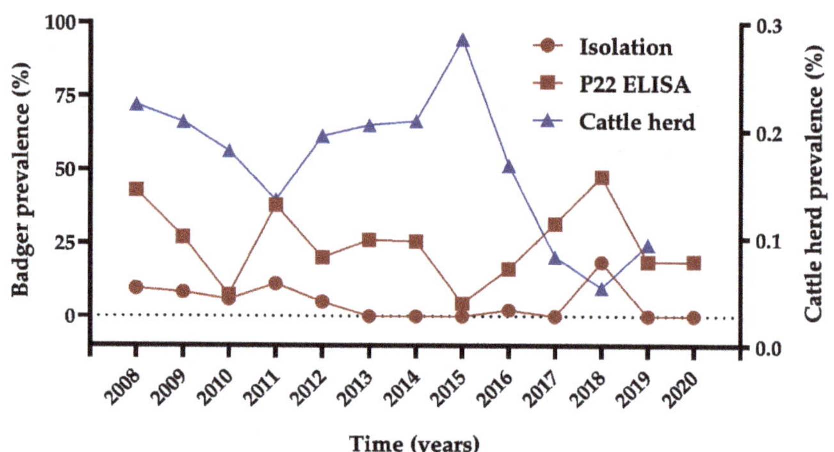

Figure 2. Prevalence of tuberculosis (TB) among badgers (red lines) and cattle (blue line) by year (2008–2020). Red circles refer to results based on bacterial isolation; red squares refer to results based on the P22 ELISA. TB prevalence based on single intradermal tuberculin test data and bacterial isolation was available for cattle herds as part of the Spanish National Program for the Eradication of Bovine TB [2].

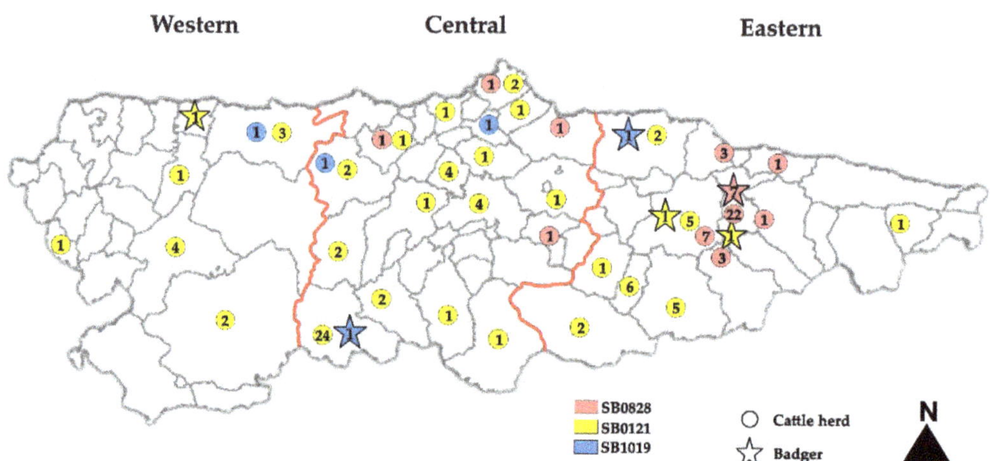

Figure 3. Shared *Mycobacterium bovis* spoligotypes (SB) identified in badgers and cattle herds in the three regions of Asturias (Atlantic Spain) from 2008 to 2020. The numbers within the color circles or stars indicate the number of badgers or cattle herds, respectively, positive to that SB during the 13-year follow-up. Red lines within the map establish the three regions of Asturias.

The rates of MTC positivity were 13.27% (13/98) among trapped badgers and 2.59% (14/541) (chi square = 23.37; $p < 0.0001$) among those killed in road traffic accidents (Figure 4, Supplementary Material 2: Trapped and RTA Badgers). The rate of positivity non-significantly varied with geographic region: 5 of 75 badgers (6.67%) from the western region were positive, compared to 4 of 209 (1.91%) from the central area and 18 of 355 (5.07%) from the eastern area. The rate of positivity was 3.81% (11/289) among males and 5.05% (16/317) among females. It was 4.40% (7/159) among subadults and 4.47% (20/447) among adults. Badgers with undetermined sex or age were all negative. No significant differences were found between areas (F = 1.78; p = 0.1695), sex (F = 1.12; p = 0.2898), or age class (F = 0.23; p = 0.6300) regarding MTC isolation. The only significant effects on isolation were year (F = 2.94; p = 0.0005) and origin (F = 6.01; p = 0.0145). We found significant differences in the P22 ELISA results between positive and negative animals in the bacteriological analyses (p < 0.0001).

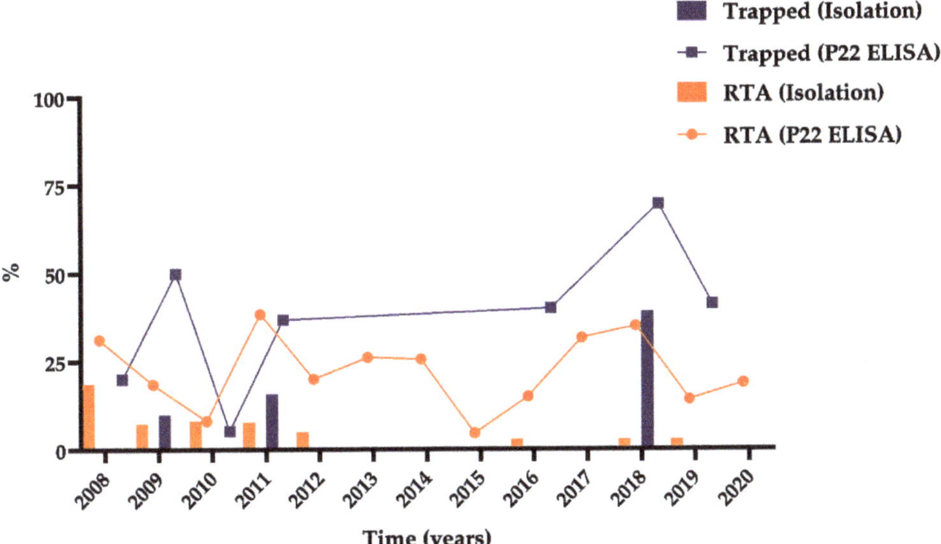

Figure 4. Prevalence of tuberculosis among trapped badgers (purple color) or badgers killed in road traffic accidents (RTA) (orange color) by year. The isolates are represented with columns for each of the two animal groups, and the lines represent seroprevalence. Circles refer to results based on P22 ELISA for the RTA, and the squares refer to results based on P22 ELISA for trapped animals. Note that trapping was not carried out in 2012–2015, 2017, or 2020, so all badgers studied in those years were RTA.

Of the 639 badgers, 28 (4.38%) were positive for MAC (Figure 5a): *Mycobacterium avium avium* (Maa, n = 14), *Mycobacterium avium hominissuis* (Mah, n = 8), or both (n = 1). Three animals positive for Maa were also positive by *M. bovis* isolation, and the spoligotypes were identified as SB0120 in two animals and SB0828 in the other. MAC isolates from five badgers could not be typed because of insufficient DNA. No differences were observed between the years 2008 and 2020 (4.76%, 5.56%, 8.82%, 4.45%, 0.00%, 0.00%, 4.65%, 4.55%, 2.17%, 0.00%, 5.56%, 4.12%, and 7.94%) (chi square = 11.70; p = 0.4699), origins (2.04% and 4.81% for trapped and road-killed, respectively) (chi square = 0.80; p = 0.3700), areas (2.67%, 5.74%, and 3.94% for western, central, and eastern, respectively) (chi square = 1.71; p = 0.4242), sex (3.46% and 5.68% for males and females, respectively) (chi square = 5.05; p = 0.0799), and age class (5.03% and 4.47% for subadults and adults, respectively) (chi square = 1.91; p = 0.3848). Badgers with undetermined sex or age class were all negative.

3.1.3. Serological Assay (P22 ELISA)

A total of 160 badgers (23.77%) were positive in the P22 ELISA, and the distribution of positives by year was as follows (Figure 2): 2008: 9 (42.86%); 2009: 10 (27.03%); 2010: 5 (7.35%); 2011: 17 (37.78%); 2012: 4 (20.00%); 2013: 6 (26.09%); 2014: 11 (25.58%); 2015:

1 (4.55%); 2016: 15 (16.30%); 2017: 18 (31.58%); 2018: 30 (47.62%); 2019: 19 (18.63%); and 2020: 15 (18.75%). The rates of positivity based on P22 ELISA were 42.86% (42/98) among trapped badgers and 20.52% (118/575) among those killed in road traffic accidents.

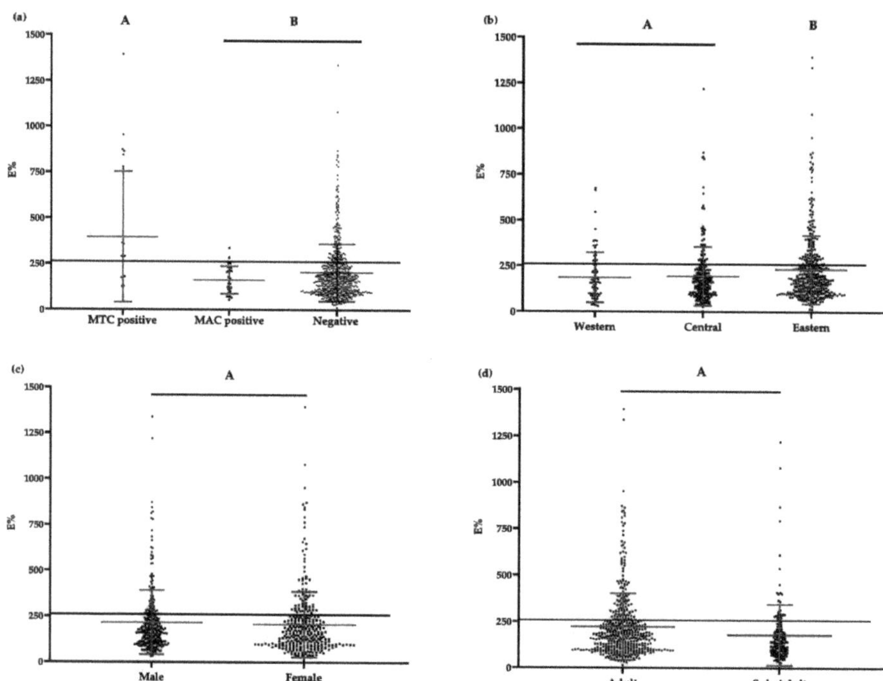

Figure 5. Proportions of badgers testing positive based on E% value in the P22 ELISA. (**a**) Positive badgers for *Mycobacterium tuberculosis* complex (MTC), positive for the *M. avium* complex (MAC), or negative for both complexes. (**b**) Positive badgers in western, central, and eastern parts of Asturias. (**c**) Male and female positive badgers. (**d**) Adult and sub-adult positive badgers. In all graphs, the horizontal black line represents the cut-off of 260% (see Section 2.3.1.3). The red lines represent means and standard deviations. Data connected with the same letter (A or B) are not significantly different based on Tukey–Kramer's test ($p > 0.05$). Results from animals in which sex or age class were unknown are not included in panels c and d.

The rate of seropositivity varied with geographic region as follows (Figure 5b): 17 of 77 badgers (22.08%) from the western region, 39 of 220 (17.73%) from the central area, and 104 of 376 (27.66%) from the eastern area were positive. The rate of seropositivity was 23.92% (72/301) among males, 22.05% (71/322) among females, and 34.00% (17/50) among animals whose sex was undetermined (Figure 5c). The rate of seropositivity was 17.79% (29/163) among subadults, 24.78% (114/460) among adults, and 34.00% (17/50) among badgers of underdetermined age class (Figure 5d).

Seropositivity significantly varied with year (F = 3.36; $p < 0.0001$), although a clear trend across years did not emerge. Seropositivity rates did not significantly differ (F = 1.93; $p = 0.1465$) between areas, although the frequency was slightly higher in the eastern area (37.04%) than in the central (30.94%) and western (29.44%) areas. However, mean E% did significantly differ according to geographical area (F = 3.91; $p = 0.0204$), with the eastern area (E% = 273.55) being higher than the western area (E% = 214.53; TK t = 2.73; $p = 0.0179$) but not than the central area (E% = 254.41; TK t = 1.31; $p = 0.3881$). The GLM identified origin (trapped or road-killed badgers) as the most influential factor on seropositivity (F = 15.86; $p < 0.0001$), with trapped animals being more positive (42.36%) than the road-killed ones (22.58%). Neither sex (F = 0.35; $p = 0.5515$) nor age class (F = 0.85; $p = 0.3568$) showed significant differences between categories (Figure 5d). Mean E% values were higher for badgers from which MTC was isolated (367.34) than for badgers from which MAC was isolated (162.35, TK t = 4.37; *p* < 0.0001) and badgers from which neither MTC nor MAC was isolated (204.91, TK t = 4.63; *p* < 0.0001). In turn, the mean E% values of culture-negative animals (neither MTC nor MAC was isolated) were higher (204.91) than those of MAC-positive animals (162.35), although they were non-significantly different (TK t = 1.30; *p* = 0.6884).

3.2. Cattle

During the study period, 362 cattle herds (2.07%) were positive for TB [2]: 84 (1.76%) were from the western area, 144 (1.74%) were from the central area, and 134 (3.04%) were from the eastern region. Prevalence remained below 1% throughout the period of study (Figure 2). The shared *M. bovis* spoligotypes identified in cattle and badgers from 2008 to 2020 in Asturias are shown in Figure 3.

The prevalence of TB in cattle tended to negatively correlate with mean E% in the P22 ELISA in badgers, but the correlation was not significant (Pearson r; *p* = 0.489). In addition, the tendency inverted over time: it was positive between 2008 and 2010 and between 2012 to 2014, but it was negative for the rest of the period (Figure 6).

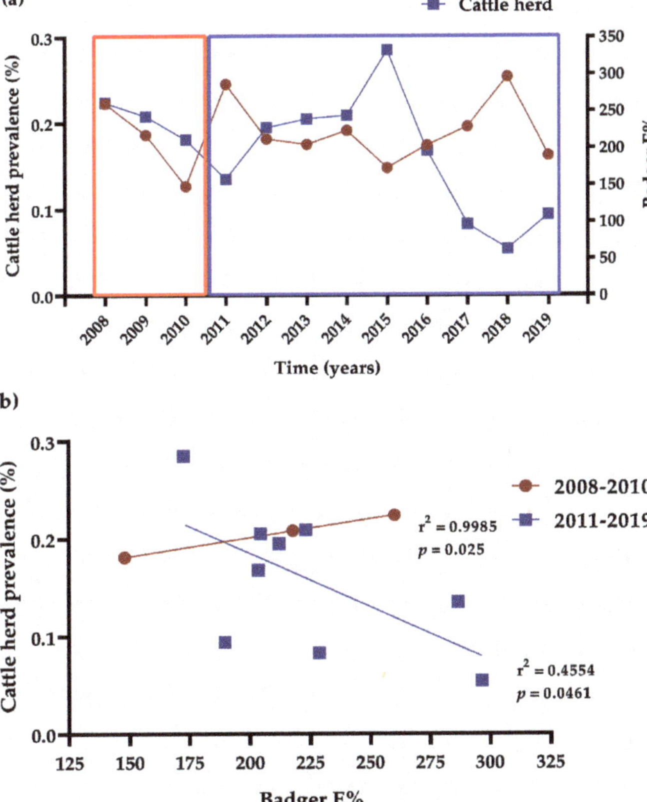

Figure 6. Correlation of tuberculosis (TB) prevalence in badgers and cattle herds in Asturias (Atlantic Spain). (**a**) Evolution of TB prevalence in badgers (red line) and cattle (blue line) by year. Prevalence in badgers was based on mean E% in the P22 ELISA. (**b**) Correlation between TB prevalence in cattle herds and mean E% in the P22 ELISA in badgers.

4. Discussion

This study examined the prevalence of TB in free-ranging Eurasian badgers in Spain for a 13-year period, which was longer than a previous seven-year study [12]. One study in England examined TB prevalence in badgers over 24 years [19], but only trapped animals were analyzed. In the present study, trapped animals and those killed in road traffic accidents were examined, with 4.23% badgers positive for MTC based on isolation and 23.77% badgers positive based on P22 ELISA. Such studies are essential for understanding pathogen transmission between species, particularly in endemic areas or hotspots, since such transmission can vary strongly from location to location [27].

P22 ELISA has provided high levels of sensitivity in previous studies in badgers [26]. Other available serological tests for diagnosing TB in badgers [28], e.g., the DPP Vet TB, the BrockTB STAT-PAK, and the chemiluminescent multiplex ELISA system have shown low-to-moderate sensitivity (from 30.77% to 58%) in naturally infected badgers [29–33]. P22 is mainly composed of MPB70 and MPB83, but there are other proteins shared with MAC or other non-tuberculous mycobacteria with a high sequence similarity to MPB70 and MPB83 [34]; that is why the cut-off is advisable to specifically define for each epidemiological situation [26]. We considered the cut-off of 260% adequate for the present study in order to avoid misclassification in an epidemiological context where MAC isolates were frequent. We found a big difference in TB prevalence using bacteriology or serology (4.32 vs. 23.77, respectively). The isolation of MTC has been considered as the gold standard method for the diagnosis of TB in wildlife [28], but sensitivity can be variable due to the lack of the active shedding of the microorganism from infected animals—hence the absence of mycobacteria in collected samples or when the number of viable microorganisms is low [35].

We detected TB-positive badgers in all parts of Asturias using both techniques, suggesting the widespread distribution of infection, mainly in trapped badgers from areas close to TB-positive farms. Our results were consistent with the idea that badgers are good hosts for TB; for example, TB-infected badgers living more than 1 km away from culled setts were detected even five years after the most recent outbreak in local cattle [36]. Indeed, TB transmission within badger subpopulations has been observed [37]. Given the widespread geographic distribution of positive badgers in our study, we suspect that badgers in Asturias have been infected on several occasions at different locations. We cannot exclude the possibility of transmission from cattle to badgers since both species shared the same or similar *M. bovis* genotypes, although this transmission is more likely when cattle have advanced disease [37] and such animals should be rare due to annual testing of all cattle in Asturias. Nevertheless, we observed a trend toward higher numbers of TB infections in cattle and badgers in the eastern part of the region, which supports the idea of cattle-to-badger transmission. This trend undoubtedly also reflects that most TB hotspots in Asturias lie in the eastern part, where several outbreaks of TB in cattle and badgers have been described, suggesting interspecies transmission and a high environmental contamination [12]. The presence of such hotspots may help explain why TB prevalence among trapped badgers in our study was more than two times the prevalence in road-killed badgers based on serology and more than four times the prevalence based on MTC isolation—trapped animals were captured in areas under active surveillance because of recent and ongoing outbreaks in cattle. Indeed, eight of the nine animals that showed gross lesions and whose infection was further confirmed by histopathology had been trapped. At the same time, we cannot exclude that at least some of the large difference in TB prevalence between the two groups of animals reflected the greater sensitivity of MTC detection in serum and tissue samples from trapped animals given that the quality of those samples may have been higher than the quality of samples from animals killed in road traffic accidents due to better preservation.

In a total geographical context, while TB prevalence among cattle herds has decreased in Asturias in recent years (reflecting the success of the official national eradication campaigns), TB prevalence among badgers has slightly increased. The increase in prevalence among badgers correlated with an increase in prevalence among cattle from 2008 to 2010 but with a decrease in prevalence among cattle from 2011 to 2019. This correlation was weaker regarding badger prevalence, as measured either by isolation or by P22 ELISA. Nevertheless, the number of positive badgers might have increased from 2016 to 2019 due to outbreaks in cattle, which itself was likely due to environmental contamination and indirect interspecies transmission [12]. This illustrates how badgers might likely transmit TB back to cattle in the future, highlighting the importance of understanding badger ecology for monitoring and controlling TB risk in cattle [12].

In our study, male badgers were marginally but not significantly more likely than female badgers to test positive for TB, which was different to studies in UK [38,39]. A study of 94 badgers killed in road traffic accidents in the UK found no sex difference in TB prevalence [40]. Studies have suggested that male badgers may transmit *M. bovis* to one another through biting during aggressive behavior [27,41]. We did not observe bite wounds in the present study, but this possibility should be explored in future work.

Adult badgers in our sample were more likely, although not significantly, to test seropositive than subadults. In our study, adult animals also showed higher E% in the P22 ELISA than subadults, probably reflecting the longer time available for animals to come into contact with *M. bovis*. The rates of MTC isolation did not significantly vary with badger age class, consistent with other studies in Europe [36,40]. Badgers in our study from which neither MAC nor MTC was isolated gave slightly higher E% in the P22 ELISA than badgers from which MAC was isolated, supporting the specificity of the assay. Our observation of intermediate E% for badgers from which neither MAC nor MTC was isolated suggests that some of these cases may have been false negatives because an immune response was apparent.

Of the 612 badgers that were TB-negative based on culture, 133 were seropositive based on the P22 ELISA, which means that the ELISA had a complementary sensitivity of 492.59% with respect to bacteriological isolation. Thus, taking the positives to either isolation or ELISA as reference, ELISA sensitivity was 84.57% compared to 15.43% for isolation. The sensitivity of culture tests might have been low because of poor sample condition, the technical limitations of the microbiological procedure, and the clinical stage of the badgers. In fact, only nine animals (a third of MTC-positive badgers) showed gross lesions, implying that most animals might have been in an early stage of infection or be MTC subclinical carriers, when MTC isolation is less sensitive [8,37]. Those early or latent stages, with infection confirmed by either immunology or bacteriological culture but with an absence of detectable pathology, are usual in badgers and have less risk of shedding greater numbers of mycobacteria than badgers in the late stages of disease [37]. We cannot exclude that other members of the MTC were present but not recovered by culture and cross-react in P22 ELISA, i.e., *Mycobacterium microti* infections have been reported in both wildlife [42,43] and domestic animals [44,45] in France, although the direct detection of that species on tissue samples was not included in this study. On the other hand, the relative specificity of each method with respect to the other remained much higher for isolation (97.45% vs. 78.27% for culture and P22 ELISA, respectively). In this regard, bacterial isolation remains the gold standard for the diagnosis of TB despite of methodological issues such as moderate sensitivity and long incubation times. Nevertheless, our results highlighted the P22 ELISA as a useful screening tool that is faster and more cost-effective and sensitive in the early or latent stages of disease than isolation for TB detection. Similarly, direct qPCR on tissue samples is also considered to be a sensitive and valuable alternative to culture that is faster than bacterial isolation [46]. Though qPCR can also detect non-viable mycobacteria, both techniques (P22 ELISA and qPCR) might be used in parallel in order to increase sensitivity and specificity of TB diagnosis in badgers.

5. Conclusions

We found that the TB status of badgers in Asturias during 2008–2020 was associated with the TB status of local cattle herds. Our results could not determine the direction of possible interspecies transmission, but they were consistent with the idea that the two hosts may exert infection pressure on each other. Adult badgers were more likely to be TB-positive than subadults. This study highlights the importance of monitoring this multi-host infection and disease in wildlife during epidemiological interventions in order to optimize outcomes under the One Health concept.

Supplementary Materials: The following are available online at https://www.mdpi.com/article/10.3390/ani11051294/s1, Supplementary Material 1: Raw Data; Supplementary Material 2: Trapped and RTA Badgers.

Author Contributions: Conceptualization, A.B.; methodology, C.B.V., T.D.B., B.R., M.Q., I.M., P.Q., J.Á.A., R.J., L.D., M.D. and A.B.; validation, A.B. and R.J.; formal analysis, C.B.V., T.D.B. and A.B.; investigation, C.B.V., T.D.B., B.R., M.Q., I.M., P.Q., J.Á.A., L.D., M.D., R.C. and A.B.; resources, A.B.; data curation, C.B.V., T.D.B., R.J. and A.B.; writing—original draft preparation, C.B.V., T.D.B. and A.B.; writing—review and editing, C.B.V., T.D.B., B.R., M.Q., I.M., P.Q., J.Á.A., R.J., L.D., M.D., R.C. and A.B. All authors have read and agreed to the published version of the manuscript.

Funding: This work was funded by the Ministerio de Ciencia, Innovación y Universidades (MCIU), the Agencia Estatal de Investigación (AEI) reference project RTI2018-096010-B-C21 (FEDER co-funded), PCTI 2018–2020 (GRUPIN: IDI2018-000237), and FEDER and Ministerio de Agricultura, Pesca y Alimentación. C.B.V. was supported by a grant from the Instituto Nacional de Investigación y Tecnología Agraria y Alimentaria (INIA), Spain. Publication costs were covered with funds from RTI2018-096010-B-C21 (FEDER co-funded). T.D.B. was supported by a fellowship from the Coordenação de Aperfeiçoamento de Pessoal de Nível Superior (CAPES; process number 88887.511077/2020-00).

Institutional Review Board Statement: The study was conducted according to the European Ethical guidelines for the experimental procedures with trapped badgers. They were approved by the Government of the Principality of Asturias Ethics Committee (reference numbers: 010/07-01-2011, PROAE 20/2015, PROAE 47/2018).

Data Availability Statement: The data published in this study are available in Supplementary Material 1: Raw Data and Supplementary Material 2: Trapped and RTA Badgers.

Acknowledgments: The authors acknowledge invaluable support from their colleagues at SERIDA (Miguel Prieto and Alberto Espí), the University of León, the Servicio de Espacios Protegidos y Conservación de la Naturaleza, the Dirección General del Medio Natural y Planificación Rural del Principado de Asturias, the Servicio de Sanidad y Producción Animal del Principado de Asturias, the Regional Animal Health Laboratory of Principado de Asturias, VISAVET and Instituto de Salud Carlos III. We thank A. Chapin Rodríguez (Creaducate Consulting GmbH) for critically reviewing the manuscript.

Conflicts of Interest: The authors declare no conflict of interest.

References

1. Olea-Popelka, F.J.; Muwonge, A.; Perera, A.; Dean, A.S.; Mumford, E.; Erlacher-Vindel, E.; Forcella, S.; Silk, B.J.; Ditiu, L.; El Idrissi, A.; et al. Zoonotic tuberculosis in humans begins caused by *Mycobacterium bovis*- a call for action. *Lancet Infect. Dis.* **2017**, *1*, e21–e25. [CrossRef]
2. MAPA. 2020. Available online: https://www.mapa.gob.es/es/ganaderia/temas/sanidad-animal-higiene-ganadera/pnetb_2020final_tcm30-523317.PDF (accessed on 8 January 2021).
3. Olea-Popelka, F.J.; Costello, E.; White, P.; McGrath, G.; Collins, J.D.; O'Keeffe, J.; Felton, D.F.; Berke, O.; More, S.; Martin, S.W. Risk factors for disclosureof additional tuberculous cattle in attested-clear herds that had one animal with a confirmed lesion of tuberculosis at slaughter during 2003 in Ireland. *Prev. Vet. Med.* **2008**, *85*, 81–91. [CrossRef]
4. Delahay, R.J.; Smith, G.C.; Barlow, A.M.; Walker, N.; Harris, A.; Clifton-Hadley, R.S.; Cheeseman, C.L. Bovine tuberculosis infection in wild mammals in the South-West region of England: A survey of prevalence and a semi-quantitative assessment of the relative risks to cattle. *Vet. J.* **2007**, *173*, 287–301. [CrossRef]
5. Zanella, G.; Duvauchelle, A.; Hars, J.; Moutou, F.; Boschiroli, M.L.; Durand, B. Patterns of lesions of bovine tuberculosis in wild red deer and wild boar. *Vet. Rec.* **2008**, *163*, 43–47. [CrossRef]
6. Payne, A.; Philipon, S.; Hars, J.; Dufour, B.; Gilot-Fromont, E. Wildlife Interactions on baited places and waterholes in a French area Infected by bovine tuberculosis. *Front. Vet. Sci.* **2017**, *3*, 122. [CrossRef]
7. Sobrino, R.; Martín-Hernando, M.P.; Vicente, J.; Aurtenetxe, O.; Garrido, J.M.; Gortázar, C. Bovine tuberculosis in a badger (*Meles meles*) in Spain. *Vet. Rec.* **2008**, *163*, 159–160. [CrossRef]

8. Balseiro, A.; Rodríguez, O.; González-Quirós, P.; Merediz, I.; Sevilla, I.A.; Davé, D.; Dalley, D.J.; Lesellier, S.; Chambers, M.A.; Bezos, J. Infection of Eurasian badgers (*Meles meles*) with *Mycobacterium bovis* and *Mycobacterium avium* complex in Spain. *Vet. J.* **2011**, *190*, 21–25. [CrossRef] [PubMed]
9. Corner, L.A.L.; Clegg, T.A.; More, S.J.; Williams, D.H.; O'Boyle, I.; Costello, E.; Sleeman, D.P.; Griffin, J.M. The effect of varying levels of population control on the prevalence of tuberculosis in badgers in Ireland. *Res. Vet. Sci.* **2008**, *85*, 238–249. [CrossRef] [PubMed]
10. Jenkins, H.E.; Woodroffe, R.; Donnelly, C.A. The duration of the effects of repeated widespread badger culling on cattle tuberculosis following the cessation of culling. *PLoS ONE* **2010**, *5*, e9090. [CrossRef] [PubMed]
11. Gortázar, C.; Delahay, R.J.; McDonald, R.A.; Boadella, M.; Wilson, G.J.; Gavier-Widen, D.; Acevedo, P. The status of tuberculosis in European wild mammals. *Mammal Rev.* **2012**, *42*, 193–206. [CrossRef]
12. Acevedo, P.; Prieto, M.; Quirós, P.; Merediz, I.; de Juan, L.; Infantes-Lorenzo, J.A.; Triguero-Ocaña, R.; Balseiro, A. Tuberculosis epidemiology and badger (*Meles meles*) spatial ecology in a hotspot-area in Atlantic Spain. *Pathogens.* **2019**, *8*, 292. [CrossRef]
13. Barasona, J.A.; Vicente, J.; Díez-Delgado, I.; Aznar, J.; Gortázar, C.; Torres, M.J. Environmental presence of *Mycobacterium tuberculosis* complex in aggregation points at the wildlife/livestock interface. *Transbound. Emerg. Dis.* **2017**, *64*, 1148–1158. [CrossRef] [PubMed]
14. Vicente, J.; Barasona, J.A.; Acevedo, P.; Ruíz-Fons, J.F.; Boadella, M.; Diez-Delgado, I.; Beltrán-Beck, B.; González-Barrio, D.; Queirós, J.; Montoro, V.; et al. Temporal trend of tuberculosis in wild ungulates from Mediterranean Spain. *Transbound. Emerg. Dis.* **2013**, *1*, 92–103. [CrossRef] [PubMed]
15. LaHue, N.P.; Baños, J.V.; Acevedo, P.; Gortázar, C.; Martínez-López, B. Spatially explicit modeling of animal tuberculosis at the wildlife-livestock interface in Ciudad Real province, Spain. *Prev. Vet. Med.* **2016**, *128*, 101–111. [CrossRef]
16. Muñoz-Mendoza, M.; Marreros, N.; Boadella, M.; Gortázar, C.; Menéndez, S.; de Juan, L.; Bezos, J.; Romero, B.; Copano, M.F.; Amado, J.; et al. Wild boar tuberculosis in Iberian Atlantic Spain: A different picture from Mediterranean habitats. *BMC Vet. Res.* **2013**, *9*, 176. [CrossRef] [PubMed]
17. Balseiro, A.; González-Quirós, P.; Rodríguez, O.; Copano, M.F.; Merediz, I.; De-Juan, L.; Chambers, M.A.; Delahay, R.J.; Marreros, N.; Royo, L.J.; et al. Spatial relationships between Eurasian badgers (*Meles meles*) and cattle infected with *Mycobacterium bovis* in Northern Spain. *Vet. J.* **2013**, *197*, 739–745. [CrossRef]
18. Byrne, A.W.; Kenny, K.; Fogarty, U.; O'Keeffe, J.J.; More, S.J.; McGrath, G.; Teeling, M.; Martin, S.W.; Dohoo, I.R. Spatial and temporal analyses of metrics of tuberculosis infection in badgers (*Meles meles*) from the Republic of Ireland: Trends in apparent prevalence. *Prev. Vet. Med.* **2015**, *122*, 345–354. [CrossRef] [PubMed]
19. Delahay, R.J.; Walker, N.; Smith, G.C.; Wilkinson, D.; Clifton-Hadley, R.S.; Cheeseman, C.L.; Tomlinson, A.J.; Chambers, M.A. Long-term temporal trends and estimated transmission rates for *Mycobacterium bovis* infection in an undisturbed high-density badger (*Meles meles*) population. *Epidemiol. Infect.* **2013**, *141*, 1445–1456. [CrossRef]
20. Ninyerola, M.; Pons, X.; Roure, J.M. *Atlas Climático Digital de la Península Ibérica. Metolodología y Aplicaciones en Bioclimatología y Geobotánica*, 1st ed.; Universidad Autónoma de Barcelona: Bellaterra, Spain, 2005.
21. Acevedo, P.; González-Quirós, P.; Etherington, T.R.; Prieto, J.M.; Gortázar, C.; Balseiro, A. Generalizing and transferring spatial models: A case study to predict Eurasian badger abundance in Atlantic Spain. *Ecol. Model.* **2014**, *275*, 1–8. [CrossRef]
22. Coetsier, C.; Vannuffel, P.; Bloondel, N.; Denef, J.F.; Cocito, C.; Gala, J.L. Duplex PCR for differential identification of *Mycobacterium bovis*, *M. avium*, and *M. avium* subsp. *paratuberculosis* in formalin-fixed paraffin-embedded tissues from cattle. *J. Clin. Microbiol.* **2000**, *38*, 3048–3054. [CrossRef]
23. Rodríguez-Campos, S.; Aranaz, A.; de Juan, L.; Sáez-Llorente, J.L.; Romero, B.; Bezos, J.; Jiménez, A.; Mateos, A.; Domínguez, L. Limitations of spoligotyping and variable number tandem repeat typing for molecular tracing of *Mycobacterium bovis* in a high diversity setting. *J. Clin. Microbiol.* **2011**, *49*, 3361–3364. [CrossRef]
24. Mycobacterium bovis Spoligotype Database Website. Available online: www.Mbovis.org (accessed on 2 October 2019).
25. Álvarez, J.; García, I.G.; Aranaz, A.; Bezos, J.; Romero, B.; de Juan, L.; Mateos, A.; Gómez-Mampaso, E.; Domínguez, L.J. Genetic diversity of *Mycobacterium avium* isolates recovered from clinical samples and from the environment: Molecular characterization for diagnostic purposes. *Clin. Microbiol.* **2008**, *46*, 1246–1251. [CrossRef]
26. Infantes-Lorenzo, J.A.; Dave, D.; Moreno, I.; Anderson, P.; Lesellier, S.; Gormley, E.; Domínguez, L.; Balseiro, A.; Gortázar, C.; Domínguez, M.; et al. New serological platform for detecting antibodies against *Mycobacterium tuberculosis* complex in European badgers. *Vet. Med. Sci.* **2019**, *5*, 61–69. [CrossRef] [PubMed]
27. Schroeder, P.; Hopkins, B.; Jones, J.; Galloway, T.; Pike, R.; Rolfe, S.; Hewinson, G. Temporal and spatial *Mycobacterium bovis* prevalence patterns as evidenced in the All Wales Badgers Found Dead (AWBFD) survey of infection 2014–2016. *Sci. Rep.* **2020**, *10*, 1–11. [CrossRef]
28. Thomas, J.; Balseiro, A.; Gortázar, C.; Risalde, M.A. Diagnostic of tuerculosis in wildlife: A systematic review. *Vet. Res.* **2021**, *52*, 31. [CrossRef]
29. Courcier, E.A.; Pascual-Linaza, A.V.; Arnold, M.E.; McCormick, C.M.; Corbett, D.M.; O'Hagan, M.J.H.; Collins, S.F.; Trimble, N.A.; McGeown, C.F.; McHugh, G.E.; et al. Evaluating the application of the dual path platform VetTB tests for badgers (*Meles meles*) in the test and vaccinate or remove (TVR) wildlife research intervention project in Northern Ireland. *Res. Vet. Sci.* **2020**, *130*, 170–178. [CrossRef] [PubMed]

30. Drewe, J.A.; Tomlinson, A.J.; Walker, N.J.; Delahay, R.J. Diagnostic accuracy and optimal use of three tests for tuberculosis in live badgers. *PLoS ONE.* **2010**, *5*, e11196. [CrossRef]

31. Buzdugan, S.N.; Chambers, M.A.; Delahay, R.J.; Drewe, J.A. Diagnosis of tuberculosis in groups of badgers: An exploration of the impact of trapping efficiency, infection prevalence and the use of multiple tests. *Epidemiol. Infect.* **2016**, *144*, 1717–1727. [CrossRef] [PubMed]

32. Chambers, M.A.; Waterhouse, S.; Lyashchenko, K.; Delahay, R.; Sayers, R.; Hewinson, R.G. Performance of TB immunodiagnostic tests in Eurasian badgers (*Meles meles*) of different ages and the influence of duration of infection on serological sensitivity. *BMC Vet. Res.* **2009**, *5*, 42. [CrossRef]

33. Aznar, I.; Frankena, K.; More, S.J.; O'Keeffe, J.; McGrath, G.; de Jong, M.C.M. Quantification of *Mycobacterium bovis* transmission in a badger vaccine field trial. *Prev. Vet. Med.* **2017**, *149*, 29–37. [CrossRef]

34. Infantes-Lorenzo, J.A.; Moreno, I.; de los Ángeles Risalde, M.; Roy, A.; Villar, M.; Romero, B.; Ibarrola, N.; de la Fuente, J.; Puentes, E.; de Juan, L.; et al. Proteomic characterisation of bovine and avian purified protein derivatives and identification of specific antigens for serodiagnosis of bovine tuberculosis. *Clin. Proteomics.* **2017**, *14*, 36. [CrossRef] [PubMed]

35. Miller, M.; Buss, P.; Hofmeyr, J.; Olea-Popelka, F.; Parsons, S.; van Helden, P. Antemortem diagnosis of *Mycobacterium bovis* infection in free-ranging African lions (*Panthera leo*) and implications for transmission. *J. Wildl. Dis.* **2015**, *51*, 493–497. [CrossRef]

36. Murphy, D.; Gormley, E.; Collins, D.M.; McGrath, G.; Sovsic, E.; Costello, E.; Corner, L.A.L. Tuberculosis in cattle herds are sentinels for *Mycobacterium bovis* infection in European badgers (*Meles meles*): The Irish Greenfield Study. *Vet. Microbiol.* **2011**, *151*, 120–125. [CrossRef] [PubMed]

37. Corner, L.A.L.; Murphy, D.; Gormley, E. *Mycobacterium bovis* infection in the Eurasian badger (*Meles meles*): The disease, pathogenesis, epidemiology and control. *J. Comp. Pathol.* **2011**, *144*, 1–24. [CrossRef]

38. Cheeseman, C.L.; Jones, G.W.; Gallagher, J.; Mallinson, P.J. The population structure, density and prevalence of tuberculosis (*Mycobacterium bovis*) in badgers (*Meles meles*) from four areas in south west England. *J. Appl. Ecol.* **1981**, *18*, 795–804. [CrossRef]

39. Bourne, F.J.; Donnelly, C.A.; Cox, D.R.; Gettinby, G.; McInerney, J.P.; Morrison, W.I.; Woodroffe, R. TB policy and the ISG's findings. *Vet. Rec.* **2007**, *161*, 633–635. [CrossRef]

40. Sandoval Barron, E.; Swift, B.; Chantrey, J.; Christley, R.; Gardner, R.; Jewell, C.; McGrath, I.; Mitchell, A.; O'Cathail, C.; Prosser, A.; et al. A study of tuberculosis in road traffic-killed badgers on the edge of the British bovine TB epidemic area. *Sci. Rep.* **2018**, *8*, 17206. [CrossRef] [PubMed]

41. Delahay, R.J.; Walker, N.J.; Forrester, G.J.; Harmsen, B.; Riordan, P.; Macdonald, D.W.; Newman, C.; Cheeseman, C.L. Demographic correlates of bite wounding in Eurasian badgers, *Meles meles* L., in stable and perturbed populations. *Anim. Behav.* **2006**, *71*, 1047–1055. [CrossRef]

42. Reveillaud, E.; Desvaux, S.; Boschiroli, M.L.; Hars, J.; Faure, E.; Fediaevsky, A.; Cavalerie, L.; Chevalier, F.; Jabert, P.; Poliak, S.; et al. Infection of wildlife by *Mycobacterium bovis* in France assessment through a national surveillance system, Sylvatub. *Front. Vet. Sci.* **2018**, *5*, 262. [CrossRef]

43. Michelet, L.; de Cruz, K.; Tambosco, J.; Hénault, S.; Boschiroli, M.L. *Mycobacterium microti* interferes with bovine tuberculosis surveillance. *Microorganisms* **2020**, *8*, 1850. [CrossRef]

44. Michelet, L.; de Cruz, K.; Phalente, Y.; Karoui, C.; Henault, S.; Beral, M.; Boschiroli, M.L. *Mycobacterium microti* infection in dairy goats, France. *Emerg. Infect. Dis.* **2016**, *22*, 569–570. [CrossRef] [PubMed]

45. Michelet, L.; de Cruz, K.; Karoui, C.; Henault, S.; Boschiroli, M.L. *Mycobacterium microti* infection in a cow in France. *Vet. Rec.* **2017**, *180*, 429. [CrossRef]

46. Lorente-Leal, V.; Liandris, E.; Pacciarini, M.; Botelho, A.; Kenny, K.; Loyo, B.; Fernández, R.; Bezos, J.; Domínguez, L.; de Juan, L.; et al. Direct PCR on tissue samples to detect *Mycobacterium tuberculosis* complex: An alternative to the bacteriological culture. *J. Clin. Microbiol.* **2021**, *59*, e01404-20. [CrossRef] [PubMed]

Article

Exposure of Free-Ranging Wild Animals to Zoonotic *Leptospira interrogans* Sensu Stricto in Slovenia

Diana Žele-Vengušt [1], Renata Lindtner-Knific [2], Nina Mlakar-Hrženjak [2], Klemen Jerina [3] and Gorazd Vengušt [1,*]

[1] Institute of Pathology, Wild Animals, Fish and Bees, Veterinary Faculty, University of Ljubljana, Gerbičeva 60, 1000 Ljubljana, Slovenia; diana.zelevengust@vf.uni-lj.si

[2] Institute of Poultry, Birds, Small Mammals and Reptiles, Veterinary Faculty, University of Ljubljana, Gerbičeva 60, 1000 Ljubljana, Slovenia; renata.lindtnerknific@vf.uni-lj.si (R.L.-K.); nina.mlakarhrzenjak@vf.uni-lj.si (N.M.-H.)

[3] Department of Forestry and Renewable Forest Resources, Biotechnical Faculty, Večna pot 83, 1000 Ljubljana, Slovenia; klemen.jerina@bf.uni-lj.si

* Correspondence: gorazd.vengust@vf.uni-lj.si; Tel.: +386-1-4779-196

Simple Summary: Wildlife can serve as a reservoir for highly contagious and deadly diseases, many of which are infectious to domestic animals and/or humans. Wildlife pathogen and disease surveillance is, thus, an essential tool that can provide valuable information on population health status and protect human health. Blood samples from 244 wild animals and 5 from carcasses were tested for specific antibodies against *Leptospira* serovars in Slovenia between 2019 and 2020 using the microscopic agglutination test. The results confirm that various wildlife species were exposed to *Leptospira interrogans* and may be used as a sentinel for leptospirosis, which is considered a significant health threat to other wildlife species and to humans.

Citation: Žele-Vengušt, D.; Lindtner-Knific, R.; Mlakar-Hrženjak, N.; Jerina, K.; Vengušt, G. Exposure of Free-Ranging Wild Animals to Zoonotic *Leptospira interrogans* Sensu Stricto in Slovenia. *Animals* **2021**, *11*, 2722. https://doi.org/10.3390/ani11092722

Academic Editors: Laila Darwich Soliva and David González-Barrio

Received: 24 July 2021
Accepted: 14 September 2021
Published: 17 September 2021

Publisher's Note: MDPI stays neutral with regard to jurisdictional claims in published maps and institutional affiliations.

Abstract: A total of 249 serum samples from 13 wild animal species namely fallow deer (*Dama dama*, $n = 1$), roe deer (*Capreolus capreolus*, $n = 80$), red deer (*Cervus elaphus*, $n = 22$), chamois (*Rupicapra rupicapra*, $n = 21$), mouflon (*Ovis musimon*, $n = 4$), brown hare (*Lepus europaeus*, $n = 2$), nutria (*Myocastor coypus*, $n = 1$), red fox (*Vulpes vulpes*, $n = 97$), stone marten (*Martes foina*, $n = 12$), European badger (*Meles meles*, $n = 2$), golden jackal (*Canis aureus*, $n = 2$) Eurasian lynx (*Lynx lynx*, $n = 2$) and grey wolf (*Canis lupus*, $n = 3$) were analysed for the presence of antibodies against *Leptospira interrogans* sensu stricto. Serum samples were examined via the microscopic agglutination test for the presence of specific antibodies against *Leptospira* serovars Icterohaemorrhagiae, Bratislava, Pomona, Grippotyphosa, Hardjo, Sejroe, Australis, Autumnalis, Canicola, Saxkoebing and Tarassovi. Antibodies to at least one of the pathogenic serovars were detected in 77 (30.9%; CI = 25–37%) sera. The proportion of positive samples varied intraspecifically and was the biggest in large carnivores (lynx, wolf and jackal; 86%), followed by mezzo predators: stone marten (67%) and red fox (34%), and large herbivores: red deer (32%), roe deer (25%), alpine chamois (10%) and mouflon (0%). Out of the 77 positive samples, 42 samples (53.8%) had positive titres against a single serovar, while 35 (45.4%) samples had positive titres against two or more serovars. The most frequently detected antibodies were those against the serovar Icterohaemorrhagiae. The present study confirmed the presence of multiple pathogenic serovars in wildlife throughout Slovenia. It can be concluded that wild animals are reservoirs for at least some of the leptospiral serovars and are a potential source of leptospirosis for other wild and domestic animals, as well as for humans.

Keywords: wildlife; *Leptospira interrogans*; microscopic agglutination test; serology; Slovenia

1. Introduction

In recent decades, international attention on wildlife diseases, including surveillance and monitoring programmes, has increased [1,2]. Wildlife diseases occur in numerous forms in a wide range of species and populations around the globe. Leptospirosis is a zoonosis of global importance, affecting many species of wild and domestic animals, as well as humans [1,3]. *Leptospira* spp. are also considered as small mammal-associated zoonotic pathogens causing diseases with potentially similar symptoms in humans [4]. It is considered as one of the most important re-emerging health threats to humans by the World Organisation for Animal Health [5,6]. Various pathogenic serovars of *Leptospira* have been serologically classified into 22 serogroups and over 300 serovars based on the microscopic agglutination test (MAT) or the cross-agglutination absorption test (CAAT), respectively [7], with each serovar tending to be maintained by a host group and capable of causing the disease [8–10]. Small rodents are the usual reservoirs of leptospires in natural herds [11–13]. Data from the Netherlands show that insectivores and rodents could serve as indicators of environmental contamination and/or wildlife contamination with *Leptospira* spp. [14]. Studies worldwide indicate that various wild ruminants [15–18], lagomorphs [15,18] and carnivores [15,19–21] are also potential sources of leptospires. Wildlife species are generally considered to be important epidemiological vectors, mainly because of their frequent reactivity to *Leptospira* serovars native to their habitat [16]. These reservoirs are thought to act as a source of infection for humans and domestic animals, who can then become a source of infection for other animals and humans [6,11]. However, data on the epidemiology of *Leptospira* infections in wildlife and the public health significance of wildlife species worldwide are lacking [20,22].

The presence of antibodies in wild animals may indicate previous or current infection, which may have occurred either by direct contact with the contaminated urine of another animal or by the consumption of infected prey [23]. Several domestic and wild animals become infected and, thus, become kidney carriers, excreting the pathogen through their urine [23] and via parent–offspring transmission [24]. Wildlife are reservoir hosts for leptospires and often show no clinical signs of disease; however, these reservoirs can serve as a source of infection for humans and domestic animals, who can then become a source of infection for other animals and humans [5,6,25]. In addition, there are few data on Leptospira antibodies and infections in wildlife, although transmission to livestock and humans often originates from or is maintained by wildlife [25].

Humans may be exposed to *Leptospira* infections directly through contact with infected material or indirectly through the contaminated environment [26,27]. Studies have shown that peak incidence of disease occurred after periods of excessive rainfall and flooding [28]. Infections with *Leptospira* spp. can affect not only people in exposed professions (i.e., veterinarians, trappers, abattoir workers, farm workers, hunters, animal shelter workers and scientists and technicians involved with animals in laboratories or in the field) but also people who work with marine mammals, fishmen, researchers, wildlife rehabilitators, animal trainers and zoological park workers [3]. A study conducted in Austria [29] reported that hunters in particular are exposed to zoonotic agents, including leptospires, probably through the direct contact of abraded skin or mucous membranes with the tissues, blood or urine of infected animals [30,31]. In some cases, leptospirosis can present as a severe disease in both animals and humans and can lead to death [11]. Human infections by *Leptospira* spp. and orthohantaviruses are almost indistinguishable in their clinical presentation [32] and can often be confused with each other [4]. In addition, coinfections with orthohantaviruses are frequently observed at sites where the prevalence of *Leptospira* spp. in small mammals exceeded 35% [4]. Therefore, long-term active surveillance and studies of wildlife reservoirs will help to understand the role of wildlife as a reservoir and as a source of leptospires and other pathogens to humans [13]. According to Podgoršek [33], the epidemiological picture in Slovenia is comparable to that in Europe. Up to 30 cases of *Leptospira* infection with the predominant serovars Grippotyphosa and Icterohaemorrhagiae and the species *L. kirschneri* and *L. interrogans* sensu stricto are reported annually [33,34]. In Slovenia, the risk of contracting leptospirosis is associated with occupational and recreational exposure [34].

MAT is considered the gold standard for sero-diagnosis of leptospirosis because of its unsurpassed diagnostic specificity [35]; however it is not sufficiently sensitive for the diagnosis of the acute phase of the disease [36]. It would be an important tool for epidemiological purposes, such as identifying infecting serovars [36]. Antigens can be detected by histological, histochemical or immunostaining techniques. Unfortunately, none of these tests are currently suitable for routine laboratory use because of technical limitations and low sensitivity [37]. Isolation of *Leptospira* from the clinical specimen is difficult because leptospires are fastidious, slow growing and require special growth media, and it is time consuming and laborious [38]. Therefore, PCR assay is very useful as a contemporary method for diagnosis in the acute phase of leptospirosis [39].

To date, studies in Slovenia have confirmed the presence of specific antibodies against 11 pathogenic *Leptospira* serovars in wild boar [40], but there is currently no information on the seroprevalence and distribution of leptospirosis in other wildlife species in Slovenia. The aim of this study was to investigate the seroprevalence of pathogenic *L. interrogans* serovars in different wild species in Slovenia.

2. Materials and Methods

2.1. Samples

During the 2019 and 2020 hunting season (May to December), blood samples were collected nationwide from a total of 244 apparently healthy, free-ranging wild animals, and 5 clotted blood samples were collected from carcasses (Table 1 and Figure 1). Licensed game wardens and hunters were invited to submit samples from animals shot during the regular annual cull or from animals found dead in nature. Prior to sample collection, hunters were instructed on procedures and were provided with field sample kits. Immediately after animal death, blood samples were collected from the jugular vein or heart. As a part of the national passive health surveillance of wildlife in Slovenia, carcasses of wild large predators (Eurasian lynx and grey wolf) found dead in their habitats were sent for necropsy to the Veterinary Faculty, University of Ljubljana. Clotted blood samples were taken from the heart. The authors declare that no animals were killed for the purpose of this study and that all procedures contributing to this work met the ethical standards of the relevant national and European regulations on the care and use of animals (Directive 2010/63/EC).

Table 1. Samples from 249 free-ranging wild animals, harvested or found dead (*).

Species Common Name	Latin Name	No. of Animals
Alpine chamois	*Rupicapra rupicapra*	21
Brown hare	*Lepus europaeus*	2
European badger	*Meles meles*	2
European mouflon	*Ovis musimon*	4
Fallow deer	*Dama dama*	1
Golden jackal	*Canis aureus*	2
Nutria	*Myocastor coypus*	1
Red deer	*Cervus elaphus*	22
Red fox	*Vulpes vulpes*	97
Roe deer	*Capreolus capreolus*	80
Stone marten	*Martes foina*	12
* Eurasian lynx	*Lynx lynx*	2
* Grey wolf	*Canis lupus*	3
Total samples		249

2.2. Laboratory Methods

After field collection, blood samples were transported to the Veterinary Faculty, University of Ljubljana, within 24 h. Many of the collected samples were haemolysed and were, therefore, rejected at the pre-analysis stage ($n = 34$). Fresh blood samples and clotted blood samples were centrifuged at 4000 rpm for 15 min (LC 320) to obtain the sera. Sera were transferred with serum pipettes into sterile Eppendorf tubes and stored at $-20\ ^\circ$C before being tested for the presence of specific antibodies against pathogenic serovars of *Leptospira interrogans* sensu stricto using the MAT.

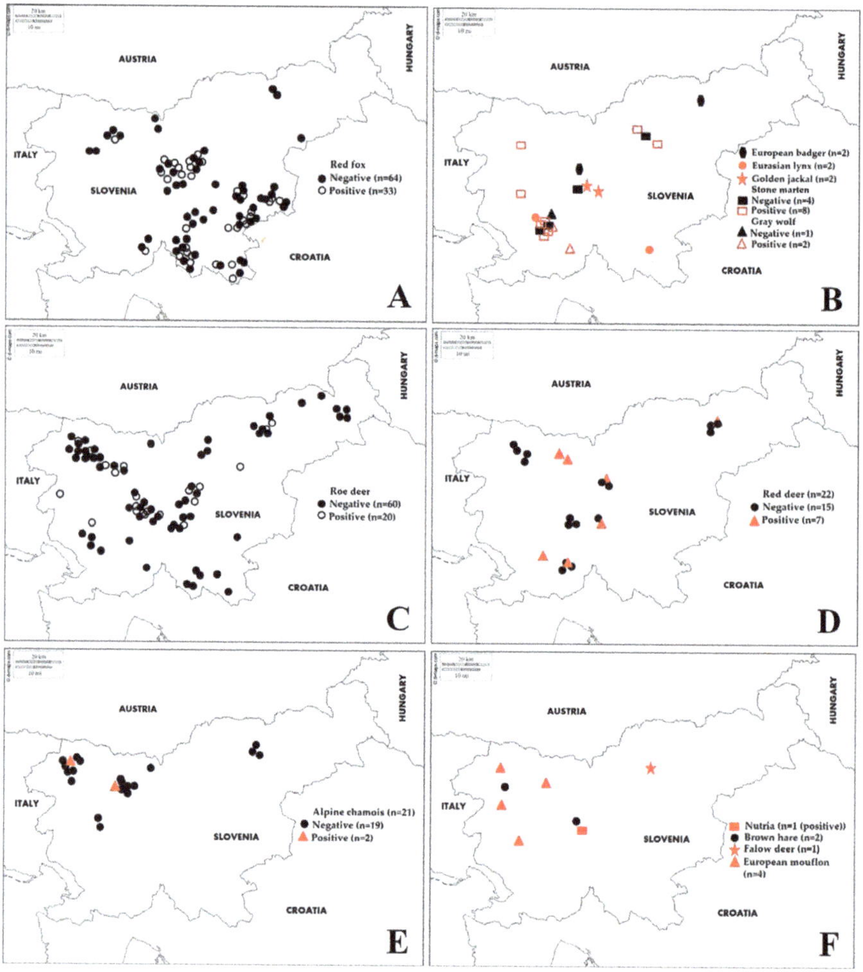

Figure 1. Geographical location of *Leptospira interrogans* antibody-negative and -positive samples of different wildlife species (**A**)—red fox; (**B**)—European badger, Eurasian lynx, golden jackal, stone marten, grey wolf; (**C**)—roe deer; (**D**)—red deer; (**E**)—chamois; (**F**)—nutria, brown hare, fallow deer, European mouflon) detected by MAT in Slovenia from 2019 to 2020.

Live cultures of different serovars were used as antigens: Grippotyphosa, strain Moskva V; Sejroe, strain Mallerdorf 84; Pomona, strain Pomona; Tarassovi, strain Mitis Johnson; Copenhageni (serological group: Icterohaemorrhagiae), strain Wijnberg; Canicola, strain Hond Utrecht IV; Australis, strain Ballico; Autumnalis, strain Akyami A; Bataviae, strain Van Tienen; Saxkoebing, strain MUS 24; Bratislava, strain Jež Bratislava; and Hardjo, strain Hardjo Bovis. The MAT was performed according to the accredited method in accordance with the protocol standard operating procedure (SOP 120) in the laboratory for leptospirosis at the Veterinary Faculty in Ljubljana and was carried out in two phases. In the first phase (pretest), the presence of specific antibodies for the serovars used in the test was determined, while in the second phase, a twofold titration of positive sera, starting with the dilution 1:50, was performed. Phosphate buffer (PBS; Dulbecco's phosphate-buffered saline, Sigma-Aldrich, Burlington, MA, USA) was used for serum dilutions. Results were read using a darkfield microscope with a magnification of 160, and the endpoint was estimated as 50% agglutination or the lysis of leptospires in the microscopic field. Samples that had titres of $\geq$50 against one or more serovars were considered positive.

2.3. Statistical Analyses

Estimates and confidence intervals (CIs, for $p = 0.05$) of the estimated proportions of individuals exposed to *L. interrogans* for each of the species studied were calculated, taking into account the binomial distribution of the exposure outcome (yes/no). Confidence intervals were estimated only for the species with adequate sample size. Differences in the extent of exposure to *L. interrogans* among the animal species studied were also investigated. These differences were evaluated with chi-square tests of homogeneity, first for all species together, and in the second phase, for all combinations of pairs of species. Test of homogeneity provides reliable results if the theoretical frequency in each cross-section of the levels of tested variables is larger than 1. To reach this condition, species with very small sample size ($n < 4$) were either excluded from analysis (fallow deer, badger, hare and nutria) or, in the case of species with similar biology and thus expected similar prevalence for examined disease, data from several species were pulled together in a wider group, e.g., species Eurasian lynx, grey wolf and golden jackal were joined in new group named "large carnivores".

3. Results

Examination of 249 blood sera from wild animals revealed antibodies to at least one of the pathogenic serovars in 77 sera (30.9%; CI 25.2–36.7%) (Table 2 and Figure 1). Of the 77 positive samples, 42 samples (53.8%) had positive titres against a single serovar, while 35 (45.4%) samples had positive titres against two or more serovars. Of all positive reactions, the highest antibody seroprevalence was found for serovar Australis in red fox and stone marten; serovar Icterohaemorrhagiae in golden jackal, grey wolf and roe deer; serovar Pomona in red deer; serovars Icterohaemorrhagiae and Sejroe in Eurasian lynx; serovars Icterohaemorrhagiae and Tarassovi in chamois; and serovar Bratislava in nutria. Serovar Icterohaemorrhagiae showed the highest antibody titre, 1:6400 in red fox (Table 3). No antibodies were detected in fallow deer, European badger, European mouflon or brown hare.

Table 2. Number of serum samples (Total no.) collected from different wildlife species in Slovenia from 2019 to 2020, testing positive (No. pos.) to antibody against *Leptospira* serovars.

Common Name	Total No.	No. Pos.	Proportion of Positives (and CI)	Serovars								
Alpine chamois	21	2	10 (0–22)%	Ictero	Brat	Tar	-	-	-	-	-	-
Brown hare *	2	0	-	-	-	-	-	-	-	-	-	-
European badger *	2	0	-	-	-	-	-	-	-	-	-	-
European mouflon *	4	0	-	-	-	-	-	-	-	-	-	-
Fallow deer *	1	0	-	-	-	-	-	-	-	-	-	-
Red deer	22	7	32 (12–51)%	Ictero	Brat	Pom	Grip	Sejroe	Aut	Can	-	-
Roe deer	80	20	25 (16–34)%	Ictero	Brat	Pom	Grip	Sejroe	-	-	-	-
European badger *	2	0	-	-	-	-	-	-	-	-	-	-
Eurasian lynx	2	2	Large carnivores 86 (60–100)%	Ictero	Sejroe	-	-	-	-	-	-	-
Golden jackal	2	2		Ictero	Pom	Hardjo	Sejroe	Sax	-	-	-	-
Gray wolf	3	2		Ictero	Grip	-	-	-	-	-	-	-
Red fox	97	33	34 (25–43)%	Ictero	Brat	Pom	Grip	Sejroe	Aus	Aut	Sax	Can
Stone marten	12	8	67 (40–93)%	Ictero	Brat	Pom	Aus	Sax	-	-	-	-
	249	77	30.9 (25.2–36.7)%									

* The proportion and CI of positive samples are not presented due to unreliability resulting from small sample size. Abbreviations: Icterohaemorrhagiae (Ictero), Bratislava (Brat), Pomona (Pom), Grippotyphosa (Grip), Australis (Aus), Autumnalis (Aut), Canicola (Can), Saxkoebing (Sax) and Tarassovi (Tar).

Table 3. Serovars and antibody titres against *Leptospira* serovars in tested wild animals.

Common Name	Serovars	Titre					
		50	100	200	400	800	$\geq$1600
Alpine chamois	Ictero	0	1	0	0	0	0
	Tarassovi	1	0	0	0	0	0
Eurasian lynx	Ictero	0	1	0	0	0	0
	Sejroe	1	0	0	0	0	0
Golden jackal	Ictero	0	1	1	0	0	0
	Pomona	0	0	0	1	0	0
	Hardjo	0	1	0	0	0	0
	Sejroe	0	0	0	0	1	0
	Saxkoebing	0	0	0	1	0	0
Gray wolf	Ictero	1	1	0	0	0	0
	Grippo	0	1	0	0	0	0
Nutria	Bratislava	0	1	0	0	0	0
Red fox	Ictero	3	2	3	0	1	2
	Bratislava	2	4	7	2	1	0
	Pomona	2	1	3	2	0	0
	Grippo	1	0	0	0	0	0
	Sejroe	2	4	2	1	0	0
	Australis	7	1	6	0	2	1
	Autumnalis	0	0	1	0	0	0
	Canicola	0	1	0	1	0	0
	Saxkoebing	0	2	3	1	0	0
Red deer	Ictero	0	1	0	0	0	0
	Bratislava	0	1	0	0	0	0
	Pomona	1	1	0	1	0	0
	Grippo	1	0	0	0	0	0
	Sejroe	1	0	0	0	0	0
	Autumnalis	0	1	0	0	0	0
	Canicola	1	1	0	0	0	0
	Bratislava	2	1	3	0	0	0
	Pomona	0	0	1	0	0	0
	Australis	2	1	1	1	2	1
	Saxkoebing	1	0	0	0	0	0
Roe deer	Ictero	8	4	1	0	0	0
	Bratislava	4	0	0	0	0	0
	Pomona	1	0	0	1	0	0
	Grippo	1	0	0	0	0	0
	Sejroe	1	0	0	0	0	0
Stone marten	Ictero	0	2	0	1	0	0

A test of homogeneity shows that there are differences in the proportion of positive cases between different animal species (Pearson chi-square = 25.6; df = 7; p = 0.00059). Overall, the largest proportion of positive cases was detected in the large carnivores group (86%; CI = 60–100%; n = 7), followed by two medium-sized predator species: stone marten (67%; CI = 40–93%; n = 12) and red fox (34% CI = 25–43%; n = 97), and in declining order the following four species of large herbivores: roe deer (25%; CI = 16–34%; n = 80), red deer (32%; CI = 12–51%; n = 22), alpine chamois (10%; CI = 0–22%; n = 21) and as the last European mouflon with no detected positive cases (0%; n = 4). Other species had too low a sample size for statistical analysis. The single analysed sample of nutria was positive and the sample of fallow deer was negative for the presence of specific antibodies against *Leptospira* serovars; moreover, both results of both tested samples of brown hare and European badgers were negative.

Paired comparisons between species proved differences between large carnivores and different species of large herbivores, between stone marten and species of large herbivores, and between red fox and alpine chamois (p < 0.05), but it is noteworthy that outcomes of the formal statistical test depend on sample size and are prone to type I error in cases of small samples (Table 4).

Table 4. Outcome of chi-square homogeneity tests of differences in proportion of detected antibodies for *L. interrogans* between pairs of species. Numbers in the table are *p*-values; significant differences are in bold. The proportion of positive cases is given in parentheses.

Species (Proportion of Positive Cases)	Mouflon (0%)	Alpine Chamois (10%)	Roe Deer (25%)	Red Deer (32%)	Red Fox (34%)	Stone Marten (67%)	Large Carnivores (86%)
Mouflon (0%)	1.000	0.526	0.255	0.199	0.158	**0.037**	**0.027**
Alpine chamois (10%)	0.526	1.000	0.129	0.080	**0.028**	**0.002**	**0.001**
Roe deer (25%)	0.255	0.129	1.000	0.522	0.194	**0.004**	**0.001**
Red deer (32%)	0.199	0.080	0.522	1.000	0.844	0.059	**0.019**
Red fox (34%)	0.158	**0.028**	0.194	0.844	1.000	0.030	**0.008**
Stone marten (67%)	**0.037**	**0.002**	**0.004**	0.059	0.030	1.000	0.376
Large carnivores (86%)	**0.027**	**0.001**	**0.001**	**0.019**	**0.008**	0.376	1.000

4. Discussion

Wildlife and domestic animals play an important role as the reservoir for particular *Leptospira* serovars. Environmental characteristics, topography, meteorology, human presence and species interactions can influence the occurrence and density of *Leptospira* species [41]. In wild animals, specific climatic, edaphic and hydrological factors also determine the incidence of leptospirosis in different habitats [42]. Differences in prevalence and serovars between studies in wildlife could be due to inconsistencies in cutoff titres, serovars tested, sampling site characteristics, climate or geographic location and timing of the year of the study [43].

In this study, blood samples were taken at random from animals shot during the regular annual cull or found dead in the wild without any knowledge of possible infection or its duration. Determination of antibody titre by MAT has been used as a tool for leptospirosis diagnosis. Different diagnostic tests that can be used to detect leptospirosis have advantages and disadvantages, and laboratory diagnosis of leptospirosis is challenging. A positive culture of biological samples (blood, urine, tissue) is the definitive proof of infection, but culturing leptospires is laborious and fastidious. The bacterium requires special growth media, and incubation can last for months [44]. Histopathological examination of the kidneys is not indicated to replace the serological diagnosis of leptospirosis, and may be used only as a complementary examination [45]. In the early stages of the disease, the only sensitive and specific test is PCR. Its limitation is that it does not detect DNA in the blood during the first 5–10 days after the onset of the disease and until the 15th day [37]. MAT is the most commonly used serological test in the diagnosis of leptospirosis, despite being a technically demanding and laborious procedure [46]. MAT can be positive from the 10th to 12th day after disease onset and can detect both class M and class G antibodies [37]. MAT has a sensitivity of 41% in the first week, 82% in the second to fourth week and 96% after the fourth week of illness [47]. MAT is considered the gold standard for serodiagnosis of leptospirosis due to its unsurpassed diagnostic specificity [35], but it is not sufficiently sensitive for diagnosis of the acute phase of the disease [36]. Another limitation of serology is that it cannot distinguish between current, recent or previous infections [37]. In the present study, paired samples to confirm acute or convalescent infection were not available.

The analyses performed in the present study showed that most of the carnivores studied (with the exception of the European badger) and the most abundant wild ruminants in Slovenia were frequently exposed to *L. interrogans*. The significant percentage of seropositive red fox, stone marten, grey wolf and other carnivores tested may indicate that those species are reliable sentinels for epidemiological monitoring in Slovene forest habitats, which can also explain positive titres to *L. interrogans* in roe deer and red deer sharing the same biotope. Infection by multiple serogroups was confirmed, suggesting that multiple epidemiological cycles exist in the Slovenian region. The results of our study confirmed antibodies against 10 pathogenic *Leptospira* serovars in carnivores, 8 in wild ruminants and 1 in nutria. The observed seroprevalence of leptospiral antibodies in the tested wildlife species could not be extrapolated to the whole population level in Slovenia due to the statistically insufficient number of samples but could be a good indicator of the importance of these wildlife species in leptospirosis transmission.

The red fox (*Vulpes vulpes*) is widely distributed in European countries [48] and is the most widespread mesopredator in Slovenia. Since 2013, the hunting bag of red fox increased from 10.400 to 15.715 in 2019 [49,50]. Several serological studies using MAT have shown that red foxes are frequently exposed to different serovars of *Leptospira* spp. [19,21,51–53]. The results of our study confirm that interactions between different *Leptospira* serovars are also common in the Slovenian red fox population. The seroprevalence of antibodies to leptospiral serovars (34%) found in this study was lower than the seroprevalence reported in red foxes from Spain (47.1%) [19] but higher than that in other European countries, such as Poland (26.3%) [21], Croatia (31.25%) [33], Norway (9.9%) [51] and Germany (1.9%) [52], all using MAT. The different results from the different countries are difficult to explain because research on the seroprevalence of *Leptospira* spp. in foxes requires extensive ecological knowledge of fox population dynamics and must include considerations of juvenile fall migrations, home range, population density, litter size, yearly accession, mortality rate and hunting pressure [53]. It is also very important to consider epidemiological data on leptospiral archaic foci, reservoirs, maintenance hosts and serovar distribution [42]. Antibodies against serovar Australis (51.5%) were detected most frequently, followed by those against serovar Bratislava (48.5%) and Icterohaemorrhagiae (33.3%). Antibodies against serovar Australis were also the most frequently detected antibodies in red foxes in Croatia [54]. According to data from studies in other European countries, the most common serovar in red foxes is Icterohaemorrhagiae [19,51], while the Bratislava serovar is less common in Europe. The exposure of foxes to this serovar is not surprising because rodents, an important food source for foxes, are probably the most important host for a variety of *Leptospira* serovars in rural and urban environments [55,56].

Among all European members of the marten family, the stone marten is the only species whose population is increasing, and it is one of the most widespread mustelids in the Eurasian region [57]. The current population size is unknown. The results of our study confirmed the presence of specific antibodies against various serovars of *Leptospira* in the stone marten population. *Leptospira* antibodies were found in eight animals, a seroprevalence of 66.6% (8/12). The high seroprevalence in stone martens in Slovenia is comparable to that in Croatia (4/7; 62.50%) [42], while in Spain and France, all samples (*n* = 8) were negative against *L. interrogans* serovars using MAT [18,58]. In Slovenia, antibodies against serovar Australis (66.66%) were most frequently detected in stone martens, followed by those against serovar Bratislava (50%) and Icterohaemorrhagiae (25%).

The golden jackal (*Canis aureus*) is one of the most widespread canid species [59] and has also established territories in Slovenia [60]. A rough estimate of the population size in Slovenia is about 1000 individuals [61]. The results of our study confirmed the presence of specific antibodies against different serovars of *Leptospira* in the golden jackal population. *Leptospira* antibodies were found in two animals with a seroprevalence of 100% (2/2). The study data indicated that golden jackals could transmit different *Leptospira* serovars. They showed titres against Icterohaemorrhagiae, Pomona, Hardjo, Sejroe and Saxkoebing serovars. The seroprevalence of antibodies against leptospiral serovars found in this study was comparable to the seroprevalence reported in golden jackals from Ukraine (100%; 9/9). Both golden jackals tested had titres against five serovars [62].

The grey wolf (*Canis lupus*) is the largest wild member of the dog family (*Canidae*). Its population in Slovenia is increasing and includes over 100 individuals [63]. *Leptospira* antibodies were found in two animals with a seroprevalence of 66.6% (2/3). Serological reactions for the Grippotyphosa, Pomona and Icterohaemorrhagiae serogroups were detected in a grey wolf in Italy [64]. Few studies have been conducted in the United States of America, where seroprevalence in grey wolves ranged from 1% to 11% [65,66], and the most commonly detected serovar was Grippotyphosa.

The Eurasian lynx (*Lynx lynx*) is the third-largest predator in Europe after the brown bear and grey wolf. The population in Slovenia is estimated at only about 15 individuals [67]. *Leptospira* antibodies were found in two animals with a seroprevalence of 100% (2/2). The study data in Eurasian lynx showed low titres of antibodies against serovars Icterohaemorrhagiae and Sejroe. Seroprevalence in Iberian lynx in Spain [19] and Quebec in wild lynx (*Lynx canadensis*) was 32% (7/22) and 1% (1/97), respectively. In Spain, the most frequently detected serovars were Icterohaemorrhagiae, and in Quebec, in one case, Pomona and Bratislava. According to Labelle et al. [68], the low seroprevalence of antibodies to *L. interrogans* in lynx is unexpected because rodents, one of the main food sources for these animals, are known reservoirs of *L. interrogans*. Throughout Europe and in Slovenia, the diet of lynx usually consists of European roe deer, which is clearly the preferred prey of Eurasian lynx [69]. In this study, the seroprevalence of antibodies to *L. interrogans* in roe deer was 25%. We believe that roe deer may also serve as a source of leptospirosis for lynx.

The European roe deer (*Capreolus capreolus*) is the most common and widespread deer species in Europe [70]. The rough estimate of roe deer population size in Slovenia is about 110,000 individuals [71]. In recent decades, the population size and, at the same time, the hunting bag of roe deer have greatly increased in most parts of Europe [72]. Roe deer is one of the most important game species and a crucial prey of large carnivores in Europe [70,73]. In roe deer, a seroprevalence of 25% (20/80) was observed. Animals showed titres against five serovars. The majority of the positive samples had positive titres against a single serovar. Antibodies against serovar Icterohaemorrhagiae (65%) were most frequently detected. The seroprevalence of antibodies against leptospiral serovars found in this study was considerably higher than that in Croatia (6.0%) [42] or in Poland, which showed an overall seroprevalence in deer (roe deer, red deer and fallow deer) of 4.8% [74], or in Germany (2%) [75]. No positive serological reactions were found in roe deer (*n* = 66) in Italy [17].

The red deer (*Cervus elaphus*) is the second most abundant deer species in almost all of Europe [76]. The rough estimate of the population size is 10,000–14,000 animals [77]. A seroprevalence of 31.8% (7/22) was found in red deer. The animals showed titres against seven serovars. Antibodies against serovar Pomona (42.8%) were detected most frequently. The seroprevalence of antibodies against leptospiral serovars found in this study was significantly higher than that in Italy (6.33%) [17] or in Poland, which showed a total seroprevalence in deer (roe deer, red deer and fallow deer) of 4.8% [74], or in Croatia (19.02%) [42].

Chamois (*Rupicapra rupicapra*) is a habitat-specialised ungulate inhabiting "continental archipelagos" with fragmented rocky habitats, often restricted to high altitudes [78]. The estimated number of chamois in Slovenia is 10,000 [77]. *Leptospira* antibodies were found in two animals, a seroprevalence of 9.52% (2/21). Study data revealed that chamois had antibodies against serovars Icterohaemorrhagiae and Tarassovi. In Italy, no positive serological reactions for *Leptospira* serovars were found in chamois (*n* = 138) [17]. To our knowledge, our study is the first to report positive samples in chamois for *Leptospira* antibodies. It is, however, noteworthy that the proportion of positive samples in chamois was among the lowest between all analysed species in Slovenia being followed only by mouflon (0/4).

The coypu, also known as the nutria (*Myocastor coypus*), is a semiaquatic rodent and significant carrier of pathogenic *Leptospira* in Europe [79]. The current population size is unknown. We tested one animal and detected antibodies against serovar Bratislava. Several researchers in Europe have presented information on antibodies to leptospiral serovars in nutria, ranging from 11.5% in Italy [80] to 76% in France [81] with the predominance of the Icterohaemorrhagiae serogroup. These results are consistent with the idea that nutria should be considered a risk factor for leptospirosis in humans and domestic animals and should be taken into account by public health decision makers, especially with regard to prevention and population control [79].

Unlike clinical disease seen in canines and humans, the health impact of leptospirosis in wildlife is unclear [43]. Necropsy of fresh carcasses of animals killed in traffic accidents to look for renal lesions [19] and collection of tissue for immunohistochemistry would help in confirming the disease [82].

The risk of contracting leptospirosis is associated with occupational and recreational hazards [34]. Strategies for leptospirosis prevention are, therefore, based on education about the epidemiology and transmission mechanisms of leptospirosis [83]. Education of occupationally exposed workers about contact with contaminated water or infected animals is particularly important. Personal protective measures should also be taken for workers in high-risk occupations. The risk of infection can be reduced by increasing awareness of the routes of infection, avoiding contact with high risk water sources and using of prophylaxis during high-risk activities [84]. Increased efforts should be made to identify and treat infected animals at an early stage and to raise awareness of immunisation options for domestic and farm animals [83].

5. Conclusions

Our data confirm that large and medium-sized carnivores are frequently exposed to the pathogenic serovars of *L. interrogans* and may play the role of sentinel for leptospirosis. Data on *L. interrogans*-specific antibody-positive wild ruminants suggest that these species, although less infected, may still be a potential source of leptospirosis for humans, with the risk of infection particularly high for veterinarians, butchers, people working in forested areas and, frequently overlooked, hunting dogs. Due to the small number of samples tested, further investigation of the prevalence of infection in wild animals in Slovenia is needed to clarify the epidemiological significance of wild animals for leptospirosis transmission.

Author Contributions: Conceptualisation, D.Ž.V., G.V. and R.L.-K.; methodology, D.Ž.V., K.J., R.L.-K. and G.V.; software, K.J. and R.L.-K.; validation, R.L.-K. and N.M.-H.; formal analysis, R.L.-K., K.J. and N.M.-H.; investigation, G.V., R.L.-K., N.M.-H. and D.Ž.V.; data curation, G.V.; writing—original draft preparation, G.V. and D.Ž.V.; writing—review and editing, D.Ž.V., R.L.-K., N.M.-H., K.J. and G.V.; visualisation, D.Ž.V., G.V. and R.L.-K.; supervision, G.V. and R.L.-K.; funding acquisition, G.V., R.L.-K. and D.Ž.V. All authors have read and agreed to the published version of the manuscript.

Funding: This research was funded by the Slovenian Research Agency (research core funding No. P4-0092), the Administration of the Republic of Slovenia for Food Safety, Veterinary Sector and Plant Protection and the Slovenian Hunting Association.

Institutional Review Board Statement: No ethical approval was required for the sample types collected in this study. Animals were harvested by hunters either during the regular annual cull or when they were taken from the wild due to disease as decided by the game warden. The authors declare that no animals were killed for the purpose of this study and that all procedures that contributed to this work meet the ethical standards of the relevant national and European regulations on the care and use of animals (Directive 2010/63/EC).

Informed Consent Statement: Not applicable.

Data Availability Statement: The data presented in this study are available upon request from the corresponding author.

Acknowledgments: The authors thank all hunters for their participation in the study through collecting samples. We would also like to thank the Administration of the Republic of Slovenia for Food Safety, Veterinary Service and Plant Protection; the Slovenian Research Agency and the Hunting Association of Slovenia for supporting the research.

Conflicts of Interest: The authors declare no conflict of interest.

References

1. Mörner, T.; Obendorf, D.L.; Artois, M.; Woodford, M.H. Surveillance and monitoring of wildlife diseases. *Rev. Sci. Tech.* **2002**, *21*, 67–76. [CrossRef]
2. Guberti, V.; Stancampiano, L.; Ferrari, N. Surveillance, monitoring and surveys of wildlife diseases: A public health and conservation approach. *Hystrix. It J. Mamm* **2014**, 3–8.
3. Cilia, G.; Bertelloni, F.; Albini, S.; Fratini, F. Insight into the Epidemiology of Leptospirosis: A Review of. *Animals* **2021**, *11*, 191. [CrossRef]
4. Jeske, K.; Jacob, J.; Drewes, S.; Pfeffer, M.; Heckel, G.; Ulrich, R.G.; Imholt, C. Hantavirus-Leptospira coinfections in small mammals from central Germany. *Epidemiol. Infect.* **2021**, *149*, e97. [CrossRef]
5. Bharti, A.R.; Nally, J.E.; Ricaldi, J.N.; Matthias, M.A.; Diaz, M.M.; Lovett, M.A.; Levett, P.N.; Gilman, R.H.; Willig, M.R.; Gotuzzo, E.; et al. Leptospirosis: A zoonotic disease of global importance. *Lancet Infect. Dis.* **2003**, *3*, 757–771. [CrossRef]
6. Bengis, R.G.; Leighton, F.A.; Fischer, J.R.; Artois, M.; Mörner, T.; Tate, C.M. The role of wildlife in emerging and re-emerging zoonoses. *Rev. Sci. Tech.* **2004**, *23*, 497–511. [PubMed]
7. Picardeau, M. Virulence of the zoonotic agent of leptospirosis: Still terra incognita? *Nat. Rev. Microbiol.* **2017**, *15*, 297–307. [CrossRef]
8. Adler, B.; Peña, D.L.; Moctezuma, A. Leptospira. In *Pathogenesis of Bacterial Infections of Animals*; Gyles, C., Prescott, J., Songer, G., Thoen, C., Eds.; Blackwell Publishing: Hoboken, NJ, USA, 2004; pp. 385–396.
9. Vincent, A.T.; Schiettekatte, O.; Goarant, C.; Neela, V.K.; Bernet, E.; Thibeaux, R.; Ismail, N.; Mohd Khalid, M.K.N.; Amran, F.; Masuzawa, T.; et al. Revisiting the taxonomy and evolution of pathogenicity of the genus Leptospira through the prism of genomics. *PLoS Negl. Trop. Dis.* **2019**, *13*, e0007270. [CrossRef] [PubMed]
10. Guglielmini, J.; Bourhy, P.; Schiettekatte, O.; Zinini, F.; Brisse, S.; Picardeau, M. Genus-wide Leptospira core genome multilocus sequence typing for strain taxonomy and global surveillance. *PLoS Negl. Trop. Dis.* **2019**, *13*, e0007374. [CrossRef] [PubMed]
11. Levett, P.N. Leptospirosis. *Clin. Microbiol. Rev.* **2001**, *14*, 296–326. [CrossRef]
12. Cvetnic, Z.; Margaletic, J.; Toncic, J.; Turk, N.; Milas, Z.; Spicic, S.; Lojkic, M.; Terzic, S.; Jemersic, L.; Humski, A.; et al. A serological survey and isolation of leptospires from small rodents and wild boars in the Republic of Croatia. *Vet. Med. Czech.* **2003**, *11*, 321–329. [CrossRef]
13. Zhang, C.; Xu, J.; Zhang, T.; Qiu, H.; Li, Z.; Zhang, E.; Li, S.; Chang, Y.F.; Guo, X.; Jiang, X.; et al. Genetic characteristics of pathogenic Leptospira in wild small animals and livestock in Jiangxi Province, China, 2002-2015. *PLoS Negl. Trop. Dis.* **2019**, *13*, e0007513. [CrossRef] [PubMed]
14. Krijger, I.M.; Ahmed, A.A.A.; Goris, M.G.A.; Cornelissen, J.B.W.J.; Groot Koerkamp, P.W.G.; Meerburg, B.G. Wild rodents and insectivores as carriers of pathogenic Leptospira and Toxoplasma gondii in The Netherlands. *Vet. Med. Sci* **2020**, *6*, 623–630. [CrossRef] [PubMed]
15. Hartskeerl, R.A.; Terpstra, W.J. Leptospirosis in wild animals. *Vet. Q.* **1996**, *18* (Suppl. 3), S149–S150. [CrossRef]
16. Espí, A.; Prieto, J.M.; Alzaga, V. Leptospiral antibodies in Iberian red deer (Cervus elaphus hispanicus), fallow deer (Dama dama) and European wild boar (Sus scrofa) in Asturias, Northern Spain. *Vet. J.* **2010**, *183*, 226–227. [CrossRef]
17. Andreoli, E.; Radaelli, E.; Bertoletti, I.; Bianchi, A.; Scanziani, E.; Tagliabue, S.; Mattiello, S. Leptospira spp. infection in wild ruminants: A survey in Central Italian Alps. *Vet. Ital.* **2014**, *50*, 285–291. [CrossRef]
18. Ayral, F.; Djelouadji, Z.; Raton, V.; Zilber, A.L.; Gasqui, P.; Faure, E.; Baurier, F.; Vourc'h, G.; Kodjo, A.; Combes, B. Hedgehogs and Mustelid Species: Major Carriers of Pathogenic Leptospira, a Survey in 28 Animal Species in France (20122015). *PLoS ONE* **2016**, *11*, e0162549. [CrossRef]
19. Millán, J.; Candela, M.G.; López-Bao, J.V.; Pereira, M.; Jiménez, M.A.; León-Vizcaíno, L. Leptospirosis in wild and domestic carnivores in natural areas in Andalusia, Spain. *Vector Borne Zoonotic Dis.* **2009**, *9*, 549–554. [CrossRef]
20. Fornazari, F.; Langoni, H.; Marson, P.M.; Nóbrega, D.B.; Teixeira, C.R. Leptospira reservoirs among wildlife in Brazil: Beyond rodents. *Acta Trop.* **2018**, *178*, 205–212. [CrossRef]
21. Żmudzki, J.; Arent, Z.; Jabłoński, A.; Nowak, A.; Zębek, S.; Stolarek, A.; Bocian, Ł.; Brzana, A.; Pejsak, Z. Seroprevalence of 12 serovars of pathogenic Leptospira in red foxes (Vulpes vulpes) in Poland. *Acta Vet. Scand.* **2018**, *60*, 34. [CrossRef]
22. Vieira, A.S.; Pinto, P.S.; Lilenbaum, W. A systematic review of leptospirosis on wild animals in Latin America. *Trop. Anim. Health Prod.* **2018**, *50*, 229–238. [CrossRef]
23. Sharma, S.; Vijayachari, P.; Sugunan, A.P.; Sehgal, S.C. Leptospiral carrier state and seroprevalence among animal population—A cross-sectional sample survey in Andaman and Nicobar Islands. *Epidemiol. Infect.* **2003**, *131*, 985–989. [CrossRef]
24. Mwachui, M.A.; Crump, L.; Hartskeerl, R.; Zinsstag, J.; Hattendorf, J. Environmental and Behavioural Determinants of Leptospirosis Transmission: A Systematic Review. *PLoS Negl. Trop. Dis.* **2015**, *9*, e0003843. [CrossRef] [PubMed]
25. Cirone, S.M.; Riemann, H.P.; Ruppanner, R.; Behymer, D.E.; Franti, C.E. Evaluation of the hemagglutination test for epidemiologic studies of leptospiral antibodies in wild mammals. *J. Wildl. Dis.* **1978**, *14*, 193–202. [CrossRef] [PubMed]
26. Bierque, E.; Thibeaux, R.; Girault, D.; Soupé-Gilbert, M.E.; Goarant, C. A systematic review of Leptospira in water and soil environments. *PLoS ONE* **2020**, *15*, e0227055. [CrossRef] [PubMed]
27. Haake, D.A.; Dundoo, M.; Cader, R.; Kubak, B.M.; Hartskeerl, R.A.; Sejvar, J.J.; Ashford, D.A. Leptospirosis, water sports, and chemoprophylaxis. *Clin. Infect. Dis.* **2002**, *34*, e40–e43. [CrossRef]

28. Lau, C.L.; Smythe, L.D.; Craig, S.B.; Weinstein, P. Climate change, flooding, urbanisation and leptospirosis: Fuelling the fire? *Trans. R Soc. Trop. Med. Hyg.* **2010**, *104*, 631–638. [CrossRef]

29. Deutz, A.; Fuchs, K.; Schuller, W.; Müller, M.; Kerbl, U.; Klement, C. Studies on the seroseroprevalence of antibodies against *Leptospira interrogans* in hunters and wild boar from south-eastern Austria. *Z. Jagdwiss* **2002**, *48*, 60–65.

30. Mason, R.J.; Fleming, P.J.; Smythe, L.D.; Dohnt, M.F.; Norris, M.A.; Symonds, M.L. Leptospira interrogans antibodies in feral pigs from New South Wales. *J. Wildl. Dis.* **1998**, *34*, 738–743. [CrossRef]

31. Ko, A.I.; Goarant, C.; Picardeau, M. Leptospira: The dawn of the molecular genetics era for an emerging zoonotic pathogen. *Nat. Rev. Microbiol.* **2009**, *7*, 736–747. [CrossRef]

32. Sunil-Chandra, N.P.; Clement, J.; Maes, P.; De Silva, H.J.; Van Esbroeck, M.; Van Ranst, M. Concomitant leptospirosis-hantavirus co-infection in acute patients hospitalized in Sri Lanka: Implications for a potentially worldwide underestimated problem. *Epidemiol. Infect.* **2015**, *143*, 2081–2093. [CrossRef] [PubMed]

33. Podgoršek, D. *Phenotypic and Genotypic Identification of Leptospira spp. and Evaluation of Microbiological Methods for Diagnosis of Infection*; Biotehnical Faculty: Ljubljana, Slovenija, 2017.

34. Bedernjak, J. *Leptospiroza pri nas in v Svetu (Leptospirosis in Our Country and in the World)*; Bedernjak, J., Ed.; Pomurska Založba: Murska Sobota, Slovenija, 1993; p. 136.

35. Goris, M.G.; Hartskeerl, R.A. Leptospirosis serodiagnosis by the microscopic agglutination test. *Curr. Protoc. Microbiol.* **2014**, *32*, Unit 12E.15. [CrossRef] [PubMed]

36. Niloofa, R.; Fernando, N.; de Silva, N.L.; Karunanayake, L.; Wickramasinghe, H.; Dikmadugoda, N.; Premawansa, G.; Wickramasinghe, R.; de Silva, H.J.; Premawansa, S.; et al. Diagnosis of Leptospirosis: Comparison between Microscopic Agglutination Test, IgM-ELISA and IgM Rapid Immunochromatography Test. *PLoS ONE* **2015**, *10*, e0129236. [CrossRef]

37. Musso, D.; La Scola, B. Laboratory diagnosis of leptospirosis: A challenge. *J. Microbiol. Immunol. Infect.* **2013**, *46*, 245–252. [CrossRef]

38. Katz, A.R. Quantitative polymerase chain reaction: Filling the gap for early leptospirosis diagnosis. *Clin. Infect. Dis.* **2012**, *54*, 1256–1258. [CrossRef] [PubMed]

39. Mullan, S.; Panwala, T.H. Polymerase Chain Reaction: An Important Tool for Early Diagnosis of Leptospirosis Cases. *J. Clin. Diagn Res.* **2016**, *10*, DC08–DC11. [CrossRef]

40. Vengušt, G.; Lindtner-Knific, R.; Žele, D.; Bidovec, A. Leptospira antibodies in wild boars (Sus scrofa) in Slovenia. *Eur. J. Wildl. Res.* **2008**, *54*, 749–752. [CrossRef]

41. Soberon, J.; Peterson, A.T. Interpretation of Models of Fundamental Ecological Niches and Species' Distributional Areas. *Biodivers Inform.* **2005**, *2*, 1–10. [CrossRef]

42. Slavica, A.; Cvetnić, Ž.; Milas, Z.; Janicki, Z.; Turk, N.; Konjevi, D.; Severin, K.; Tončić, J.; Lipej, Z. Incidence of leptospiral antibodies in different game species over a 10-year period (1996–2005) in Croatia. *Eur. J. Wildl.* **2008**, *54*, 305–311. [CrossRef]

43. Grimm, K.; Rivera, N.A.; Fredebaugh-Siller, S.; Weng, H.Y.; Warner, R.E.; Maddox, C.W.; Mateus-Pinilla, N.E. Evidence of Leptospira Serovars in Wildlife and Leptospiral DNA in Water Sources in a Natural Area in East Central Illinois, USA. *J. Wildl. Dis.* **2020**, *56*, 316–327. [CrossRef]

44. Di Azevedo, M.I.N.; Pires, B.C.; Libonati, H.; Pinto, P.S.; Cardoso Barbosa, L.F.; Carvalho-Costa, F.A.; Lilenbaum, W. Extra-renal bovine leptospirosis: Molecular characterization of the Leptospira interrogans Sejroe serogroup on the uterus of non-pregnant cows. *Vet. Microbiol.* **2020**, *250*, 108869. [CrossRef]

45. Magalhães, G.M.; Alvarenga, P.B.D.; Medeiros-Ronchi, A.A.; Moreira, T.d.A.; Gundim, L.F.; Gomes, D.O.; Lima, A.M.C. Leptospirosis in slaughtered cows in the Triangulo Mineiro, Minas Gerais: Prevalence, serological profile and renal lesions. *Biosci. J.* **2020**, *36*, 539–545. [CrossRef]

46. Levett, P.N. Usefulness of serologic analysis as a predictor of the infecting serovar in patients with severe leptospirosis. *Clin. Infect. Dis.* **2003**, *36*, 447–452. [CrossRef] [PubMed]

47. Sehgal, S.C.; Vijayachari, P.; Sharma, S.; Sugunan, A.P. LEPTO Dipstick: A rapid and simple method for serodiagnosis of acute leptospirosis. *Trans. R. Soc. Trop. Med. Hyg.* **1999**, *93*, 161–164. [CrossRef]

48. Eckert, J.; Conraths, F.J.; Tackmann, K. Echinococcosis: An emerging or re-emerging zoonosis? *Int. J. Parasitol.* **2000**, *30*, 1283–1294. [CrossRef] [PubMed]

49. Slovenia-Forest-Service. Slovenia Forest Service Report on Forest for Year 2019. Available online: www.zgs.si/zavod/publikacije/letna_porocila/index.html (accessed on 7 June 2021).

50. Slovenia-Forest-Service. Slovenia Forest Service Report on Forest for Year 2013. Available online: www.zgs.si/zavod/publikacije/letna_porocila/index.html (accessed on 7 June 2021).

51. Akerstedt, J.; Lillehaug, A.; Larsen, I.L.; Eide, N.E.; Arnemo, J.M.; Handeland, K. Serosurvey for canine distemper virus, canine adenovirus, Leptospira interrogans, and Toxoplasma gondii in free-ranging canids in Scandinavia and Svalbard. *J. Wildl. Dis.* **2010**, *46*, 474–480. [CrossRef] [PubMed]

52. Müller, H.; Winkler, P. [Results of serological studies of Leptospira antibodies in foxes]. *Berl. Munch Tierarztl. Wochenschr.* **1994**, *107*, 90–93.

53. Milas, Z.; Turk, N.; Janicki, Z.; Slavica, A.; Starešina, V.; Barbić, L.J.; Lojkić, M.; Modrić, Z. Leptospiral antibodies in red foxes (*Vulpes vulpes*) in northwest Croatia. *Vet. Arhiv.* **2006**, *76*, 51–57.

54. Slavica, A.; Dezdek, D.; Konjevic, D.; Cvetnic, Z.; Sindicic, M.; Stanin, D.; Habus, J.; Turk, N. Prevalence of leptospiral antibodies in the red fox (Vulpes vulpes) population of Croatia. *Vet. Med. (Praha)* **2011**, *56*, 209–213.

55. Dupouey, J.; Faucher, B.; Edouard, S.; Richet, H.; Kodjo, A.; Drancourt, M.; Davoust, B. Human leptospirosis: An emerging risk in Europe? *Comp. Immunol. Microbiol. Infect. Dis.* **2014**, *37*, 77–83. [CrossRef] [PubMed]

56. Boey, K.; Shiokawa, K.; Rajeev, S. Leptospira infection in rats: A literature review of global prevalence and distribution. *PLoS Negl. Trop. Dis.* **2019**, *13*, e0007499. [CrossRef]

57. Genovesi, P.; Secchi, M.; Boitani, L. Diet of stone martens: An example of ecological flexibility. *J. Zool* **1996**, *238*, 545–555. [CrossRef]

58. Millán, J.; Velarde, R.; Chirife, A.D.; León-Vizcaíno, L. Carriage of pathogenic Leptospira in carnivores at the wild/domestic interface. *Pol. J. Vet. Sci.* **2019**, *22*, 589–598. [CrossRef] [PubMed]

59. Arnold, J.; Humer, A.; Heltai, M.; Murariu, D.; Spassov, N.; Hackländer, K. Current status and distribution of golden jackals *Canis aureus* in Europe. *Mamm. Rev.* **2011**, *42*, 1–11. [CrossRef]

60. Krofel, M. Confirmed presence of territorial groups of golden jackals (*Canis aureus*) in Slovenia. *Nat. Slov.* **2009**, *11*, 65–68.

61. Potočnik, H.; Pokorny, B.; Flajšman, K.; Kos, I. *Evrazijski Šakal (Eurasian Jackal)*; Leskovic, B., Ed.; Lovska zveza Slovenije: Ljubljana, Slovenia, 2019; p. 248.

62. Nakonechnyi, I.V.; Perots'ka, L.V.; Pyvovarova, I.V.; Chornyi, V.A. Ecological and epizootic roles of Golden jackal, genus Canis aureus in the Northwest of Black Sea coast. *Sci. Messenger LNU Vet. Med. Biotechnol.* **2019**, *21*, 37–43. [CrossRef]

63. Simčič, G.; Fležar, U.B.; Bartol, M.; Černe, R.; Berce, T.; Krofel, M.; Potočnik, H.; Jelenčič, M.; Kuralt, Ž. Spremljanje varstvenega stanja volkov v Sloveniji v letih 2018 in 2019 (Monitoring of wolf population in Slovenija in seasons 2018 and 2019). *Lovec* **2020**, *103*, 328–331.

64. Bregoli, M.; Pesaro, S.; Ustulin, M.; Vio, D.; Beraldo, P.; Galeotti, M.; Cocchi, M.; Lucchese, L.; Bertasio, C.; Boniotti, M.B.; et al. Environmental Exposure of Wild Carnivores to Zoonotic Pathogens. *Int. J. Environ. Res. Public Health* **2021**, *18*, 2512. [CrossRef]

65. Zarnke, R.L.; Ballard, W.B. Serologic survey for selected microbial pathogens of wolves in Alaska, 1975-1982. *J. Wildl. Dis.* **1987**, *23*, 77–85. [CrossRef]

66. Khan, M.A.; Goyal, S.M.; Diesch, S.L.; Mech, L.D.; Fritts, S.H. Seroepidemiology of leptospirosis in Minnesota wolves. *J. Wildl. Dis* **1991**, *27*, 248–253. [CrossRef]

67. Žerjav, S.; Černe, R.; Stergar, M.; Krofel, M.; Majić Skrbinšek, A.; Skrbinšek, T.; Berce, T.; Bartol, M.; Potočnik, H.; Kos, I.; et al. Project LIFE Lynx - Preventing the Extinction of the Dinaric-SE Alpine Lynx Population. *Lovec* **2017**, *9*, 428–429.

68. Labelle, P.; Mikaelian, I.; Martineau, D.; Beaudin, S.; Blanchette, N.; Lafond, R.; St-Onge, S. Seroprevalence of leptospirosis in lynx and bobcats from Quebec. *Can. Vet. J.* **2000**, *41*, 319–320.

69. Krofel, M.; Huber, D.; Kos, I. Diet of Eurasian lynx Lynx lynx in the northern Dinaric Mountains (Slovenia and Croatia). *Acta Theriol.* **2011**, *56*, 315–322. [CrossRef]

70. Melis, C.; Nilsen, E.B.; Panzacchi, M.; Linnell, J.D.; Odden, J. Roe deer face competing risks between predators along a gradient in abundance. *Ecosphere* **2013**, *4*, 1–12. [CrossRef]

71. Jerina, K.; Stergar, M.; Jelenko, I.; Pokorny, B. *Spatial Distibution, Fitness, and Population Dynamics of Ungulates in Slovenia: Studies on the Effects of Spatially Explicit Habitat and Species-Specific Factors and Predicting Future Trends*; Biotechnical Faculty of the University of Ljubljana: Ljubljana, Slovenia, 2010; p. 48.

72. Burbaitė, L.; Csányi, S. Roe deer population and harvest changes in Europe. *Estonian J. Ecol* **2009**, *58*, 169–180. [CrossRef]

73. Linnell, J.D.C.; Zachos, F. Status and distribution patterns of European ungulates: Genetics, population history and conservation. In *Ungulate Management in Europe—Problems and Practices*; Putman, R., Apollonio, M., Andersen, R., Eds.; Cambridge University Press: Cambridge, UK, 2011; pp. 12–53.

74. Zmudzki, J.; Jablonski, A.; Arent, Z.; Zebek, S.; Nowak, A.; Stolarek, A. Parzeniecka-Jaworska, M. First report of Leptospira infections in red deer, roe deer, and fallow deer in Poland. *J. Vet.* **2016**, *60*, 257–260. [CrossRef]

75. Witt, W.; Dedek, J.; Loepelmann, H. Occurrence of Leptospira antibodies in red, roe and fallow deer and mouflon. *Monatsh Veterinarmed* **1988**, *43*, 65–68.

76. Burbaité, L.; Csányi, S. Red deer population and harvest changes in Europe. *Acta Zool Lit.* **2010**, *20*, 179–188. [CrossRef]

77. Adamič, M.; Jerina, K. Ungulates and their management in Slovenia. In *European Ungulates and Their Management in the 21st Century*; Apollonio, M., Andersen, R., Putman, R., Eds.; Cambridge University Press: Cambridge, UK, 2010; pp. 507–527.

78. Bužan, E.; Bryja, J.; Zemanová, B.; Kryštufek, B. Population genetics of chamois in the contact zone between the Alps and the Dinaric mountains: Uncovering the role of habitat fragmentation and past management. *Cons Genet.* **2013**, *14*, 401–412. [CrossRef]

79. Ayral, F.; Kodjo, A.; Guédon, G.; Boué, F.; Richomme, C. Muskrats are greater carriers of pathogenic Leptospira than coypus in ecosystems with temperate climates. *PLoS ONE* **2020**, *15*, e0228577. [CrossRef]

80. Bollo, E.; Pregel, P.; Gennero, S.; Pizzoni, E.; Rosati, S.; Nebbia, P.; Biolatti, B. Health status of a population of nutria (Myocastor coypus) living in a protected area in Italy. *Res. Vet. Sci.* **2003**, *75*, 21–25. [CrossRef]

81. Vein, J.; Leblond, A.; Belli, P.; Kodjo, A.; Berny, P.J. The role of the coypu (Myocastor coypus), an invasive aquatic rodent species, in the epidemiological cycle of leptospirosis: A study in two wetlands in the East of France. *Eur. J. Wildl. Res.* **2014**, *60*, 125–133. [CrossRef]

82. Shearer, K.E.; Harte, M.J.; Ojkic, D.; Delay, J.; Campbell, D. Detection of Leptospira spp. in wildlife reservoir hosts in Ontario through comparison of immunohistochemical and polymerase chain reaction genotyping methods. *Can. Vet. J.* **2014**, *55*, 240–248. [PubMed]

83. Poeppl, W.; Orola, M.J.; Herkner, H.; Müller, M.; Tobudic, S.; Faas, A.; Mooseder, G.; Allerberger, F.; Burgmann, H. High prevalence of antibodies against Leptospira spp. in male Austrian adults: A cross-sectional survey, April to June 2009. *Euro Surveill.* **2013**, *18*. [CrossRef] [PubMed]

84. Monahan, A.M.; Miller, I.S.; Nally, J.E. Leptospirosis: Risks during recreational activities. *J. Appl. Microbiol.* **2009**, *107*, 707–716. [CrossRef] [PubMed]

Brief Report

Presence of *Helicobacter pylori* and *H. suis* DNA in Free-Range Wild Boars

Francisco Cortez Nunes [1,2,3], Teresa Letra Mateus [4,5], Sílvia Teixeira [1,2,3], Patrícia Barradas [5,6,7], Chloë de Witte [8], Freddy Haesebrouck [9], Irina Amorim [1,2,3,*] and Fátima Gärtner [1,2,3]

1 Institute of Biomedical Sciences Abel Salazar (ICBAS), University of Porto, 4050-313 Porto, Portugal; franciscojvcnunes@gmail.com (F.C.N.); silvia.goncalves.teixeira@gmail.com (S.T.); fgartnes@ipatimup.pt (F.G.)
2 Institute for Research and Innovation in Health (i3S), University of Porto, 4200-135 Porto, Portugal
3 Institute of Molecular Pathology and Immunology of the University of Porto (IPATIMUP), 4200-135 Porto, Portugal
4 CISAS—Centre for Research and Development in Agrifood Systems and Sustainability, Escola Superior Agrária, Instituto Politécnico de Viana do Castelo, 4900-347 Viana do Castelo, Portugal; tlmateus@esa.ipvc.pt
5 Epidemiology Research Unit (EPIUnit), Institute of Public Health, University of Porto, 4050-091 Porto, Portugal; patricia.barradas@ispup.up.pt
6 Escola Superior Agrária, Instituto Politécnico de Viana do Castelo, 4900-347 Viana do Castelo, Portugal
7 Vasco da Gama Research Center (CIVG), Vasco da Gama University School (EUVG), 3020-210 Coimbra, Portugal
8 Department of Pharmaceutics, Faculty of Pharmaceutical Sciences, Ghent University, 9820 Merelbeke, Belgium; chloe.dewitte@ugent.be
9 Department of Pathology, Bacteriology and Avian Diseases, Faculty of Veterinary Medicine, Ghent University, 9820 Merelbeke, Belgium; Freddy.Haesebrouck@UGent.be
* Correspondence: iamorim@ipatimup.pt

Citation: Cortez Nunes, F.; Letra Mateus, T.; Teixeira, S.; Barradas, P.; de Witte, C.; Haesebrouck, F.; Amorim, I.; Gärtner, F. Presence of *Helicobacter pylori* and *H. suis* DNA in Free-Range Wild Boars. *Animals* **2021**, *11*, 1269. https://doi.org/10.3390/ani11051269

Academic Editor: David González Barrio

Received: 23 March 2021
Accepted: 26 April 2021
Published: 28 April 2021

Publisher's Note: MDPI stays neutral with regard to jurisdictional claims in published maps and institutional affiliations.

Simple Summary: *Helicobacter pylori* and *H. suis* are associated with gastric pathologies in humans. To obtain better insights into the potential role of wild boars as reservoirs of these pathogens, gastric samples of 14 animals were tested for the presence of *H. pylori* and *H. suis* DNA. Two wild boars were found PCR-positive for *H. pylori* and one for *H. suis*. This indicates that these microorganisms may colonize the stomach of wild boars.

Abstract: *Helicobacter pylori* (*H. pylori*) is a Gram-negative bacterium that infects half of the human population worldwide, causing gastric disorders, such as chronic gastritis, gastric or duodenal ulcers, and gastric malignancies. *Helicobacter suis* (*H. suis*) is mainly associated with pigs, but can also colonize the stomach of humans, resulting in gastric pathologies. In pigs, *H. suis* can induce gastritis and seems to play a role in gastric ulcer disease, seriously affecting animal production and welfare. Since close interactions between domestic animals, wildlife, and humans can increase bacterial transmission risk between species, samples of gastric tissue of 14 free range wild boars (*Sus scrofa*) were evaluated for the presence of *H. pylori* and *H. suis* using PCR. Samples from the antral gastric mucosa from two animals were PCR-positive for *H. pylori* and another one for *H. suis*. These findings indicate that these microorganisms were able to colonize the stomach of wild boars and raise awareness for their putative intervention in *Helicobacter* spp. transmission cycle.

Keywords: one health; wildlife; zoonosis; *Helicobacter* spp.; PCR; *Sus scrofa*

1. Introduction

The number of infectious diseases has been increasing in humans, with about 60% of those being zoonotic [1]. Of these emerging zoonoses, about 72% are transmitted from wildlife animals [2].

Wild boars are known to be reservoirs of a considerable number of zoonotic bacteria, viruses, and parasites [3], and these infections can be bi-directional (wild/domestic). The exposure to wild boars' pathogens can happen through different pathways: Direct contact, meat consumption, or indirect intake of contaminated water, food, or through the environment [4]. Changes in human living habits, increased hunting activities, and consumption of wild boar meat play a role in the risk of human exposure to infectious agents [3].

Helicobacter species are Gram-negative, spiral-shaped motile bacteria that colonize the gastrointestinal tract of both humans and animals [5–7], and have been studied over the years for their association with gastrointestinal diseases [8]. In humans, *Helicobacter pylori* (*H. pylori*) is the most common gastric pathogen, affecting more than half of the world's population, being responsible for development of gastritis, gastroduodenal ulcers, gastric adenocarcinoma, mucosa-associated lymphoid tissue (MALT) lymphoma, and extra digestive diseases [5,9,10]. In addition, *Helicobacter suis* (*H. suis*) is the most prevalent human gastric non-*Helicobacter pylori Helicobacter* (NHPH) and has been associated with a range of gastric pathologies, including MALT lymphoma, and possibly also extra digestive diseases. Recent reports reinforce that these infections most likely originate from pigs, emphasizing their zoonotic potential [5,11–14].

In pigs, *H. suis* mainly colonizes the fundic and pyloric gland zone of the stomach [15]. The bacterium presents tropism for the gastric acid-producing parietal cells [16]. Its prevalence appears to be very low prior to weaning, but increases rapidly thereafter, being very high at slaughter age (77%) and in adults (>90%) [15–17]. *H. suis* infection causes gastritis, decreased daily weight gain, and plays a role in induction of gastric ulcers, clearly affecting animal production and welfare [18–20].

In pigs, there is a report of a natural infection by a *H. pylori*-like bacterium, described as attached to the mucosa of the cardiac and antral portions of the stomach of two out of four healthy young pigs, without gross lesions associated [6,21]. This agent seemed to be morphologically similar to, but antigenically different from *H. pylori*, and its exact identity is not clear [18].

Until now, studies trying to assess the presence of *H. suis* in wild boars (*Sus scrofa*) have been mainly unsuccessful, as described by Flahou et al. [14] and others [22,23].

Helicobacter spp. have been described to have zoonotic potential and the close contact between humans, domestic animals, and wild animals deserves more consideration [14,24]. Although reservoirs of wild and domestic animals can be considered as important sources of emerging infectious diseases, it is the human impact on ecological systems that determines the level of risk at the human/animal interface upon the occurrence of emerging zoonotic diseases [14,24].

From an eco-epidemiological perspective, wild boars have an important role in spread of several pathogens [3,4]; thus, the aim of this study was to screen different regions of the stomach (*Pars oesophagea*, fundic, and pyloric gland zone) collected from wild boars for the presence of *H. pylori* and *H. suis*.

2. Materials and Methods

2.1. Samples Collection

Samples were collected using convenience sampling from fourteen hunted animals during two national campaigns, one in the north and other in the center of Portugal (Vila Real and Coimbra districts, respectively) (Figure 1). All the sampled animals were older than 9 months according to teeth assessment [25,26]. From each animal, gastric samples were collected from three different gastric regions: *Pars oesophagea*, fundic gland zone, and pyloric gland zone (gastric *antrum*) using a Kruuse® Biopsy punch 8 mm. After collection, samples were stored at −20 °C until DNA extraction.

The animals were not slaughtered or euthanized in order to carry out this study, and the fresh gastric tissue specimens were obtained as sub-products derived from the normal activity associated with the meat inspection procedures occurring during these conventional campaigns. None of the actions was performed solely for research purposes and the researchers had no influence on the campaign organization, nor in the meat inspection actions.

Figure 1. Illustration of the geographical origin of the samples collected in Portugal and evaluated in this study. Samples were collected from 10 wild boars during a national campaign in Vila Real (VR10) and 4 wild boars during a national campaign in Coimbra (C4).

2.2. DNA Extraction, PCR Conditions and Sequencing

DNA was extracted from 8-mm gastric frozen tissue samples, using EXTRACTME® DNA tissue kit (BLIRT, Poland), according to the instructions provided by the supplier.

All the samples were tested for *H. pylori* and *H. suis* by conventional PCR, according to previously described protocols (Table 1).

Table 1. Primer sequences used for detection of *H. pylori* and *H. suis* and thermo cycling conditions.

Primer	Sequence	Region	Amplicon Size	Cycle			Ref.
				Nr. Cycles	Temp. (°C)	Time	
BFHsuis_F1	5′-AAA ACA MAG GCG ATC GCC CTG TA-3′	*ureA* gene	150bp	40	95	20 s	[5]
BFHsuis_R1	5′-TTT CTT CGC CAG GTT CAA AGC G-3′	*ureA* gene			60	30 s	
					72	30 s	
BFHpyl_F1	5′-AAA GAG CGT GGT TTT CAT GGC G-3′	*ureAB* gene	217bp	45	94	30 s	[27]
BFHpyl_R1	5′-GGG TTT TAC CGC CAC CGA ATT TAA-3′	*ureAB* gene			59	30 s	
					72	1 min	

Aliquots of each PCR product were electrophoresed on 1.5% agarose gel, stained with Xpert Green Safe DNA gel stain (GRISP, Porto, Portugal) and examined for the presence of a specific fragment under UV light. DNA fragment size was compared with the standard molecular weight, 100bp DNA ladder (GRISP, Porto, Portugal), and the molecular weight of the positive controls (*H. pylori* with 217 and *H. suis* with 150 bp) (Figure S1). As a negative control, distilled water was used. As positive controls, DNA was extracted from pure cultures of *H. pylori* strain 26695 and *H. suis* strain HS1.

To exclude false positive samples, the amplicons from each positive sample were sequenced. Bidirectional sequencing was performed with Sanger method at the genomics core facility of the Institute of Molecular Pathology and Immunology of the University of Porto, Portugal. Sequence editing and multiple alignments were performed with the MegaX Molecular Evolutionary Genetic Analysis version 10.1.8. The sequences obtained were subject to the basic local alignment search tool (BLAST) using the non-redundant nucleotide database (http://blast.ncbi.nlm.nih.gov/Blast.cgi, accessed on 6 January 2021) [28,29].

3. Results and Discussion

A total of 42 samples, collected from 14 wild boars, were analyzed.

Based on PCR results, two samples corresponding to two distinct animals were *H. pylori* PCR-positive and another, also corresponding to a different animal, was *H. suis* PCR-positive. These positive samples all originated from the pyloric gland zone (gastric *antrum)* of these animals.

The bidirectional sequencing and BLAST analysis of consensus sequences of partial *ureA* and *ureB* genes showed a homology of 100% with *H. pylori* (GenBank® accession no. AF507994 and AY368264) for two of them and the other positive sample showed a homology of 95.39% with *H. suis* (GenBank® accession no. EF204592).

In the current study, *H. pylori* DNA was detected in the pyloric gland zone of the stomach (gastric *antrum)* from two animals and *H. suis* DNA in the same gastric region of another animal. This region is also a preferential colonization site of *H. suis* in domesticated pigs [17]. These findings might indicate that wild boars are occasionally colonized by *H. pylori* and *H. suis*. It can, however, not be excluded that presence of DNA of these microorganisms might be a consequence of recent contamination; for instance, through contact with domesticated pigs in the case of *H. suis*, or contaminated water or environments for both species. Both *H. pylori* PCR-positive cases did not present any macroscopic gastric alterations. The wild boar stomach where *H. suis* DNA was detected showed signs of mild antral inflammation consisting of mild erythema and congestion at gross examination. Unfortunately, the tissue preservation conditions prevented their detailed microscopic evaluation and the presence of microorganisms with *Helicobacter*-like morphology could not be examined. Further studies using a larger sample and including immunohistochemical and histopathological analysis of gastric tissue from wild boars are absolutely relevant to confirm our findings.

To the authors' knowledge, this is the first report of *H. pylori* DNA detected in gastric samples of free-ranging wild boars and the first report of *H. pylori* and *H. suis* DNA detection in free-range wild boars in Portugal.

Previously, gastric *antrum* specimens of 17 free-ranging wild boars from Poland were tested, using a *Helicobacter* genus-specific 16S rRNA PCR and sequence analysis of positive samples [23]. In one sample, DNA was detected of a microorganism related to the group of NHPH mainly associated with dogs and cats (previously referred to as *Helicobacter heilmannii* type 2) [23]. More recently, a novel *Helicobacter* species, *H. apri*, was described in wild boars, but this is an enterohepatic *Helicobacter* species [22]. In another survey, very low numbers of *H. suis* were detected in the cardiac gland zone and *Pars oesophagea* from 2 out of 9 wild boars from Belgium. These gastric regions are not the preferential *Helicobacter* colonization sites in the porcine stomach, possibly indicating recent contamination for instance through contact with domesticated pigs or their excretes [14].

Wild boars are known to be reservoirs for several agents for important infectious diseases transmissible to other wild animals, domestic animals, and humans [3,4]. Depending on the pathogen properties and on wild boar density and management, the eco-epidemiological role of these animals can vary from a dead-end over spill over, up to maintenance host [3,4]. Contacts between wild boars and outdoor domestic pigs should be considered a risk for transmission of these pathogens that can directly affect the swine production and animal welfare, since wild boars and domestic pigs belong to the same species (*Sus scrofa*) [3,4,17,19,30].

Hunting dogs and humans can be exposed to wild boars' pathogens either from fresh carcass contact, handling or consumption of raw, undercooked meat, or indirect contact from contaminated water or environment, and therefore, high risk exposure would include game wardens, hunters, butchers, and other wildlife professional duties [3,4]. Another aspect to consider is the increasing wild boar population and its adaptability to urban areas that can prompt wild boar contact with humans in these areas, making humans more vulnerable to be exposed to wild boars' pathogens [3,4,31]. The current findings raise *One health* concerns regarding the impact of *H. suis* and *H. pylori* in wild boar welfare, its role and impact on bacterial spread and transmission to the environment, and also to wild and domestic animals, and, ultimately, to humans [3,13,24,31].

4. Conclusions

H. pylori and *H. suis* DNA was detected in the stomach of free-range wild boars, which might indicate that these animals were colonized by these microorganisms. It can be hypothesized that wild boars might act as reservoirs and contribute to the spread of the *H. pylori* and *H. suis* in the environment, raising a public health concern.

Supplementary Materials: The following are available online at https://www.mdpi.com/article/10.3390/ani11051269/s1, Figure S1: Example of agarose gel electrophoresis of PCR products of *H. pylori* gene fragments. M, molecular marker; +C, DNA extracted from pure culture of 26695 strain was used as a positive control; WB, wild boar DNA; -C, negative control consisting solely of mix solution (NC).

Author Contributions: Conceptualization, F.C.N., T.L.M. and I.A.; methodology, F.C.N., T.L.M., S.T. and I.A.; formal analysis, F.C.N., P.B., T.L.M. and I.A.; investigation, F.C.N.; resources, F.C.N., T.L.M. and I.A.; data curation, F.C.N., T.L.M. and I.A.; writing—original draft preparation, F.C.N., T.L.M. and I.A.; writing—review and editing, S.T., T.L.M., I.A., P.B., C.d.W., F.H. and F.G.; supervision, T.L.M., I.A., C.d.W., F.H. and F.G. All authors have read and agreed to the published version of the manuscript.

Funding: This research received no external funding.

Institutional Review Board Statement: Not applicable.

Data Availability Statement: The data presented in this study are available on request from the corresponding author

Acknowledgments: We thank Madalena Vieira-Pinto (University of Trás-dos-Montes e Alto Douro) and Carlos Venâncio (University of Trás dos-Monte e Alto Douro) for facilitating the sample collection during the hunting campaign. F.C. Nunes (SFRH/BD/147761/2019) and S. Teixeira (SFRH/BD/139669/2018) acknowledge the FCT—Portuguese Foundation for Science and Technology for financial support.

Conflicts of Interest: The authors declare no conflict of interest.

References

1. Jones, K.E.; Patel, N.G.; Levy, M.A.; Storeygard, A.; Balk, D.; Gittleman, J.L.; Daszak, P. Global trends in emerging infectious diseases. *Nat. Cell Biol.* **2008**, *451*, 990–993. [CrossRef] [PubMed]
2. Vorou, R.M.; Papavassiliou, V.G.; Tsiodras, S. Emerging zoonoses and vector-borne infections affecting humans in Europe. *Epidemiol. Infect.* **2007**, *135*, 1231–1247. [CrossRef] [PubMed]
3. Meng, X.J.; Lindsay, D.S.; Sriranganathan, N. Wild boars as sources for infectious diseases in livestock and humans. *Philos. Trans. R Soc. Lond. Ser. B Biol. Sci.* **2009**, *364*, 2697–2707. [CrossRef] [PubMed]
4. Meier, R.; Ryser-Degiorgis, M. Wild boar and infectious diseases: Evaluation of the current risk to human and domestic animal health in Switzerland: A review. *Schweiz. Arch. Tierheilkd* **2018**, *160*, 443–460. [CrossRef] [PubMed]
5. Hahadori, A.; De Witte, C.; Agin, M.; De Bruyckere, S.; Smet, A.; Tümgör, G.; Guven Gokmen, T.; Haesebrouck, F.; Koksal, F. Presence of gastric Helicobacter species in children suffering from gastric disorders in Southern Turkey. *Helicobacter* **2018**, *23*, e12511. [CrossRef] [PubMed]
6. Haesebrouck, F.; Pasmans, F.; Flahou, B.; Chiers, K.; Baele, M.; Meyns, T.; Decostere, A.; Ducatelle, R. Gastric Helicobacters in Domestic Animals and Nonhuman Primates and Their Significance for Human Health. *Clin. Microbiol. Rev.* **2009**, *22*, 202–223. [CrossRef] [PubMed]
7. Choi, Y.K.; Han, J.H.; Joo, H.S. Identification of Novel Helicobacter Species in Pig Stomachs by PCR and Partial Sequencing. *J. Clin. Microbiol.* **2001**, *39*, 3311–3315. [CrossRef] [PubMed]
8. Mladenova-Hristova, I.; Grekova, O.; Patel, A. Zoonotic potential of *Helicobacter* spp. *J. Microbiol. Immunol. Infect.* **2017**, *50*, 265–269. [CrossRef] [PubMed]
9. Lopo, I.; Libânio, D.; Pita, I.; Dinis-Ribeiro, M.; Pimentel-Nunes, P. Helicobacter pylori antibiotic resistance in Portugal: Systematic review and meta-analysis. *Helicobacter* **2018**, *23*, e12493. [CrossRef]
10. Guevara, B.; Cogdill, A.G. Helicobacter pylori: A Review of Current Diagnostic and Management Strategies. *Dig. Dis. Sci.* **2020**, *65*, 1917–1931. [CrossRef]

11. Augustin, A.D.; Savio, A.; Nevel, A.; Ellis, R.J.; Weller, C.; Taylor, D.; Tucker, R.M.; Ibrahim, M.A.A.; Bjarnason, I.; Dobbs, S.M.; et al. Helicobacter suis Is Associated With Mortality in Parkinson's Disease. *Front. Med.* **2019**, *6*. [CrossRef] [PubMed]

12. Bosschem, I.; Flahou, B.; Van Deun, K.; De Koker, S.; Volf, J.; Smet, A.; Ducatelle, R.; Devriendt, B.; Haesebrouck, F. Species-specific immunity to Helicobacter suis. *Helicobacter* **2017**, *22*, e12375. [CrossRef] [PubMed]

13. Joosten, M.; Flahou, B.; Meyns, T.; Smet, A.; Arts, J.; De Cooman, L.; Pasmans, F.; Ducatelle, R.; Haesebrouck, F. Case Report:Helicobacter suisInfection in a Pig Veterinarian. *Helicobacter* **2013**, *18*, 392–396. [CrossRef]

14. Flahou, B.; Rossi, M.; Bakker, J.; Langermans, J.A.; Heuvelman, E.; Solnick, J.V.; E Martin, M.; O'Rourke, J.; Ngoan, L.D.; Hoa, N.X.; et al. Evidence for a primate origin of zoonotic Helicobacter suis colonizing domesticated pigs. *ISME J.* **2017**, *12*, 77–86. [CrossRef] [PubMed]

15. De Witte, C.; Ducatelle, R.; Haesebrouck, F. The role of infectious agents in the development of porcine gastric ulceration. *Veter J.* **2018**, *236*, 56–61. [CrossRef]

16. Zhang, G.; Ducatelle, R.; Mihi, B.; Smet, A.; Flahou, B.; Haesebrouck, F. Helicobacter suis affects the health and function of porcine gastric parietal cells. *Veter Res.* **2016**, *47*, 101. [CrossRef]

17. Hellemans, A.; Chiers, K.; De Bock, M.; Decostere, A.; Haesebrouck, F.; Ducatelle, R.; Maes, D. Prevalence of "Candidatus Helicobacter suis" in pigs of different ages. *Vet. Rec.* **2007**, *161*, 189–192. [CrossRef]

18. De Witte, C.; Devriendt, B.; Flahou, B.; Bosschem, I.; Ducatelle, R.; Smet, A.; Haesebrouck, F. Helicobacter suis induces changes in gastric inflammation and acid secretion markers in pigs of different ages. *Veter Res.* **2017**, *48*, 1–13. [CrossRef]

19. De Bruyne, E.; Flahou, B.; Chiers, K.; Meyns, T.; Kumar, S.; Vermoote, M.; Pasmans, F.; Millet, S.; Dewulf, J.; Haesebrouck, F.; et al. An experimental Helicobacter suis infection causes gastritis and reduced daily weight gain in pigs. *Veter Microbiol.* **2012**, *160*, 449–454. [CrossRef] [PubMed]

20. Baele, M.; Decostere, A.; Vandamme, P.; Ceelen, L.; Hellemans, A.; Mast, J.; Chiers, K.; Ducatelle, R.; Haesebrouck, F. Isolation and characterization of *Helicobacter suis* sp. nov. from pig stomachs. *Int. J. Syst. Evol. Microbiol.* **2008**, *58*, 1350–1358. [CrossRef] [PubMed]

21. Krakowka, S.; Ringler, S.S.; Flores, J.; Kearns, R.J.; Eaton, K.A.; Ellis, J.A. Isolation and preliminary characterization of a novel Helicobacter species from swine. *Am. J. Veter Res.* **2005**, *66*, 938–944. [CrossRef]

22. Zanoni, R.G.; Piva, S.; Florio, D.; Bassi, P.; Mion, D.; Cnockaert, M.; Luchetti, A.; Vandamme, P. *Helicobacter apri* sp. nov., isolated from wild boars. *Int. J. Syst. Evol. Microbiol.* **2016**, *66*, 2876–2882. [CrossRef] [PubMed]

23. Fabisiak, M.; Sapierzyński, R.; Salamaszyńska-Guz, A.; Kizerwetter-Swida, M. The first description of gastric Helicobacter in free-ranging wild boar (*Sus scrofa*) from Poland. *Pol. J. Veter Sci.* **2010**, *13*, 171–174.

24. Sleeman, J.M.; Richgels, K.L.D.; White, C.L.; Stephen, C. Integration of wildlife and environmental health into a One Health approach. *Rev. Sci. Tech.* **2019**, *38*, 91–102. [CrossRef] [PubMed]

25. Magnell, O.; Carter, R. The Chronology of tooth development in wild boars—A guide to age determination of linear ennamel hypoplasia in prehistoric and medival pigs. *Vet. Med. Zoot* **2007**, *40*, 43–48.

26. Animal and Plant Health Inspection Service; U.S. Department of Agriculture. Available online: https://www.aphis.usda.gov/wildlife_damage/feral_swine/pdfs/tech-note-aging-feral-swine.pdf (accessed on 23 April 2020).

27. Altschul, S.F.; Gish, W.; Miller, W.; Myers, E.W.; Lipman, D.J. Basic local alignment search tool. *J. Mol. Biol.* **1990**, *215*, 403–410. [CrossRef]

28. Benson, D.A.; Karsch-Mizrachi, I.; Lipman, D.J.; Ostell, J.; Rapp, B.A.; Wheeler, D.L. GenBank. *Nucleic Acids Res.* **2002**, *30*, 17–20. [CrossRef]

29. Flahou, B.; Van Deun, K.; Pasmans, F.; Smet, A.; Volf, J.; Rychlik, I.; Ducatelle, R.; Haesebrouck, F. The local immune response of mice after Helicobacter suis infection: Strain differences and distinction with Helicobacter pylori. *Veter Res.* **2012**, *43*, 75. [CrossRef]

30. De Witte, C.; Demeyere, K.; De Bruyckere, S.; Taminiau, B.; Daube, G.; Ducatelle, R.; Meyer, E.; Haesebrouck, F. Characterization of the non-glandular gastric region microbiota in Helicobacter suis-infected versus non-infected pigs identifies a potential role for Fusobacterium gastrosuis in gastric ulceration. *Veter Res.* **2019**, *50*, 39. [CrossRef]

31. Fernández-Aguilar, X.; Gottschalk, M.; Aragon, V.; Càmara, J.; Ardanuy, C.; Velarde, R.; Galofré-Milà, N.; Castillo-Contreras, R.; López-Olvera, J.R.; Mentaberre, G.; et al. Urban Wild Boars and Risk for ZoonoticStreptococcus suis, Spain. *Emerg. Infect. Dis.* **2018**, *24*, 1083–1086. [CrossRef]

Communication

A Freedom of *Coxiella burnetii* Infection Survey in European Bison (*Bison bonasus*) in Poland

Michał K. Krzysiak [1,2,*], Martyna Puchalska [3], Wanda Olech [4] and Krzysztof Anusz [3]

1   Białowieża National Park, Park Pałacowy 11, 17-230 Białowieża, Poland
2   Institute of Forest Sciences, Faculty of Civil Engineering and Environmental Sciences,
     Białystok University of Technology, Wiejska 45 E, 15-351 Białystok, Poland
3   Department of Food Hygiene and Public Health Protection, Institute of Veterinary Medicine,
     Warsaw University of Life Sciences—SGGW, Nowoursynowska 159, 02-776 Warsaw, Poland;
     martyna_puchalska@sggw.edu.pl (M.P.); krzysztof_anusz@sggw.edu.pl (K.A.)
4   Department of Animal Genetics and Conservation, Institute of Animal Science,
     Warsaw University of Life Sciences—SGGW, Ciszewskiego 8, 02-786 Warsaw, Poland;
     wanda_olech@sggw.edu.pl
*   Correspondence: michal.krzysiak@bpn.com.pl

Simple Summary: Q fever is one of the important diseases transmissible from animals to humans. The source of infection can be numerous species of animals including mammals, birds, reptiles, amphibians as well as ticks. The role of wildlife in its epidemiology is poorly understood. Therefore, we examined 523 sera samples obtained from European bison for the presence of specific antibodies to assess whether infection occurs in this species and whether European bison may be an important source of infection in the natural environment as suggested by historical reports. The antibodies were found only in one free-living bull, while two other samples were doubtful. The results suggest the transmission of infection to the European bison was rather accidental and its role as an important source of infection nowadays is unlikely.

Citation: Krzysiak, M.K.; Puchalska, M.; Olech, W.; Anusz, K. A Freedom of *Coxiella burnetii* Infection Survey in European Bison (*Bison bonasus*) in Poland. *Animals* **2021**, *11*, 651. https://doi.org/10.3390/ani11030651

Academic Editor: David González-Barrio

Received: 1 February 2021
Accepted: 24 February 2021
Published: 1 March 2021

Abstract: Q fever is an important zoonosis caused by the intracellular Gram-negative bacteria *Coxiella burnetii*. The source of infection are numerous species of mammals, birds, reptiles and amphibians, as well as ticks. The disease is widespread throughout Europe, but the role of wildlife in its epidemiology is poorly understood. The European bison (*Bison bonasus*) population has been growing European-wide quite dynamically over the last few years. The aim of this study was to determine whether *C. burnetii* infection occurs in European bison and whether it can be considered an important bacterial reservoir in the natural environment. Five hundred and twenty three samples of European bison sera originating from 14 (out of the 26 existing) Polish populations were examined for the presence of specific antibodies using an ID Screen Q Fever Indirect Multispecies ELISA test. Only one (0.19%) serum sample was positive in ELISA, and two other samples were doubtful. The only seropositive animal found in this study was a free-living bull. It suggests possible transmission from domestic cattle by sharing pastures. The transmission of *C. burnetii* into the European bison was rather accidental in the country and its role as an important wild reservoir is unlikely. Since no tests are available for wildlife ruminants there is a need for the adaptation of the available tests.

Keywords: *Coxiella burnetii*; Q fever; serology; epidemiology; wildlife; European bison

Publisher's Note: MDPI stays neutral with regard to jurisdictional claims in published maps and institutional affiliations.

1. Introduction

Q fever is a widely distributed reproductive disease of ruminants caused by intracellular bacteria *Coxiella burnetii*. The pathogen is commonly transmitted to humans and remains an emerging concern for public health. The source of infection are numerous species of mammals, birds, reptiles and amphibians. Ticks are important vectors of

C. burnetii transmission; therefore, the climate changes observed in the recent decades may provoke the shift of those arthropods into new locations, increasing the risk of tick-borne infections to occur [1]. In humans, infection as a result of a tick bite is rare, while potential sources of infection may be dusty feces of infected ticks, or exposure to dust with dried secretions of the reproductive system spread by the wind up to a distance of 2 km. Free-living animals may be a reservoir and source of *C. burnetii* infection for both humans and domestic animals. Q fever is widely spread all over Europe, however the role of wildlife in its transmission is still poorly understood. Dairy cattle are considered the main reservoir in Poland with the animal-level and herd-level seroprevalence around 25% [2]. *C. burnetii* DNA was detected in 3% of wildlife tested in a survey in the north-west of the country, which included roe deer, red deer and wild boar [3]. In the same study, the presence of bacteria was also confirmed in the ticks *Ixodes ricinus* from the same area. Other reports also confirmed the infection in wild ruminants, rabbits, rodents, foxes and ticks associated with their sylvatic ecosystems [4,5]. Moreover, specific antibodies were found in wild ranging cervids and mouflon [6,7].

European bison are the largest herbivores in Europe. Their conservation strategies should also include the monitoring of zoonotic diseases as an element of the One Health concept. However, Q fever studies have been neglected for the last two decades. The first report on the occurrence of *C. burnetii* in European bison dates back to the nineteen-eighties, when the exposure of forty-seven free-ranging European bison was investigated from Borecka forest (54°5′18.09″ N 21°55′21.467″ E) in connection to a Q fever epidemic in local cattle herds. The presence of specific antibodies was then confirmed in 76% of animals by the microagglutination test and complement fixation test (CFT) [8]. Furthermore, earlier in 1980–1983, a natural foci of Q fever could have been located in the Białowieska forest (52°42′9.861″ N 23°51′4.52″ E) [9], where the conservation of the endangered species after their complete extinction from the wild was initiated. Additionally, transmission to humans was suspected as 10% of European bison caretakers from Białowieża National Park (52°42′9.861″ N 23°51′4.52″ E) became infected [8]. Further studies excluded exposure of European bison to *C. burnetii* by serological testing in 122 European bison from Białowieska forest in years 1991–2001 [10,11]. The population of the species, which remains listed by the International Union for Conservation (IUCN) of Nature's Red List of Threatened Species increases successfully thanks to many national and international projects, reaching already \ total of over 8400 individuals. They are most often reared in the enclosures but wild European bison population with frequent contact with people and domestic animals significantly increases [12]. In 2019, 74% of European bison were free-living [12].

Taking into account abovementioned, the aim of this study was to determine whether *C. burnetii* infection occurs in the European bison and whether it can be considered an important bacterial reservoir in the natural environment.

2. Materials and Methods

2.1. Study Design

To calculate the representative number of European bison in each of the populations to substantiate a prevalence of 1.5% or below with the overall α error set to 0.05, the FreeCalc software of Epitools [13,14] was used. The mean population size was determined by European bison Pedigree Books (2013–2018). The targeted population size number increased from 1196 animals in 2013 to 1478 in 2018 with the mean value of 1377 individuals [12]. Considering the design prevalence at 1%, test sensitivity and specificity of about 100% and desired type I and II errors at 0.05, the targeted sample size for the whole population tested was 267. However, for the smaller herds, all the animals or all the available samples were included in the study.

The samples were collected during the monitoring of European bison health in the years 2011–2017 as a part of the scientific cooperation of National Veterinary Research Institute in Puławy (NVRI) with herd/population managers based on: the opinion of the Minister of the Environment of 27 October 2014 (ZOP/06-061/51/2014); the Decision of the General Director for Environmental Protection of 31 December 2014 (DZP-WG.6401.06.23.2014.km.2); individual permits of the Director of the Białowieża National Park of 20 December 2013 (PN/061/22/2013) and 15 May 2017 (PN/061/14/2017) and permits of the Kobiór Forest District (ZG-7326 (1) -9/2014). Since 2017, the sample collection has been carried out as part of the project "Complex project of European bison conservation by State Forests" financed by the Forest Fund (Poland) in accordance with contract No. OR.271.3.10.2017.

A total of 523 serum samples were collected from European bison from 14 (out of 26 existing) different populations spread across the country including wild-ranging European bison from Białowieska (including Białowieża National Park) (n = 170), Borecka (n = 37) and Knyszyńska (n = 52) forests and several other herds kept in captivity between 2013 and 2018. The distribution and numbers of the studied European bison are presented in Figure 1. The samples originated both from female (n = 286) and male (n = 219) European bison aged between 5 days and 29 years. Samples were taken from pharmacologically immobilized (for placing collars with telemetric transmitters or diagnostic reasons), fallen or euthanized due to poor health individuals in accordance with the corresponding decisions of the Minister of the Environment and the General Director for Environmental Protection. The blood was collected immediately after immobilization or culling through the puncture of the external jugular vein (*vena iugularis externa*), less often from the tail vein (*vena caudalis mediana*). Blood from dead, necropsied animals was collected in the form of a clot from the heart or from body cavities. The decayed or extensively hemolyzed samples were excluded from the study to minimize the risk of false result. Blood was collected into the sterile 7–9 mL tubes, centrifuged within 24 h, and the obtained serum was frozen at $-70\,^{\circ}$C until analysis in the sample bank of the Department of Virology, NVRI, Poland.

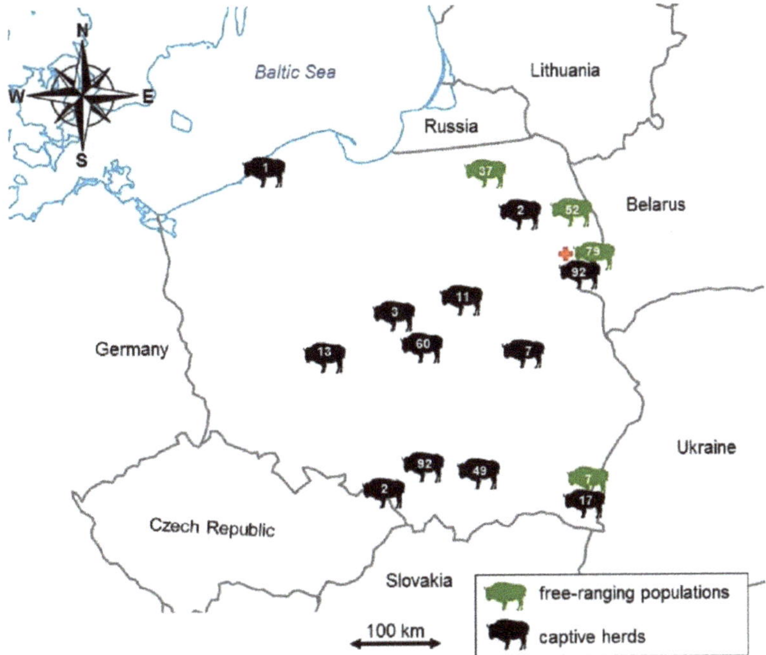

Figure 1. The map of Poland showing the numbers of European bison divided into free-ranging populations (green) and captive herds (black). The red cross indicates the location of single seropositive, 6-year-old male from wild population of Białowieska Forest.

2.2. ELISA (Enzyme-Linked Immunosorbent Assay)

Serum samples were tested for presence of antibodies to *Coxiella burnetii* using commercial ELISA test ID Screen Q Fever Indirect Multi-species (IDvet, Grabels, France). This kit is based on a mix of phase I and II antigens obtained after the purification and inactivation of a *C. burnetii* strain isolated from the placenta of an aborting cow. The test was performed according to the manufacturer's instructions. Briefly, tested and control sera (negative control—NC and positive control—PC) were diluted 1:50 in Dilution Buffer 2 at the dilution plate and then, 100 μL of the diluted sera were transferred to the test plate. The plate was incubated for 45 ± 4 min at 21 °C (± 5 °C) and washed 3 times with the washing solution. A conjugate working solution was prepared by diluting the concentrated Conjugate ($10\times$) in Dilution Buffer 3 in a ratio of 1:10. An amount of 100 μL of conjugate working solution was added to all wells. The plates were incubated for 30 ± 3 min at 21 °C (± 5 °C) and then washed 3 times with the washing fluid solution. An amount of 100 μL of Substrate solution was added to all wells and the plates were incubated in a dark place for 15 ± 2 min at 21 °C (± 5 °C). The reaction was stopped by adding 100 μL Stop Solution to each well. The optical densities (OD) were read at 450 nm using an ELISA plate reader (Epoch, BioTek, Winooski, VT). The results were interpreted by calculation of the Sample to Positive percentage (S/P%) as $S/P\% = \frac{(ODsample - ODNC)}{(ODPC - ODNC)} \times 100$ The result was considered reliable if: the average OD_{PC} value is greater than 0.350 and the ratio of the average OD_{PC} value and the average OD_{NC} value is greater than 3. If S/P% $\leq$ 40%, the result was considered negative; if 40% < S/P% $\leq$ 50%, the result was considered doubtful; if S/P% > 50%, the result was considered positive.

2.3. Statistical Analysis

For statistical evaluation, STATA software version 11 (StataCorp., College Station, TX, USA) was used.

The percentage of *C. burnetii* seropositive animals and 95% confidence interval (CI) values were calculated using binomial exact test. The S/P values were analyzed by kernel estimated frequency plot. The univariate associations between the seropositivities to *C. burnetii*, environmental (origin, population type, health status) and individual-level (sex, age) variables were estimated using Fisher's exact test.

3. Results

Only one (0.19%; 95% CI: 0.005–1.1) serum sample reacted positively in the ELISA. It derived form 6-year-old European bison bull (No 1983, born in 2007) culled due to poor body condition and severe balanoposititis from the free-living population in January 2013. Furthermore, two other samples were doubtful. These included also free-living males: a 3-month-old calf (No 2250; born in 2012) and 6-year-old bull (No 1984; born in 2007), which were also eliminated by culling from Białowieska forest in 2013. The results in relation to different variables are presented at Table 1.

Table 1. Descriptive statistics of seropositive to *Coxiella burnetii* European bison (*Bison bonasus*) with regard to their different characteristics.

	Number Positive/Examined	% (95% Confidence Interval)
Location (Global Positioning System) (*N* * = 523)		
Bałtów (51°1′3.759″ N 21° 32′30.098″ E)	0/7	0 (0–41.0)
Białowieska forest (52°42′9.861″ N 23°51′4.52″ E)	1/171 **	0.58 (0.1–3.2)
Bieszczady mountains (49°7′12.898″ N 22°45′0.782″ E)	0/24	0 (0–14.2)
Borecka forest (54°5′18.09″ N 21°55′21.467″ E)	0/37	0 (0–9.5)
Gołuchów (51°50′58.047″ N 17°55′50.863″ E)	0/12	0 (0–26.5)
Kiermusy (53°12′7.699″ N 22°42′44.245″ E)	0/2	0 (0–84.2)
Knyszyńska forest (53°15′35.036″ N 23°38′37.11″ E)	0/52	0 (0–6.8)
Niepołomice (50°1′45.67″ N 20°20′46.365″ E)	0/49	0 (0–7.2)
Pszczyna (49°58′29.284″ N 18°55′52.465″ E)	0/92	0 (0–3.9)
Smardzewice (51°28′39.975″ N 20°3′0.964″ E)	0/60	0 (0–6.0)
Strzelinko (54°31′45.236″ N 16°56′58.023″ E)	0/1	0 (0–97.5)
Ustroń (49°42′58.955″ N 18°49′59.765″ E)	0/2	0 (0–84.2)
ZOO Łódź (51°45′39.629″ N 19°24′45.333″ E)	0/3	0 (0–70.8)
ZOO Warsaw (52°15′28.9″ N 21°1′21.035″ E)	0/11	0 (0–28.5)
Population type (*N* = 523)		
free-living	1/179 **	0.56 (0.1–3.1)
captive	0/344	0 (0–1.1)
Gender (*N* = 505)		
female	0/281	0 (0–1.3)
male	1/224 **	0.44 (0.01–2.4)
Age group (*N* = 474)		
≤1 year old	0/98 ***	0 (0–3.7)
2–3 years old	0/114	0 (0 = 3.2)
≥4 years old	1/262 ***	0.38 (0.1–2.1)
Health status (*N* = 501)		
immobilized (apparently healthy)	0/348	0 (0–1.1)
eliminated (by culling)	1/134 **	0.74 (0.02–4.1)
fallen	0/15	0 (0–21.8)
traffic accident	0/4	0 (0–60.2)

* number of examined in the category (taking into account the missing data); ** two and *** one additional doubtful result was detected.

The distribution of S/P% with a single peak skewed towards negative values may be observed in Figure 2.

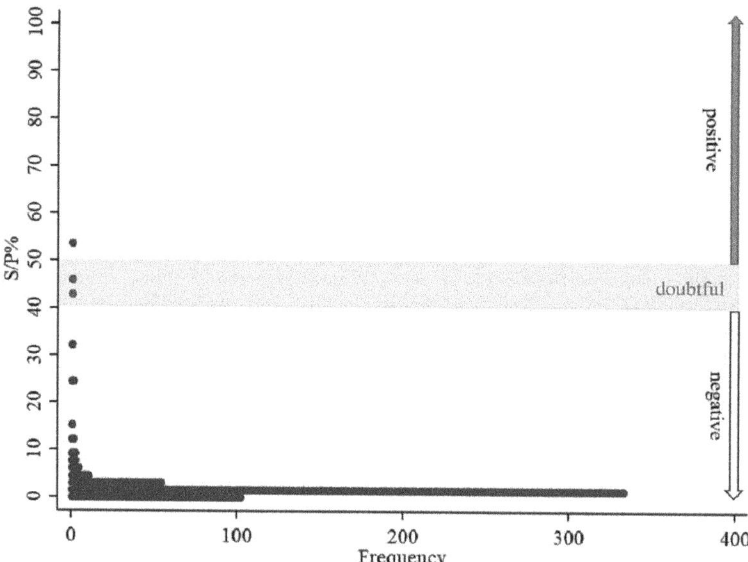

Figure 2. Dotblot distribution of Sample to Positive percentage (S/P%) values of ID Screen Q Fever Indirect Multi-species (IDvet, Grabels, France) testing for the presence of *Coxiella burnetii* antibodies in 523 sera of European bison (*Bison bonasus*) collected between 2013 and 2018 in 14 different locations in Poland.

4. Discussion

Nowadays, the increasing European bison population size and density, decreasing wild habitat and introduction strategies induces more frequent contact with other wild reservoirs and domestic animals at the shrinking interface. The population size of European bison in Poland and in Europe has tripled in the last twenty years [12]. Moreover, already three quarters of the animals are free ranging including in several new locations where the species was introduced or re-introduced after the extinction in Europe [12]. The role of the species as a reservoir of the zoonotic bacteria at present was therefore investigated using representative numbers of animals in the largest Polish populations as well as in the smaller, captive herds.

In our study, the only seropositive animal was a free-living bull. In 1980–1983 the studies of European bison culled at that time revealed antibodies to *C. burnetii* in a 2-year-old heifer. It was thought to be associated with the presence of a natural foci of Q fever in the Białowieska forest [9]. The outbreak was probably associated with the occurrence of Q fever epizootic outbreaks in cattle and sheep in neighboring villages [9]. However, despite continuous occurrence of Q fever in the domestic ruminants in 1980–1983, no transmission to European bison was observed. European bison males usually wander solitarily or in small male groups often approaching farmland and villages, where they can become into contact with domestic animals and humans. Meanwhile, females with calves and juveniles create so-called mixed groups, which rarely leave the forest, except for in winter, when they may come out onto fields or winter-feeding places in search for food. Therefore, it can be assumed that the exposure of *C. burnetii* in free-living bull in our study may have been associated with transmission of pathogen from domestic ruminants by sharing pastures. In the eighties, due to the occurrence of Q fever in domestic animals in north-eastern Poland, forty-seven free-ranging European bison from Borecka forest were examined for the presence of specific antibodies by the microagglutination test and complement fixation test (CFT). The high seroprevalence (76%) of *Coxiella burnetii* found in European bison suggested that they could be considered as a potential reservoir of Q fever [8]. Szarek et al. [15] have linked *C. burnetii* infection to the pathomorphological changes of heart and kidneys specific for Q fever observed in those studied European bison. Moreover, European bison–human transmission was suspected, since the infection was also confirmed in 10% of the employees of the Białowieża National Park (BNP) staff. However, further studies suggested that the Q fever foci were self-limited in the following years [10,11]. The results of our studies follow this trend.

The results of our studies, however, suggest that the exposure of European bison to *C. burnetii* is rather accidental if even occurring, which may be surprising, as the infections are quite common in cattle in the country [2]. Assuming that the positive result in only one free-living bull was not a false positive reaction, a transmission from cattle would be suspected. The seroprevalence of *C. burnetii* in cattle in the Podlaskie province, where two major European bison free-living populations of Białowieska and Knyszyńska forests are living reaches 48% [2]. The density of cattle in the region is also the highest in the country with the intensive milk production and increasing pasture grazing [16]. These findings suggest low susceptibility of European bison to *C. burnetii* infection despite the potential increasing risk with increasing contacts with domestic Q fever reservoirs in endemic areas.

The few reports on Q fever occurrence in wild ruminants in Europe suggest low seroprevalence of *C. burnetii*, and therefore their role in the transmission and maintaining the bacteria in the natural environment is rather doubtful. Low seroprevalences of *C. burnetii* were reported previously in wild ranging cervids and mouflon [6,7]. Higher seroprevalences were found in farmed deer, which may suggest increased exposure to *C. burnetii* connected to higher density of animals [17]. However, the factors modulating the risk of exposure to *C. burnetii* in wildlife are quite complex and should be evaluated individually [18]. Our study together with previous concerning *C. burnetii* infections in European bison [9,11] also suggest that they may be accidental hosts for the pathogen. Nevertheless, since *C. burnetii* may cause large reproductive loses to domestic ungulates, we should be cautious when it comes to endangered species as European bison and monitor the epidemic situation.

According to OIE [World Organisation for Animal Health; 19], at present, no gold standard technique is available in diagnosis of Q fever. Serological ELISA and direct detection and quantification by PCR should be considered as the methods of choice. Serological analyses may be carried out using ELISA, a complement fixation test (CFT) or indirect immunofluorescence assay (IFA). Among serological methods, ELISA is recommended for routine serological testing of animals for Q fever because it has a high sensitivity and a good specificity, and it is convenient for large-scale screening. The relative sensitivity is the lowest for CFT [19]. An antibody ELISA was used in our research. Some authors question the high test performance in other than small ruminants [20]; however, the assay was successfully used in Q fever epidemiological studies in other species [21]. In general, ELISA also shows improved sensitivity and specificity in relation to the complement fixation test [22,23], despite its performance possibly varying between different manufacturers, different antigens used for plate coating and the species to which it is applied [24]. The report on Q fever cases in European bison in the 1980s was based on CFT [8]; therefore, it could explain why further studies, including present, have differed on *C. burnetii* exposure in the species, despite potentially increasing risk. Studies on detecting antibodies against *C. burnetii* demonstrate only previous exposure to the pathogen, not current shedding of bacteria but they can be useful for epidemiological analysis of endemic areas [2,25]. Moreover, seropositivity without detection of pathogen in the sera may be indicative of chronic exposures to *C. burnetii* [26]. Moreover, some shedders of *C. burnetii* may be seronegative [27]. Therefore, some discrepancies in the results of serological versus PCR tests carried out on the same animals may be observed. Knap et al. [25] have detected the presence of specific antibodies in the sera of 60 out of 150 (40.4%) animals (cattle, sheep), while the presence of pathogen DNA was confirmed only in 14 out of 150 (9.3%) blood samples. Furthermore, Bellabidi et al. [21] found antibodies to *C. burnetii* in 75.54% of tested samples of camel sera and no DNA of the pathogen in sera of 138 seropositive animals. It is noteworthy that there are differences in the shedding of bacteria in different animal species. Cattle shed the bacteria almost exclusively in milk, goats mostly in milk with a minority shedding it in vaginal mucous or feces and sheep in feces, vaginal mucous and milk [27].

Since no tests dedicated to specific species, except for cattle, are recommended, and no tests are available for wildlife ruminants, the adaptation and optimization of the available test should be considered in further studies [5,18,23,24,28]. Since Q fever may be a potential emerging disease in the areas where contacts between wildlife and humans or domestic animals increase, the need for a reliable test for monitoring of wildlife is growing. Frosinski et al. [29] have also discussed the need for cut-off adaptation, which may be beneficial to obtaining good quality prevalence data. In our study, some samples gave slightly higher S/P% value approaching the cut-off value for doubtful results. However, optimizing cut-off value would require testing a higher number of known *C. burnetii* seropositive samples collected from the species which are not available at the moment (Figure 2). Further studies are needed.

5. Conclusions

The only seropositive European bison found in this study was a free-living bull which suggests possible transmission from domestic cattle by sharing pastures; however, a false positive result may be also suspected. The transmission of *C. burnetii* into the European bison was rather accidental in the country and its role as an important wild reservoir is questionable. Since no tests are available for wildlife ruminants there is a need for the adaptation of the available tests in order to monitor possible *C. burnetii* circulation in the sylvatic environment.

Author Contributions: Conceptualization, M.K.K. and M.P.; formal analysis, M.K.K.; investigation, M.P. and M.K.K.; resources, M.K.K.; writing—original draft preparation, M.K.K. and M.P.; writing— review and editing, K.A., M.K.K. and M.P.; funding acquisition, W.O. and M.K.K. All authors have read and agreed to the published version of the manuscript.

Funding: This work is part of the project "Complex project of European bison conservation by State Forests", which is financed by the Forest Found (Poland), contract number OR.271.3.10.2017.

Institutional Review Board Statement: Not applicable.

Data Availability Statement: Raw data are available upon request from the corresponding author.

Acknowledgments: The authors want to thank Magdalena Larska, National Veterinary Research Institute, Puławy, Poland for providing samples for testing and statistical analysis.

Conflicts of Interest: The authors declare no conflict of interest. The funders had no role in the design of the study; in the collection, analyses, or interpretation of data; in the writing of the manuscript, or in the decision to publish the results.

References

1. Yon, L.; Duff, J.P.; Ågren, E.O.; Erdélyi, K.; Ferroglio, E.; Godfroid, J.; Hars, J.; Hestvik, G.; Horton, D.; Kuiken, T.; et al. Recent Changes in Infectious Diseases in European Wildlife. *J. Wildl. Dis.* **2019**, *55*, 3–43. [CrossRef]
2. Szymańska-Czerwińska, M.; Jodełko, A.; Niemczuk, K. Occurrence of *Coxiella burnetii* in Polish dairy cattle herds based on serological and PCR tests. *Comp. Immunol. Microbiol. Infect. Dis.* **2019**, *67*, 101377. [CrossRef]
3. Bielawska-Drózd, A.; Cieślik, P.; Żakowska, D.; Głowacka, P.; Wlizło-Skowronek, B.; Zięba, P.; Zdun, A. Detection of *Coxiella burnetii* and Francisella tularensis in Tissues of Wild-living Animals and in Ticks of North-west Poland. *Pol. J. Microbiol.* **2018**, *67*, 529–534. [CrossRef] [PubMed]
4. González, J.; González, M.G.; Valcárcel, F.; Sánchez, M.; Martín-Hernández, R.; Tercero, J.M.; Olmeda, A.S. Prevalence of *Coxiella burnetii* (Legionellales: Coxiellaceae) Infection Among Wildlife Species and the Tick *Hyalomma lusitanicum* (Acari: Ixodidae) in a Meso-Mediterranean Ecosystem. *J. Med. Èntomol.* **2019**, *57*, 551–556. [CrossRef] [PubMed]
5. Meredith, A.L.; Cleaveland, S.C.; Denwood, M.J.; Brown, J.K.; Shaw, D.J. *Coxiella burnetii* (Q-Fever) Seroprevalence in Prey and Predators in the United Kingdom: Evaluation of Infection in Wild Rodents, Foxes and Domestic Cats Using a Modified ELISA. *Transbound. Emerg. Dis.* **2014**, *62*, 639–649. [CrossRef]

6. Kazimírová, M.; Hamšíková, Z.; Špitalská, E.; Minichová, L.; Mahríková, L.; Caban, R.; Sprong, H.; Fonville, M.; Schnittger, L.; Kocianová, E. Diverse tick-borne microorganisms identified in free-living ungulates in Slovakia. *Parasites Vectors* **2018**, *11*, 495. [CrossRef] [PubMed]

7. López-Olvera, J.R.; Vidal, D.; Vicente, J.; Pérez, M.; Luján, L.; Gortázar, C. Serological survey of selected infectious diseases in mouflon (*Ovis aries musimon*) from south-central Spain. *Eur. J. Wildl. Res.* **2008**, *55*, 75–79. [CrossRef]

8. Ciecierski, H.; Anusz, K.; Borko, K.; Anusz, Z.; Tsakalidis, S. Occurrence of antibodies against *Coxiella burnetii* in wild animals in the foci of Q fever in 1985–1988. *Med. Weter.* **1988**, *44*, 652–654.

9. Kita, J.; Anusz, K. Serologic Survey for Bovine Pathogens in Free-Ranging European Bison from Poland. *J. Wildl. Dis.* **1991**, *27*, 16–20. [CrossRef]

10. Rypuła, K.; Krasińska, M.; Kita, J.; Płoneczka-Janeczko, K.; Kapuśniak, W. The prevalence of specific antibody to selected viral and bacterial infections in wild ruminants in Poland. *Cent. Eur. J. Immunol.* **2011**, *36*, 180–183.

11. Salwa, A.; Anusz, K.; Arent, Z.; Paprocka, G.; Kita, J. Seroprevalence of selected viral and bacterial pathogens in free-ranging European bison from the Białowieża Primeval Forest (Poland). *Pol. J. Leterinary Sci.* **2007**, *10*, 19–23.

12. Raczyński, J. European Bison Pedigree Books. Białowieża National Park Webpage 2013–2018. Available online: https://bpn.com.pl/index.php?option=com_content&task=view&id=1133&Itemid=213 (accessed on 27 September 2020).

13. Cameron, A.R. *Survey Toolbox for Livestock Diseases—A Practical Manual and Software Package for Active Surveillance of Livestock Diseases in Developing Countries*; Australian Centre for International Agricultural Research: Canberra, Australia, 1999.

14. Cameron, A.R.; Baldock, F. A new probability formula for surveys to substantiate freedom from disease. *Prev. Vet. Med.* **1998**, *34*, 1–17. [CrossRef]

15. Szarek, J.; Rotkiewicz, T.; Anusz, Z.; Khan, M.Z.; Chishti, M.A. Pathomorphological Studies in European Bison (*Bison bonasus* Linnaeus, 1758) with Seropositive Reaction to *Coxiella burnetii*. *J. Vet. Med. Ser. B* **1994**, *41*, 618–624. [CrossRef]

16. Statistical Office in Białystok, Agriculture in Podlaskie Voivodship in 2019. 2020. Available online: https://bialystok.stat.gov.pl/download/gfx/bialystok/en/defaultaktualnosci/693/1/9/1/agriculture_in_podlaskie_voivodship_in_2019.pdf (accessed on 19 January 2021).

17. Ruiz-Fons, F.; Óscar, R.; Torina, A.; Naranjo, V.; Gortázar, C.; De La Fuente, J. Prevalence of *Coxiella burnetti* infection in wild and farmed ungulates. *Vet. Microbiol.* **2008**, *126*, 282–286. [CrossRef] [PubMed]

18. González-Barrio, D.; Ruiz-Fons, F. *Coxiella burnetii* in wild mammals: A systematic review. *Transbound. Emerg. Dis.* **2019**, *66*, 662–671. [CrossRef] [PubMed]

19. OIE. *Manual of Diagnostic Tests and Vaccines for Terrestrial Animals*, 8th ed.; OIE: Paris, France, 2019; pp. 560–577.

20. Burns, R.J.L.; Douangngeun, B.; Theppangna, W.; Khounsy, S.; Mukaka, M.; Selleck, P.W.; Hansson, E.; Wegner, M.D.; Windsor, P.A.; Blacksell, S.D. Serosurveillance of Coxiellosis (Q-fever) and Brucellosis in goats in selected provinces of Lao People's Democratic Republic. *PLOS Neglected Trop. Dis.* **2018**, *12*, e0006411. [CrossRef] [PubMed]

21. Bellabidi, M.; Benaissa, M.H.; Bissati-Bouafia, S.; Harrat, Z.; Brahmi, K.; Kernif, T. *Coxiella burnetii* in camels (*Camelus dromedarius*) from Algeria: Seroprevalence, molecular characterization, and ticks (Acari: Ixodidae) vectors. *Acta Trop.* **2020**, *206*, 105443. [CrossRef]

22. Jóźwik, A.; Jakubowski, T.; Kaba, J.; Jurkowski, W.; Witkowski, L.; Nowicki, M.; Frymus, T. Evaluation of the agreement of ELISA and complement fixation test in the diagnostics of Q fever in cattle. *Med. Weter.* **2007**, *63*, 655–657.

23. Szymańska-Czerwińska, M.; Niemczuk, K.; Jodełko, A. Evaluation of qPCR and phase I and II antibodies for detection of *Coxiella burnetii* infection in cattle. *Res. Vet. Sci.* **2016**, *108*, 68–70. [CrossRef]

24. Horigan, M.W.; Bell, M.M.; Pollard, T.R.; Sayers, A.R.; Pritchard, G.C. Q fever diagnosis in domestic ruminants. *J. Vet. Diagn. Investig.* **2011**, *23*, 924–931. [CrossRef] [PubMed]

25. Knap, N.; Žele, D.; Biškup, U.G.; Avšič-Županc, T.; Vengušt, G. The prevalence of *Coxiella burnetii* in ticks and animals in Slovenia. *BMC Vet. Res.* **2019**, *15*, 1–6. [CrossRef] [PubMed]

26. Dhaka, P.; Malik, S.S.; Yadav, J.P.; Kumar, M.; Baranwal, A.; Barbuddhe, S.B.; Rawool, D.B. Seroprevalence and molecular detection of coxiellosis among cattle and their human contacts in an organized dairy farm. *J. Infect. Public Health* **2019**, *12*, 190–194. [CrossRef] [PubMed]

27. Barlozzari, G.; Sala, M.; Iacoponi, F.; Volpi, C.; Polinori, N.; Rombolà, P.; Vairo, F.; Macrì, G.; Scarpulla, M. Cross-sectional serosurvey of *Coxiella burnetii* in healthy cattle and sheep from extensive grazing system in central Italy. *Epidemiol. Infect.* **2020**, *148*, e9. [CrossRef] [PubMed]

28. Wood, C.; Muleme, M.; Tan, T.; Bosward, K.; Gibson, J.; Alawneh, J.; McGowan, M.; Barnes, T.S.; Stenos, J.; Perkins, N.; et al. Validation of an indirect immunofluorescence assay (IFA) for the detection of IgG antibodies against *Coxiella burnetii* in bovine serum. *Prev. Vet. Med.* **2019**, *169*, 104698. [CrossRef] [PubMed]

29. Frosinski, J.; Hermann, B.P.; Maier, K.; Boden, K. Enzyme-linked immunosorbent assays in seroprevalence studies of Q fever: The need for cut-off adaptation and the consequences for prevalence data. *Epidemiol. Infect.* **2016**, *144*, 1148–1152. [CrossRef] [PubMed]

Article

Investigating the Role of Micromammals in the Ecology of *Coxiella burnetii* in Spain

David González-Barrio [1,2,3,*], Isabel Jado [4], Javier Viñuela [1], Jesús T. García [1], Pedro P. Olea [5,6], Fernando Arce [7] and Francisco Ruiz-Fons [1,*]

1   Instituto de Investigación en Recursos Cinegéticos IREC (CSIC-UCLM-JCCM), Ronda de Toledo 12, 13071 Ciudad Real, Spain; Javier.Vinuela@uclm.es (J.V.); jesusgarcia.irec@gmail.com (J.T.G.)
2   Parasitology Reference and Research Laboratory, Spanish National Centre for Microbiology, Health Institute Carlos III, Ctra. Majadahonda-Pozuelo Km 2, Majadahonda, 28220 Madrid, Spain
3   Viral Hepatitis Reference and Research Laboratory, Spanish National Centre for Microbiology, Health Institute Carlos III, Ctra. Majadahonda-Pozuelo Km 2, Majadahonda, 28220 Madrid, Spain
4   Special Pathogens Reference and Research Laboratory, Spanish National Centre for Microbiology, Health Institute Carlos III, Ctra. Majadahonda-Pozuelo Km 2, Majadahonda, 28220 Madrid, Spain; ijado@isciii.es
5   Departamento de Ecología, Universidad Autónoma de Madrid (UAM), 28049 Madrid, Spain; pedrop.olea@uam.es
6   Centro de Investigación en Biodiversidad y Cambio Global (CIBC-UAM), Universidad Autónoma de Madrid, 28049 Madrid, Spain
7   School of Natural Sciences, University of Tasmania, Hobart, TAS 7004, Australia; fernando.arcegonzalez@utas.edu.au
*   Correspondence: dgonzalezbarrio@gmail.com (D.G.-B.); josefrancisco.ruiz@uclm.es (F.R.-F.)

Citation: González-Barrio, D.; Jado, I.; Viñuela, J.; García, J.T.; Olea, P.P.; Arce, F.; Ruiz-Fons, F. Investigating the Role of Micromammals in the Ecology of *Coxiella burnetii* in Spain. *Animals* **2021**, *11*, 654. https://doi.org/10.3390/ani11030654

Academic Editors: Julie Arsenault and Marsilio Fulvio

Received: 27 January 2021
Accepted: 23 February 2021
Published: 2 March 2021

Publisher's Note: MDPI stays neutral with regard to jurisdictional claims in published maps and institutional affiliations.

Simple Summary: *Coxiella burnetii*, the causal agent of human Q fever and animal Coxiellosis, is a zoonotic infectious bacterium with a complex ecology that replicates in multiple host species. However, the role of wildlife in its transmission is poorly understood. We examined 816 spleen samples obtained from ten species of micromammals and 130 vaginal swabs from *Microtus arvalis* females to detect the presence of *C. burnetii* DNA by qPCR. Our aim was assessing whether infection occurs in micromammals in Spain and what species could be relevant hosts in pathogen maintenance. The 9.7% of the spleen samples were qPCR positive. The infection prevalence level was highest (10.8%) in *Microtus arvalis* and also one vaginal swab was PCR positive. Positive samples were also found in *Apodemus sylvaticus* (8.7%), *Crocidura russula* (7.7%), and *Rattus rattus* (6.4%). A genotype II+ strain was identified in one of the positive samples from *M. arvalis*. The results of the study are consistent with previous findings suggesting susceptibility of micromammals to *C. burnetii* infection. We also provide further support to consider micromammals when tracing the origin of human Q fever cases in Europe as one of the authors probably got infected while handling *M. arvalis*.

Abstract: *Coxiella burnetii*, the causal agent of human Q fever and animal Coxiellosis, is a zoonotic infectious bacterium with a complex ecology that results from its ability to replicate in multiple (in)vertebrate host species. Spain notifies the highest number of Q fever cases to the ECDC annually and wildlife plays a relevant role in *C. burnetii* ecology in the country. However, the whole picture of *C. burnetii* hosts is incomplete, so this study seeks to better understand the role of micromammals in *C. burnetii* ecology in the country. Spleen samples from 816 micromammals of 10 species and 130 vaginal swabs from *Microtus arvalis* were analysed by qPCR to detect *C. burnetii* infection and shedding, respectively. The 9.7% of the spleen samples were qPCR positive. The highest infection prevalence (10.8%) was found in *Microtus arvalis*, in which *C. burnetii* DNA was also detected in 1 of the 130 vaginal swabs (0.8%) analysed. Positive samples were also found in *Apodemus sylvaticus* (8.7%), *Crocidura russula* (7.7%) and *Rattus rattus* (6.4%). Positive samples were genotyped by coupling PCR with reverse line blotting and a genotype II+ strain was identified for the first time in one of the positive samples from *M. arvalis*, whereas only partial results could be obtained for the rest of the samples. Acute Q fever was diagnosed in one of the researchers that participated in the study, and it was presumably linked to *M. arvalis* handling. The results of the study are consistent with previous findings suggesting that micromammals can be infected by *C. burnetii*. Our findings additionally suggest that micromammals may be potential sources to trace back the origin of human Q fever and animal Coxiellosis cases in Europe.

Keywords: micromammals; *Coxiella burnetii*; Q fever; zoonosis

1. Introduction

Coxiella burnetii is a multi-host bacterium that causes Q fever in humans, a zoonosis that is emerging worldwide [1]. In humans, Q fever is associated with a multiple clinical spectrum, from asymptomatic to fatal disease. A low percentage of acute cases, especially patients with previous valvulopathy and, to a lesser extent, immunocompromised persons and pregnant women, develop chronic disease that may present with endocarditis, vascular alterations, chronic hepatitis, chronic pulmonary infections, or the so-called post-Q fever fatigue syndrome [2].

It is assumed that domestic ruminants are the main reservoir of *C. burnetii* for humans. Nonetheless, the origin of several human Q fever cases remains unclarified [3] and human–wildlife interaction has been suggested as a risk factor for human infection with *C. burnetii* [4]. The current changes in the patterns of wildlife–human interactions caused by variations in human and wildlife population dynamics and behaviour imply an increased risk of *C. burnetii* inter-species transmission [4]. The ecology of *C. burnetii* in wildlife is still poorly understood and the influence of host, environmental and pathogen factors is almost unknown [4]. *C. burnetii* infection has been neglected in wildlife despite the evidence of particular wild species behaving as true *C. burnetii* reservoirs [4]. Indeed, scientific publications focused on *C. burnetii* in livestock outnumber those in wildlife tenfold. The circulation of *C. burnetii* in wild vertebrates in the sylvatic cycle may perhaps be also enhanced by tick-borne transmission [5]. Wild terrestrial small mammals such as rodents are thought to constitute maintenance hosts of infection in the domestic cycle of *C. burnetii* [6–9].

Within Europe, Spain has reported the highest number of human Q fever cases annually since 2016 (Q fever is of mandatory notification in Spain since 2015). In 2018, Spain accounted for more than a third of the overall number of cases with 418 reported notifications [10]. Livestock (mainly cattle, sheep and goats) is an important reservoir of *C. burnetii* for humans in Spain [11]. However, the geographical location of the country (between Mediterranean and Atlantic oceanic climates) and its orography account for a wide diversity of habitats and biotopes that make Spain a European biodiversity hotspot. As *C. burnetii* is a multi-host pathogen by evolution, the implication of wild reservoirs in its life cycle was expected and already reported in different studies [12–14]. The role of wild micromammals such as rodents and insectivores in *C. burnetii* ecology is currently poorly known. A recent review of studies performed in wild mammals suggests that several micromammal species worldwide may be relevant hosts for *C. burnetii* [4]. It would be expected that the large diversity of micromammals in Spain would add up with joint effects favouring the proliferation of *C. burnetii* because of the higher host availability. Furthermore, the current expansion of some 'pest' rodent species (e.g., *Microtus arvalis*) would have an impact in the sylvatic cycle of the bacterium and therefore in the incidence of human Q fever and animal coxiellosis in the country. A large area in Spain is occupied by farming areas that are in line with the intensification trend of the agriculture in Europe, and it has suffered tremendous transformations in the last 2–3 decades [15]. This transformation is behind the massive spatial expansion and the cyclic population outbreaks of the common vole (*M. arvalis*) that is considered a severe agricultural pest at European level, a matter of intersectoral conflict and a risk for human and animal health [15,16]. Furthermore, the sustained human migration from rural to urban areas over the last four decades in Spain has notably contributed to re-wilding of Iberian forests [17], consequently bearing a re-colonization of lost areas by wildlife, including wild forest micromammals. In both agricultural and natural (forested) landscapes in Spain, the (direct and indirect) interaction with humans, grazing livestock and other wildlife may constitute a risk factor for the exchange of specific strains of *C. burnetii* among different hosts.

According to these premises and to the hypothesis of a relevant implication of wild micromammals in *C. burnetii* ecology in Spain (and beyond), the objectives of this study were to estimate the presence and prevalence of the bacterium in different wild micromammal species and phylogenetically characterize the *C. burnetii* genotypes present in these animals as a first stage to estimate the implication of wild micromammals in the epidemiology of Q fever.

2. Materials and Methods

2.1. Sampling

Between 2003 and 2014, and in the framework of different studies, samples from wild micromammals were collected in 16 locations in mainland Spain using LFATDG Sherman Live Traps (7.62 cm × 8.89 cm × 22.86 cm, H. B. Sherman Traps, Inc., Tallahassee, FL, USA) (Figure 1 and Table 1). Capture and handling procedures for sampling were approved by the UCLM Ethics Committee (reference number CEEA: PR20170201) and were in accordance with the Spanish and European policy for animal protection and experimentation. The researchers and technicians involved in the captures only employed gloves as personal protective equipment. Some of the individuals captured were randomly selected, sedated with an intramuscular injection of a solution containing Ketamin (10 mg/kg) and medetomidine (1 mg/kg) and thereafter humanely euthanised by cervical dislocation. These animals were transported refrigerated to our labs where a detailed necropsy was performed under biosafety 2 containment in cabinets, and tissue samples were collected and preserved frozen at −20 °C. Vaginal swabs (Aluminium + viscose AMIES swabs, Deltalab, Spain) were collected from live female *M. arvalis* captured in northwestern Spain along 2012. The swabs were thereafter preserved frozen at −20 °C until DNA purification. Some species (*Arvicola terrestris*, *Sciurus vulgaris* and *Eliomys quercinus*; Table 2) were surveyed after being found dead close to trap capture sites or by environment agents and brought to the lab for necropsy.

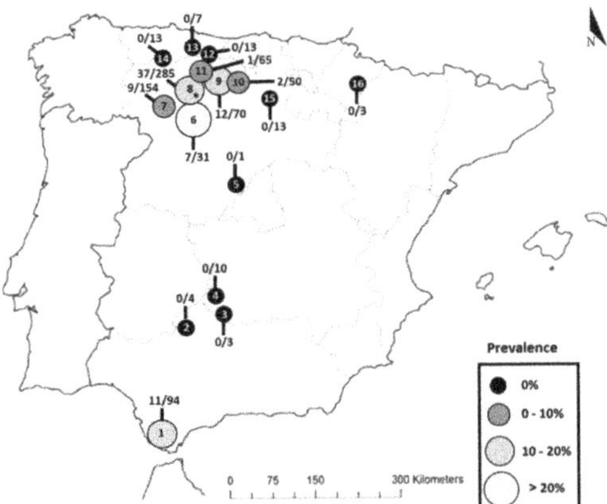

Figure 1. Spatial distribution and prevalence of *Coxiella burnetii* DNA in spleen samples from micromammals. Each dot (overall sample size included) represents a surveyed population of micromammals. The numbers shown per location indicate the number of positive samples with respect to local sample size (positives/total). The size and color of the dots show population prevalence of *C. burnetii* infection as detailed in the legend. The asterisk (*) in a dot indicates that a *C. burnetii* genotype was obtained in this population.

Table 1. Species of micromammals surveyed per study location as shown in Figure 1: *Apodemus flavicollis* (Af), *Apodemus sylvaticus* (Ap), *Arvicola terrestris* (At), *Crocidura russula* (Cr), *Eliomys quercinus* (Eq), *Microtus arvalis* (Ma), *Mus musculus* (Mm), *Mus spretus* (Ms), *Rattus rattus* (Rr), *Sciurus vulgaris* (Sv). The PCR-positive (p) vs. the total number (n) of samples per micromammal species (p/n) and location is shown. In addition, the year(s) and month(s) of sampling, the habitat type, the presence of other co-existing animal species as well as the existence of evidence of previous detection of C. burnetii in non-micromammal species in the location are included. n.a. = data not available.

Location Reference	Micromammal Species Surveyed	Sampling Period	Habitat Type	Co-Habitation with Other Animals	Previous DNA Detection of *C. burnetii* in Other Species
1	As (7/33), Cr (1/8), Eq (0/2), Ms (0/4), Rr (3/47)	2013 (January, June, July, December)	Natural Mediterranean scrubland with large areas of irrigated prairies.	Wildlife	Yes [13,14]
2	As (0/1), Cr (0/2), Ms (0/1)	2003 (April)	Natural Mediterranean scrubland with Savannah-like areas	Wildlife	Yes [13,18]
3	As (0/1), Mm (0/2)	2003 (June)	Natural Mediterranean scrubland	Wildlife	Yes [18]
4	As (0/5), Ms (0/5)	2004 (April)	Natural Mediterranean scrubland	Wildlife	n.a.
5	Sv (0/1)	2008 (December)	Natural Mediterranean scrub with pinelands	Wildlife	Yes [19]
6	As (0/1), Ma (7/30)	2013 (April) 2014 (May)	Agricultural areas	Occasionally sheep	n.a.
7	As (0/18), Ma (9/134), Ms (0/2)	2012 (March–July, October)	Agricultural areas	Occasionally sheep	n.a.
8	As (4/45), Cr (0/5), Ma (33/232), Ms (0/3)	2012 (January–November) 2013 (March–May) 2014 (May)	Agricultural areas	Occasionally sheep	Yes [19]
9	As (1/13), Cr (0/2), Ma (11/51), Ms (0/4)	2011 (November–December) 2012 (January–April)	Agricultural areas	Occasionally sheep	n.a.
10	As (0/4), Cr (0/1), Ma (2/44), Ms (0/1)	2012 (September)	Agricultural areas	Occasionally sheep	n.a.
11	As (0/2), At (0/1), Cr (1/3), Ma (0/59)	2012 (August, September, November)	Agricultural areas	Occasionally sheep	n.a.
12	As (0/1), Ma (0/12)	2012 (October)	Agricultural areas	Wildlife and extensive cattle breeding	n.a.
13	Af (0/2), As (0/4), Mm (0/1),	2003 (July) 2013 (July)	Atlantic forest interspersed with scrublands and prairies	Wildlife and extensive cattle breeding	Yes [19]
14	Cr (0/3), Ma (0/10)	2012 (August)	Atlantic forest interspersed with scrublands and prairies	Wildlife and extensive cattle breeding	n.a.
15	As (0/10), Cr (0/1), Mm (0/2),	2003 (June)	Atlantic forest interspersed with scrublands and prairies	Wildlife	n.a.
16	Cr (0/1), Ms (0/2)	2012 (December)	Steppe and Mediterranean vegetation ("Bardenas Reales")	Wildlife	Yes [13]

Table 2. qPCR results by micromammal species object of survey in this study for each location. Species are displayed along with sample size (*n*) and the number of qPCR positive samples (PCR positives). In addition, 95% exact confidence intervals are shown within brackets.

Location Reference	Species	*n*	PCR Positives	% PCR Positive
1	*Apodemus sylvaticus*	33	7	21.2 (10.7–37.8)
	Crocidura russula	8	1	12.5 (2.2–47.1)
	Eliomys quercinus	2	0	0.0 (0.0–65.8)
	Mus spretus	4	0	0.0 (0.0–48.9)
	Rattus rattus	47	3	6.38 (2.2–17.2)
2	*Apodemus sylvaticus*	1	0	0.0 (0.0–79.3)
	Crocidura russula	2	0	0.0 (0.0–65.8)
	Mus spretus	1	0	0.0 (0.0–79.3)
3	*Apodemus sylvaticus*	1	0	0.0 (0.0–79.3)
	Mus musculus	2	0	0.0 (0.0–65.8)
4	*Apodemus sylvaticus*	5	0	0.0 (0.0–43.4)
	Mus spretus	5	0	0.0 (0.0–43.4)
5	*Sciurus vulgaris*	1	0	0.0 (0.0–79.3)
6	*Apodemus sylvaticus*	1	0	0.0 (0.0–79.3)
	Microtus arvalis	30	7	23.3 (11.8–40.9)
7	*Apodemus sylvaticus*	18	0	0.0 (0.0–17.6)
	Microtus arvalis	134	9	6.7 (3.6–12.3)
	Mus spretus	2	0	0.0 (0.0–65.8)
8	*Apodemus sylvaticus*	45	4	8.9 (3.5–20.7)
	Crocidura russula	5	0	0.0 (0.0–43.4)
	Microtus arvalis	232	33	14.2 (10.3–19.3)
	Mus spretus	3	0	0.0 (0.0–56.1)
9	*Apodemus sylvaticus*	13	1	7.7 (13.7–33.3)
	Crocidura russula	2	0	0.0 (0.0–65.8)
	Microtus arvalis	51	11	21.6 (12.5–34.6)
	Mus spretus	4	0	0.0 (0.0–48.9)
10	*Apodemus sylvaticus*	4	0	0.0 (0.0–48.9)
	Crocidura russula	1	0	0.0 (0.0–79.3)
	Microtus arvalis	44	2	4.5 (1.3–15.1)
	Mus spretus	1	0	0.0 (0.0–79.3)
11	*Apodemus sylvaticus*	2	0	0.0 (0.0–65.8)
	Arvicola terrestris	1	0	0.0 (0.0–79.3)
	Crocidura russula	3	1	33.3 (6.1–79.2)
	Microtus arvalis	59	0	0.0 (0.0–6.1)
12	*Apodemus sylvaticus*	1	0	0.0 (0.0–79.3)
	Microtus arvalis	12	0	0.0 (0.0–24.2)
13	*Apodemus flavicollis*	2	0	0.0 (0.0–65.8)
	Apodemus sylvaticus	4	0	0.0 (0.0–48.9)
	Mus musculus	1	0	0.0 (0.0–79.3)
14	*Crocidura russula*	3	0	0.0 (0.0–56.1)
	Microtus arvalis	10	0	0.0 (0.0–27.7)
15	*Apodemus sylvaticus*	10	0	0.0 (0.0–27.7)
	Crocidura russula	1	0	0.0 (0.0–79.3)
	Mus musculus	2	0	0.0 (0.0–65.8)
16	*Crocidura russula*	1	0	0.0 (0.0–79.3)
	Mus spretus	2	0	0.0 (0.0–65.8)

2.2. DNA Extraction, Coxiella burnetii DNA Detection and Genotyping

Spleen samples were the target tissue to estimate the occurrence of infection with *C. burnetii* because of the presence of *C. burnetii* DNA in this organ could only be the consequence of a generalized infection. DNA from spleen samples and swabs was extracted by using the DNeasy Blood and Tissue Kit (QIAGEN, Hilden, Germany). Around 25 mg of spleen from each animal was cut into small pieces on a sterile glass plate with a disposable scalpel blade before being disrupted in 180 μL of ATL buffer with a homogenizer (TissueLyser II, QIAGEN, Hilden, Germany). After disruption, samples were incubated at 56 °C for 1 h with 20 μL of Proteinase K. Later on, samples were vortexed for 15 s and, after adding 200 μL of AL buffer, the manufacturer's blood extraction protocol was followed (http://mvz.berkeley.edu/egl/inserts/DNeasy_Blood_&_Tissue_Handbook. pdfaccessedon14October2020). The swabs were incubated at 56 °C for 30 min in 200 μL of AL buffer containing 20 μL of proteinase K in sterile 1.5 mL nuclease-free tubes. Swabs were then vortexed for 15 s and carefully removed after squeezing out the liquid contained in them with a sterile glass rod into the tube. The remaining solution was incubated at 56 °C for 30 min. The manufacturer's blood extraction protocol was thereafter used. The DNA concentration in aliquots was quantified (NanoDrop 2000c/2000 spectrophotometer; Thermo Scientific, Waltham, MA, USA) and, if above 50 ng/μL, they were homogenised to that concentration with RNase/DNase free water (Promega, Madison, WI, USA). DNA aliquots were preserved frozen at −20 °C until the PCR was performed. Sample cross-contamination during DNA extraction was excluded by including negative controls (nuclease-free water; Promega, Madison, WI, USA) every ten samples that were also tested by PCR. A screening assay was selected for the detection of *C. burnetii* DNA in samples based on the IS1111-based PCR. Positive samples were further analysed by coupling the PCR with hybridization with a specific probe by reverse line blotting (RLB) [19,20]. The resulting genotypes were further analyzed with InfoQuest™FP 4.50 (BioRad, Hercules, CA, USA). Cluster analyses used the binary coefficient (Jaccard) and UPGMA (Unweigthed Pair Group Method Using Arithmetic Averages) to infer the phylogenetic relationships (Supplementary Figure S1).

3. Results and Discussion

We detected *C. burnetii* DNA—positive if qPCR cycle threshold (Ct) values were <40.0—in 79 of 816 spleen samples analysed (Table 2). In general, and considering overall all locations, *Microtus arvalis* was the species displaying the highest ratio of infection with *C. burnetii* (10.8%; 62/572) followed by *A. sylvaticus* (8.7%; 12/138), *Crocidura russula* (7.7%; 2/26) and *R. rattus* (6.4%; 3/47). The other micromammal species sampled were negative. One of the 130 vaginal swabs collected from *M. arvalis* females was qPCR positive (0.8%; 95%CI: 0.1–4.2). Ten positive samples with Ct values < 35.0 were analysed by RLB hybridization. Only a genotype II+ strain could be obtained in one *M. arvalis* from Northwest Spain. The acute disease antigen A (*adaA*) gene—present in some *C. burnetii* strains causing acute Q fever in humans [21]—was present in this genotype II+ strain. Genotype II+ has been previously reported in ticks (Slovak Republic), sheep (Germany, Spain) and humans (Italy) and it appears to be the most widely distributed (RLB) genotype in Europe [19]; nevertheless, the absence of molecular epidemiology studies in wildlife and in particular in micromammals makes the understanding of potential cross-species transmission difficult. Typing pathogens circulating in healthy wildlife could be partly constrained by pathogen burden in tissues [18,19]. This may occur more frequently in enzootic pathogens that have a large history of co-evolution with their hosts and replicate at a lower ratio in hosts than epidemic pathogens [22]. This, however, does not detract these hosts from playing relevant roles in the life cycle of pathogens. Currently, no clinical consequence of *C. burnetii* infection has been reported or noticed in infected micromammals; none of the micromammals surveyed in this study had symptoms compatible with Coxiellosis. A significant number of the infected individuals did indeed display very low levels of *C. burnetii* DNA in the spleen (69 of the 79 qPCR-positives had Ct values close to the negative threshold), perhaps indicative of past or subclinical infections. We did indeed obtain partial RLB typing results for a major part of the analysed samples, but only one could be completely typed.

In recent decades, zoonotic emerging infectious diseases have advanced positions to become one of the most worrisome threats to human, livestock and wildlife health [23,24]; SARS-CoV-2 has an animal origin, and it has been able to cross the inter-species barrier to emerge as the most devastating human pandemic of our time [25]. This may be owed to changes in the patterns of interaction between domestic animals, wildlife and humans [4] that are most probably occurring due to human influences on habitats, biodiversity and the climate. These changing patterns may also be behind the re-emergence of enzootic zoonoses such as Q fever that, although with a lower pandemic potential, may become a serious health problem. Our study contributes to unravelling the potential future threats of the re-emergence of *C. burnetii* infections of wildlife origin by informing about the potential implication of micromammals in the interspecific exchange of the pathogen. It also highlights the relevance of opportunistic sample collection in providing basic descriptive information useful to design future epidemiological studies. Findings reveal the occurrence of infections by *C. burnetii* in different species of wild micromammals in Spain as well as the presence of genotypes shared with humans, ticks and domestic animals and reported in different European countries [18,19].

In previous European studies, *C. burnetti* DNA was found in spleen samples of *R. rattus, R. norvegicus, Mus musculus, A. flavicollis* and *A. sylvaticus* from Cyprus, Germany, The Netherlands, Italy, Slovakia and Spain (Table 3). Prevalence ratio in medium-to-large sized studies ranged 0.6–23.5%. Other micromammal species such as the bank vole, the common vole and the common shrew that were object of medium-to-large surveys did not show infection with *C. burnetii*.

Table 3. Review of current evidences of *Coxiella burnetii* DNA detection in European micromammal species.

Common Name	Scientific Name	Country	Pos/N (Prev)	Reference
Bank vole	*Myodes glareolus*	Austria	0/40 (0.0)	[26]
		Croatia	0/43 (0.0)	[27]
		Italy	0/42 (0.0)	[28]
		Slovakia	0/23 (0.0)	[29]
			0/239 (0.0)	[30]
		Spain	0/6 (0.0)	[31]
		Czech Republic/Germany	0/78 (0.0)	[32]
Black rat	*Rattus rattus*	Netherlands	5/166 (3.0)	[6]
		Spain	3/47 (6.4)	This study
Brown rat	*Rattus norvegicus*	Germany	7/524 (1.3)	[8]
		Netherl.	8/164 (4.8)	[6]
Brown/black rat	*Rattus* spp.	Cyprus	32/136 (23.5)	[33]
		Spain	3/3 (100.0)	[20]
Common vole	*Microtus arvalis*	Austria	0/15 (0.0)	[26]
		Croatia	0/4 (0.0)	[27]
		Germany	0/109 (0.0)	[34]
		Slovakia	0/3 (0.0)	[29]
			0/19 (0.0)	[30]
		Czech Republic/Germany	0/148 (0.0)	[32]
		Spain	62/572 (10.9)	This study
Eurasian Harvest Mouse	*Micromys minutus*	Slovakia	0/1 (0.0)	[30]
European Pine Vole	*Microtus subterraneus*	Slovakia	0/1 (0.0)	[30]
European Water Vole	*Arvicola terrestris*	Spain	0/1 (0.0)	This study
		Germany	0/3 (0.0)	[34]
Field vole	*Microtus agrestis*	Croatia	0/1 (0.0)	[27]
		Czech Republic/Germany	0/1 (0.0)	[32]

Table 3. *Cont.*

Common Name	Scientific Name	Country	Pos/N (Prev)	Reference
Hazel dormouse	*Muscardinus avellanarius*	Croatia	0/1 (0.0)	[27]
House mouse	*Mus musculus*	Spain	2/28 (7.1)	[31]
		Spain	8/61 (13.1)	[35]
		Spain	0/10 (0.0)	This study
Long-tailed field mouse	*Apodemus sylvaticus*	Austria	0/26 (0.0)	[26]
		Croatia	0/3 (0.0)	[27]
		Italy	2/101 (19.8)	[28]
		Slovakia	0/3 (0.0)	[29]
			0/3 (0.0)	[30]
		Czech Republic/Germany	0/6 (0.0)	[32]
		Germany	0/2 (0.0)	[34]
		Spain	1/162 (0.6)	[31]
		Spain	12/138 (8.7)	This study
Striped Field Mouse	*Apodemus agrarius*	Croatia	0/54 (0.0)	[27]
		Czech Republic/Germany	0/2 (0.0)	[32]
Yellow-necked field mouse	*Apodemus flavicollis*	Austria	0/29 (0.0)	[26]
		Croatia	0/131 (0.0)	[27]
		Slovakia	1/38 (2.6)	[29]
			0/401 (0.0)	[30]
		Czech Republic/Germany	0/48 (0.0)	[32]
		Germany	0/3 (0.0)	[34]
		Spain	0/3 (0.0)	[31]
		Spain	0/2 (0.0)	This study
Common Shrew	*Sorex araneus*	Czech Republic/Germany	0/30 (0.0)	[32]
Crowned Shrew	*Sorex coronatus*	Czech Republic/Germany	0/7 (0.0)	[32]
Eurasian Pygmy Shrew	*Sorex minutus*	Czech Republic/Germany	0/1 (0.0)	[32]
White-toothed Shrew	*Crocidura russula*	Spain	2/26 (7.7)	This study

European studies on *C. burnetii* have mainly focused on the most widespread and abundant micromammal species in Europe such as *A. sylvaticus*, *Myodes glareolus*, *A. flavicollis*, *M. arvalis* and *Rattus* sp. The genus *Rattus* has been found consistently positive to *C. burnetii* DNA in all the studies performed in Europe, so rats are currently considered as true *C. burnetii* reservoirs [6,8,9]. *Coxiella burnetii* DNA was also detected in *R. rattus* in southern Spain in this study, so rats can be relevant hosts for *C. burnetii* in the Iberian Peninsula as well. These may play also a relevant role in the exchange of *C. burnetii* at the wildlife–livestock–human interface because rats live in natural, peridomestic and urban environments. In most of the studies in which *C. burnetii* DNA has been detected in micromammals, these were captured in peridomestic areas where domestic ruminants (known *C. burnetii* reservoirs) were present. The micromammals surveyed for this study came from wild environments where the direct interaction with domestic ruminants is low or inexistent. Wild ungulates and other proven wild reservoirs of *C. burnetii* [13,14] may be frequent in some of these areas but have been found to host mainly strains of the genotypes I and VII [19]. Our results cannot confirm that all the qPCR positive micromammal species found in this study are reservoirs for *C. burnetii*, but qPCR positive uterus samples from common voles were previously found [19] and a positive vaginal swab from *M. arvalis* was also found in this study, therefore demonstrating that *M. arvalis* is able to replicate and shed *C. burnetii*. An additional observation supporting the potential reservoir role of *M. arvalis* was the diagnosis of acute Q fever in one of the researchers (and author in this study) that participated in the survey of common voles. This person presented to a local medical practitioner in April 2012 with high fever and malaise a week after finishing a vole survey in northwestern Spain, and it was presumably diagnosed with flu. Some days later, the patient had persistent high fever and visited a hospital in northern Spain where Q fever was confirmed by detecting high titres (1/320) of specific *C. burnetii* IgG antibodies in an indirect immunofluorescent assay (IFA) and by a positive result in a specific IgM ELISA test for *C. burnetii*. The patient was negative to brucellosis, Lyme disease, tularaemia, bartonellosis and hepatitis B in a differential diagnosis approach of the probable infectious causes of persistent fever in a patient exposed to wildlife. This person had no history of exposure to livestock and other wildlife environments and had spent several weeks surveying voles before the onset of the symptoms. Whether this case was related to vole handling or was the consequence of exposure to contaminated aerosols from livestock or other wildlife shedding *C. burnetii* could not be determined.

The potential implication of Spanish micromammal species in the ecology of *C. burnetii* is based on some of the findings (in this and in other studies) such as the detection of specific *C. burnetii* antibodies in a wide diversity of micromammal species, including common vole, rats, house mouse, wood mouse and yellow-necked field mouse [4] that demonstrate susceptibility to pathogen infection. Further support comes from finding *C. burnetii* DNA in spleen samples of some micromammal species that indicates that a bacteraemia following replication did occur. In addition, *C. burnetii* DNA was found in the reproductive tract and vaginal secretions of *M. arvalis*, further supporting a potential efficient role in *C. burnetii* replication and transmission. The detection of *C. burnetii* DNA in rat faeces in other studies [36] also points out that (at least) rats also allow replication of *C. burnetii* and shedding.

The common vole experiences cyclic population density peaks under highly favourable environmental conditions that do indeed drive the exchange of zoonotic multi-host pathogens at the wildlife–human interface, e.g., *Francisella tularensis* [16]. However, in contrast to the spill-over role that *M. arvalis* plays in the transmission of *F. tularensis*, it may play a true reservoir role for *C. burnetii* and maintain it independently from the co-occurrence of other relevant hosts, e.g., lagomorphs or ungulates. Tularaemia is an emerging disease that currently is present only in the northern half of mainland Iberian Peninsula where it affects humans, wild lagomorphs and micromammals [37]. In this study, we observed in common voles from this area a slight increase in *C. burnetii* infection prevalence (unpublished data) from 2012 (low vole abundance) to 2014 (vole density peak in the study area; [38]), demonstrating persisting pathogen prevalence under conditions of contrasted host density (up to 100-fold), and supporting its role as true reservoir in clear contrast with tularaemia that practically disappears from vole populations in low abundance years [16]. During vole population outbreaks, infection rates and environmental contamination with *C. burnetii* could peak, given extremely high abundance of voles around or even inside villages, which could increase Q fever and Coxiellosis risks for humans and animals, respectively. This may be perhaps partly reflected in the observed higher incidence of Coxiellosis in sheep farms [39] during years of high vole density (2010–2011) in the region where *M. arvalis* were surveyed for this study [38]. *C. burnetii* transmission to livestock could be relevant in years or areas with high density of voles, but more precise information and longer time series should be required to confirm this possibility and any potential link among voles and domestic ruminants in *C. burnetii* exchange. The human infection case reported above indicates that frequent and tight contact with *M. arvalis* is a risk factor potentially promoting Q fever, so recommendations about avoiding contact with these animals disseminated to rural populations in vole outbreak years are highly advisable. The participation of potential coexisting micromammal reservoir species, e.g., *A. sylvaticus* and *C. russula*, would add complexity to understanding the factors shaping *C. burnetii* transmission risks in a sylvatic cycle [16]. Thus, sanitary recommendations to avoid contact with small mammals should be extended in general terms in the Iberian Peninsula.

4. Conclusions

We can conclude that this first approach provides evidence supporting the fact that there are several micromammal species that can be potential reservoirs of *C. burnetii*. Abundant and widespread species in the Iberian Peninsula, e.g., rats, wood mouse and white-toothed shrew, as well as species experiencing drastic cyclic demographic outbreaks, i.e., the common vole, might be relevant in the maintenance of wild-type *C. burnetii* strains that can be a matter of concern for animal and human health authorities.

Supplementary Materials: The following files are available online at https://www.mdpi.com/2076 -2615/11/3/654/s1, Figure S1: Dendrogram construct from hybridization data of 1 sample from this study (@), samples from González-Barrio et al. [19] and reference strains (framed text). Biological and geographic origin of the samples is displayed. Black boxes indicate the presence of the selected ORFs. *Coxiella burnetii* reference isolates used to validate the method are framed. *, Acute disease antigen A gene (*adaA*); +, Presence of *adaA*; -, Absence of *adaA*.

Author Contributions: Conceptualization, D.G.-B., J.V., J.T.G. and F.R.-F.; methodology, D.G.-B., I.J., J.V., J.T.G., P.P.O., F.A. and F.R.-F.; software, D.G.-B.; validation, D.G.-B., I.J. and F.R.-F.; formal analysis, D.G.-B. and F.R.-F.; investigation, D.G.-B. and F.R.-F.; resources, J.V., J.T.G., I.J. and F.R.-F.; data curation, D.G.-B., J.T.G., F.A. and F.R.-F.; writing—original draft preparation, D.G.-B. and F.R.-F.; writing—review and editing, D.G.-B., I.J., J.V., J.T.G., P.P.O., F.A. and F.R.-F.; supervision, D.G.-B. and F.R.-F.; project administration, D.G.-B., J.V., J.T.G. and F.R.-F.; funding acquisition, J.V., J.T.G. and F.R.-F. All authors have read and agreed to the published version of the manuscript.

Funding: This work was supported by grants CGL2011-30274 and CGL2015-71255-P of the Spanish Ministry for the Science and Innovation (MCI), and by the 'Fundación BBVA' Research Project TOPIGEPLA (2014 call). This is also a contribution to MCI-funded projects CGL2017-89866-R and E-RTA-2015-0002-C02-02. D.G.-B. was funded by MCI through Juan de la Cierva (FJCI-2016-27875) and 'Sara Borrell' (CD19CIII/00011) postdoctoral fellowships. GREFA provided partial financial support and invaluable logistic and workforce support for samplings in NW Spain, along with many students and staff from UAM.

Institutional Review Board Statement: The study was conducted according to the guidelines of the Declaration of Helsinki, and approved by UCLM Ethics Committee (reference number CEEA: PR20170201, date of approval: 15 March 2017).

Data Availability Statement: The data presented in this study are available in this article and Supplementary Material (Figure S1).

Acknowledgments: The authors extend particular thanks to Ana Benítez López, María Calero Riestra, Paqui Talavera, Iván García Egea, Jesús Herranz, Daniel Jareño Gómez, Salvador Luna, Julian Núñez Conde, Juan Oñate, Alfonso Paz Luna, Francisco Díaz Ruiz, Fernando Garcés, Carlos Cuellar, Fernando Blanca Chana, Lorena Hernández and Xurxo Piñeiro Álvarez.

Conflicts of Interest: The authors declare no conflict of interest.

References

1. Angelakis, E.; Raoult, D. Q fever. *Vet. Microbiol.* **2011**, *140*, 297–309. [CrossRef] [PubMed]
2. Maurin, M.; Raoult, D. Q fever. *Clin. Microbiol. Rev.* **1999**, *12*, 518–553. [CrossRef]
3. EFSA. The European Union summary report on trends and sources of zoonoses, zoonotic agents and food-borne outbreaks in 2012. *EFSA J.* **2014**, *12*, 3547.
4. González-Barrio, D.; Ruiz-Fons, F. Coxiella burnetii in wild mammals: A systematic review. *Transbound. Emerg. Dis.* **2019**, *66*, 662–671. [CrossRef]
5. Toledo, A.; Jado, I.; Olmeda, A.S.; Casado-Nistal, M.A.; Gil, H.; Escudero, R.; Anda, P. Detection of Coxiella burnetii in ticks collected from central Spain. *Vector Borne Zoonotic Dis.* **2009**, *9*, 465–468. [CrossRef]
6. Meerburg, B.G.; Reusken, C.B.E.M. The role of wild rodents in spread and transmission of Coxiella burnetii needs further elucidation. *Wildl. Res.* **2011**, *38*, 617–625. [CrossRef]
7. Reusken, C.; van der Plaats, R.; Opsteegh, M.; de Bruin, A.; Swart, A. Coxiella burnetii (Q fever) in Rattus norvegicus and Rattus rattus at livestock farms and urban locations in the Netherlands; could Rattus spp. represent reservoirs for (re)introduction? *Prev. Vet. Med.* **2011**, *101*, 124–130. [CrossRef] [PubMed]
8. Runge, M.; Von Keyserlingk, M.; Braune, S.; Becker, D.; Plenge-Bönig, A.; Freise, J.F.; Pelz, H.-J.; Esther, A. Distribution of rodenticide resistance and zoonotic pathogens in Norway rats in Lower Saxony and Hamburg, Germany. *Pest. Manag. Sci.* **2013**, *69*, 403–408. [CrossRef] [PubMed]
9. Izquierdo-Rodríguez, E.; Fernández-Álvarez, Á.; Martín-Carrilo, N.; Feliu, C.; Marchand, B.; Quilichini, Y.; Foronda, P. Rodents as Reservoirs of the Zoonotic Pathogens Coxiella burnetii and Toxoplasma gondii in Corsica (France). *Vector Borne Zoonotic Dis.* **2019**, *19*, 879–883. [CrossRef] [PubMed]
10. European Centre for Disease Prevention and Control. Q fever. In *ECDC. Annual Epidemiological Report for 2018*; ECDC: Stockholm, Sweden, 2019.
11. Ruiz-Fons, F.; Astobiza, I.; Barandika, J.F.; Hurtado, A.; Atxaerandio, R.; Juste, R.A.; García-Pérez, A.L. Seroepidemiological study of Q fever in domestic ruminants in semi-extensive grazing systems. *BMC Vet. Res.* **2010**, *6*, 3. [CrossRef]

12. González-Barrio, D.; Almería, S.; Caro, M.R.; Salinas, J.; Ortíz, J.A.; Gortázar, C.; Ruiz-Fons, F. Coxiella burnetii shedding by farmed red deer (Cervus elaphus). *Transbound. Emerg. Dis.* **2015**, *62*, 572–574. [CrossRef]
13. González-Barrio, D.; Velasco Ávila, A.L.; Boadella, M.; Beltrán-Beck, B.; Barasona, J.A.; Santos, J.P.V.; Queirós, J.; García-Pérez, A.L.; Barral, M.; Ruiz-Fons, F. Host and environmental factors modulate the exposure of free-ranging and farmed red deer (Cervus elaphus) to Coxiella burnetii. *Appl. Environ. Microbiol.* **2015**, *81*, 6223–6231. [CrossRef]
14. González-Barrio, D.; Maio, E.; Vieira-Pinto, M.; Ruiz-Fons, F. European rabbits as reservoir for Coxiella burnetii. *Emerg. Infect. Dis.* **2015**, *21*, 1055–1058. [CrossRef] [PubMed]
15. Jareño, D.; Viñuela, J.; Luque-Larena, J.J.; Arroyo, L.; Arroyo, B.; Mougeot, F. Factors associated with the colonization of agricultural areas by common voles Microtus arvalis in NW Spain. *Biol. Invasions* **2015**, *17*, 2315–2327. [CrossRef]
16. Luque-Larena, J.J.; Mougeot, F.; Arroyo, B.; Vidal, M.D.; Rodríguez-Pastor, R.; Escudero, R.; Anda, P.; Lambin, X. Irruptive mammal host populations shape tularemia epidemiology. *PLoS Pathog.* **2017**, *13*, e1006622. [CrossRef]
17. Navarro, L.M.; Pereira, H.M. Rewilding abandoned landscapes in Europe. *Ecosystems* **2012**, *15*, 900–912. [CrossRef]
18. González-Barrio, D.; Hagen, F.; Tilburg, J.J.H.C.; Ruiz-Fons, F. Coxiella burnetii genotypes in Iberian wildlife. *Microb. Ecol.* **2016**, *72*, 890–897. [CrossRef]
19. González-Barrio, D.; Jado, I.; Fernández-de-Mera, I.G.; Fernández-Santos, M.R.; Rodríguez-Vargas, M.; García-Amil, C.; Beltrán-Beck, B.; Anda, P.; Ruiz-Fons, F. Genotypes of Coxiella burnetii in wildlife: Disentangling the molecular epidemiology of a multi-host pathogen. *Environ. Microbiol. Rep.* **2016**, *8*, 708–714. [CrossRef]
20. Jado, I.; Carranza-Rodríguez, C.; Barandika, J.F.; Toledo, A.; García-Amil, C.; Serrano, B.; Bolaños, M.; Gil, H.; Escudero, R.; García-Pérez, A.L.; et al. Molecular method for the characterization of Coxiella burnetii from clinical and environmental samples: Variability of genotypes in Spain. *BMC Microbiol.* **2012**, *12*, 91. [CrossRef]
21. Frangoulidis, D.; Splettstoesser, W.D.; Landt, O.; Dehnhardt, J.; Henning, K.; Hilbert, A.; Bauer, T.; Antwerpen, M.; Meyer, H.; Walter, M.C.; et al. Microevolution of the chromosomal region of acute disease antigen A (adaA) in the query (Q) fever agent Coxiella burnetii. *PLoS ONE* **2013**, *8*, e53440. [CrossRef] [PubMed]
22. Casades-Martí, L.; González-Barrio, D.; Royo-Hernández, L.; Díez-Delgado, I.; Ruiz-Fons, F. Dynamics of Aujeszky's disease virus infection in wild boar in enzootic scenarios. *Transbound. Emerg. Dis.* **2020**, *67*, 388–405. [CrossRef]
23. Gortazar, C.; Reperant, L.A.; Kuiken, T.; de la Fuente, J.; Boadella, M.; Martínez-Lopez, B.; Ruiz-Fons, F.; Estrada-Peña, A.; Drosten, C.; Medley, G.; et al. Crossing the interspecies barrier: Opening the door to zoonotic pathogens. *PLoS Pathog.* **2014**, *10*, e1004129. [CrossRef] [PubMed]
24. Ruiz-Fons, F. A review of the current status of relevant zoonotic pathogens in wild swine (*Sus scrofa*) populations: Changes modulating the risk of transmission to humans. *Transbound. Emerg. Dis.* **2017**, *64*, 68–88. [CrossRef] [PubMed]
25. Latif, A.A.; Mukaratirwa, S. Zoonotic origins and animal hosts of coronaviruses causing human disease pandemics: A review. *Onderstepoort J. Vet. Res.* **2020**, *87*, e1–e9. [CrossRef] [PubMed]
26. Schmidt, S.; Essbauer, S.S.; Mayer-Scholl, A.; Poppert, S.; Schmidt-Chanasit, J.; Klempa, B.; Henning, K.; Schares, G.; Groschup, M.H.; Spitzenberger, F.; et al. Multiple infections of rodents with zoonotic pathogens in Austria. *Vector Borne Zoonotic Dis.* **2014**, *14*, 467–475. [CrossRef]
27. Tadin, A.; Tokarz, R.; Markotić, A.; Margaletić, J.; Turk, N.; Habuš, J.; Svoboda, P.; Vucelja, M.; Desai, A.; Jain, K.; et al. Molecular survey of zoonotic agents in rodents and other small mammals in Croatia. *Am. J. Trop. Med. Hyg.* **2016**, *94*, 466–473. [CrossRef] [PubMed]
28. Pascucci, I.; Domenico, M.D.; Dall'Acqua, F.; Sozio, G.; Camma, C. Detection of Lyme Disease and Q fever agents in wild rodents in Central Italy. *Vector Borne Zoonotic Dis.* **2015**, *15*, 404–411. [CrossRef] [PubMed]
29. Smetanova, K.; Schwarzová, K.; Kocianová, E. Detection of *Anaplasma phagocytophilum, Coxiella burnetii, Rickettsia* spp., and *Borrelia burgdorferi* s. l. in ticks, and wild-living animals in Western and Middle Slovakia. *Ann. N. Y. Acad. Sci.* **2006**, *1078*, 312–315. [CrossRef]
30. Minichová, L.; Hamšíková, Z.; Mahríková, L.; Slovák, M.; Kocianová, E.; Kazimírová, M.; Škultéty, L'.; Štefanidesová, K.; Špitalská, E. Molecular evidence of Rickettsia spp. in ixodid ticks and rodents in suburban, natural and rural habitats in Slovakia. *Parasites Vectors* **2017**, *10*, 158. [CrossRef]
31. Barandika, J.F.; Hurtado, A.; García-Sanmartín, J.; Juste, R.A.; Anda, P.; García-Pérez, A.L. Prevalence of tick-borne zoonotic bacteria in questing adult ticks from Northern Spain. *Vector Borne Zoonotic Dis.* **2008**, *8*, 829–835. [CrossRef] [PubMed]
32. Obiegala, A.; Jeske, K.; Augustin, M.; Król, N.; Fischer, S.; Mertens-Scholz, K.; Imholt, C.; Suchomel, J.; Heroldova, M.; Tomaso, H.; et al. Highly prevalent bartonellae and other vector-borne pathogens in small mammal species from the Czech Republic and Germany. *Parasites Vectors* **2019**, *3*, 332. [CrossRef] [PubMed]
33. Psaroulaki, A.; Chochlakis, D.; Angelakis, E.; Ioannou, I.; Tselentis, Y. Coxiella burnetii in wildlife and ticks in an endemic area. *Trans. R. Soc. Trop. Med. Hyg.* **2014**, *108*, 625–631. [CrossRef]
34. Pluta, S.; Hartelt, K.; Oehmeb, R. Prevalence of Coxiella burnetii and Rickettsia spp. In ticks and rodents in southern Germany. *Ticks Tick-Borne Dis.* **2010**, *1*, 145–147. [CrossRef] [PubMed]
35. Bolaños-Rivero, M.; Carranza-Rodríguez, C.; Rodríguez, N.F.; Gutiérrez, C.; Pérez-Arellano, J.L. Detection of Coxiella burnetii DNA in peridomestic and wild animals and ticks in an endemic region (Canary Islands, Spain). *Vector Borne Zoonotic Dis.* **2018**, *61*, 30–33.

36. Abdel-Moein, K.A.; Hamza, D.A. Rat as an overlooked reservoir for Coxiella burnetii: A public health implication. *Comp. Immunol. Microbiol. Infect. Dis.* **2017**, *17*, 630–634. [CrossRef] [PubMed]
37. Lopes de Carvalho, I.; Toledo, A.; Carvalho, C.L.; Barandika, J.F.; Respicio-Kingry, L.B.; Garcia-Amil, C.; García-Pérez, A.L.; Olmeda, A.S.; Zé-Zé, L.; Petersen, J.M.; et al. Francisella species in ticks and animals, Iberian Peninsula. *Ticks Tick Borne Dis.* **2016**, *7*, 159–165. [CrossRef] [PubMed]
38. Mougeot, F.; Lambin, X.; Rodríguez-Pastor, R.; Romairone, J.; Luque-Larena, J.J. Numerical response of a mammalian specialist predator to multiple prey dynamics in Mediterranean farmlands. *Ecology* **2019**, *100*, e02776. [CrossRef]
39. García-Seco, T. Epidemiología de la Fiebre Q en Rumiantes Domésticos en la Zona Central de la Península Ibérica. Ph.D. Thesis, Complutense University of Madrid, Madrid, Spain, 2017.

animals

Article

One Health Approach: An Overview of Q Fever in Livestock, Wildlife and Humans in Asturias (Northwestern Spain)

Alberto Espí [1,2,*], Ana del Cerro [1,2], Álvaro Oleaga [2,3], Mercedes Rodríguez-Pérez [2,4], Ceferino M. López [5], Ana Hurtado [6], Luís D. Rodríguez-Martínez [7], Jesús F. Barandika [6] and Ana L. García-Pérez [6]

1 Department of Animal Health, Regional Service for Agrofood Research and Development (SERIDA), 33394 Gijón, Spain; anadc@serida.org
2 Translational Microbiology Consolidated Group, Instituto de Investigación Sanitaria del Principado de Asturias (Health Research Institute of Asturias, ISPA), Av. del Hospital Universitario, s/n, 33011 Oviedo, Spain; alvaroleaga@yahoo.es (Á.O.); Mercedes.rodriguezp@sespa.es (M.R.-P.)
3 SERPA—Sociedad de Servicios del Principado de Asturias S.A., 33202 Gijón, Spain
4 Department of Microbiology, Central Hospital of Asturias (HUCA), 33011 Oviedo, Spain
5 Department of Animal Pathology, Animal Health, Veterinary Faculty, University of Santiago de Compostela, 27071 Lugo, Spain; c.lopez@usc.es
6 Department of Animal Health, NEIKER-Basque Institute for Agricultural Research and Development, Basque Research and Technology Alliance (BRTA), 48160 Derio, Spain; ahurtado@neiker.eus (A.H.); jbarandika@neiker.eus (J.F.B.); agarcia@neiker.eus (A.L.G.-P.)
7 Animal Health Laboratory (LSAPA), Government of Principado de Asturias, 33201 Gijón, Spain; LUISDARIO.RODRIGUEZMARTINEZ@asturias.org
* Correspondence: aespi@serida.org; Tel.: +34-653372118

Citation: Espí, A.; del Cerro, A.; Oleaga, Á.; Rodríguez-Pérez, M.; López, C.M.; Hurtado, A.; Rodríguez-Martínez, L.D.; Barandika, J.F.; García-Pérez, A.L. One Health Approach: An Overview of Q Fever in Livestock, Wildlife and Humans in Asturias (Northwestern Spain). *Animals* 2021, 11, 1395. https://doi.org/10.3390/ani11051395

Academic Editor: David González-Barrio

Received: 21 April 2021
Accepted: 10 May 2021
Published: 13 May 2021

Publisher's Note: MDPI stays neutral with regard to jurisdictional claims in published maps and institutional affiliations.

Simple Summary: We studied Q fever in an area of Spain where a significant number of human cases are diagnosed every year. Although animals are the only source of infection for people, this is the first study carried out in the autonomous community of Asturias that addresses in an integrated way the infection in domestic animals, wildlife and the environment as well as people. Our results revealed that a remarkable percentage of domestic ruminants and wild ungulates from all geographic areas of the region had been in contact with the infection's causative agent (*Coxiella burnetii*). In addition, the bacteria could be detected in the air and/or the dust of livestock farms. Finally, a statistical analysis was carried out to investigate the risk factors (age, sex, geographical area, etc.) for the human population of the region. These findings will help local health authorities to focus on the origin of the problem and facilitate applying preventive measures in the affected livestock farms.

Abstract: This study aimed to investigate the seroprevalence of *C. burnetii* in domestic ruminants, wild ungulates, as well as the current situation of Q fever in humans in a small region in northwestern Spain where a close contact at the wildlife–livestock–human interface exists, and information on *C. burnetii* infection is scarce. Seroprevalence of *C. burnetii* was 8.4% in sheep, 18.4% in cattle, and 24.4% in goats. Real-time PCR analysis of environmental samples collected in 25 livestock farms detected *Coxiella* DNA in dust and/or aerosols collected in 20 of them. Analysis of sera from 327 wild ungulates revealed lower seroprevalence than that found in domestic ruminants, with 8.4% of Iberian red deer, 7.3% chamois, 6.9% fallow deer, 5.5% European wild boar and 3.5% of roe deer harboring antibodies to *C. burnetii*. Exposure to the pathogen in humans was determined by IFAT analysis of 1312 blood samples collected from patients admitted at healthcare centers with Q fever compatible symptoms, such as fever and/or pneumonia. Results showed that 15.9% of the patients had IFAT titers ≥ 1/128 suggestive of probable acute infection. This study is an example of a One Health approach with medical and veterinary institutions involved in investigating zoonotic diseases.

Keywords: Q fever; *Coxiella burnetii*; seroprevalence; ruminants; wildlife; humans; dust; aerosols

1. Introduction

Q fever is a worldwide distributed zoonosis caused by *Coxiella burnetii*, a small intracellular bacterium belonging to γ-Proteobacteria [1,2] that infects a wide range of animal species, including mammals, birds and arthropods. People are infected through inhalation of aerosols contaminated with the bacteria expelled by infected animals during abortion or normal deliveries. Among domestic ruminants, sheep and goats are considered the main reservoirs of the infection and the principal source of human outbreaks [3,4]. *C. burnetii* has been reported in over a hundred wild mammal species that can be reservoirs for livestock and humans [5]. Reported cases of Q fever linked to exposure to wildlife can be associated with changes in the wildlife–human interactions leading to an increased risk of interspecies transmission [6]. Ticks are not essential in the domestic cycle of *C. burnetii* infection in livestock but may play a significant role in the wild cycle of transmission of coxiellosis among wild vertebrates [1,3].

Human Q fever is a public health problem worldwide [1,7]. After the outbreak in the Netherlands (2007–2010), linked to goat farms and involving more than 4000 people [2,4], the efforts devoted to studying this zoonosis have increased significantly. In Spain, the disease is considered endemic in several regions [7,8]. A systematic review recently conducted [9] showed significant differences in disease manifestations according to the geographical location. In the northern areas of Spain, pneumonia was the predominant symptom, while in the central and southern areas, isolated fever followed by hepatitis was the most frequent clinical form. In Asturias (northern Spain), pneumonia is the main clinical presentation of Q fever [10–12], and a relatively high risk of exposure to *C. burnetii* in the population in Asturias has been reported [7,9,11]. In fact, the fatality rate associated with *C. burnetii* infection in the region in the period 1997–2015 (7.69 per 100) was the highest compared to other Spanish regions [7]. There are very few studies on the exposure of wildlife to *C. burnetii*, in which red deer is the only species investigated [13,14]. In addition, little is known about the role of domestic ruminants as reservoirs of *C. burnetii* in Asturias. When dealing with zoonotic diseases like Q fever, a coordinated approach involving human and animal health professionals working together from a unique perspective (One Health) is needed to reduce the risk of infection for both humans and animals. This approach should also consider the environmental risk associated with the domestic and wild cycle of Q fever, particularly in regions of high nature tourist value like Asturias, where the human population is in close contact with nature, and consequently, with livestock and wildlife.

This study aimed to investigate the exposure to *C. burnetii* in domestic ruminants, wild ungulates and humans in northwestern Spain from a One Health perspective through the work of a multidisciplinary team integrated by microbiologists, veterinarians and epidemiologists.

2. Materials and Methods

2.1. Study Area

The study was carried out in the principality of Asturias, an autonomous community of 10,604 km^2 located in northwestern Spain with a population of 1,022,670 inhabitants [15]. The region can be divided into three different geographical areas: western, central and eastern Asturias, separated by large north-to-south oriented valleys running through the Cantabrian mountain range. The predominant climate is temperate oceanic [16], which favors developing deciduous and mixed forests interspersed with open pastures and meadows as the characteristic vegetation of this region. Livestock and wildlife are abundant in the region.

2.2. Animal and Human Population Investigated and Sample Collection

2.2.1. Livestock

Livestock activity in Asturias has a long tradition and a significant impact on the economy. The last census recorded 392,789 cattle, 46,004 sheep and 31,023 goats [17]. Beef cattle have progressively increased their census (70% of the total) at the expense of dairy cattle (30%). The vast majority of sheep and goats are meat breeds, and flocks are widely dispersed in the region, with a total of 3705 sheep and 1221 goat herds [17] holding an average of 12 and 25 animals per farm respectively. Lambing/kidding season in sheep and goats concentrates in spring, though a few intensive dairy herds can have more than one lambing season per year. Parturitions in dairy cattle can occur along the year, whereas in beef cattle, they mainly concentrate in spring and early summer but can also occur in other seasons.

The sample size was calculated to estimate the prevalence of an infection with a 95% confidence level, for an expected prevalence of 10%, an absolute error of 5% and a normal population distribution. This required 139 bovine samples, 138 for sheep and 138 for goats. Ruminant blood samples were obtained from the jugular vein in sheep and goats and from the medial coccygeal vein in cattle. Blood was collected in plain tubes without anticoagulant by the veterinarians in charge of the Livestock Official Sanitary Campaigns and then submitted to the Animal Health Laboratory of the Principality of Asturias (LSAPA), and ca. 1% of them were selected by systematic random sampling. For sheep and goats due to the annual organization of Livestock Official Sanitary Campaigns in Asturias, sera from several years had to be compiled to reach the calculated sample size and achieve geographical representation of the different areas, as follows: 2016 ($n = 60$), 2017 ($n = 74$) y 2018 ($n = 20$) for sheep, and 2015 ($n = 44$), 2016 ($n = 52$), 2017 ($n = 14$) y 2018 ($n = 25$) for goats. For cattle, samples were all collected in 2018. Finally, samples of 154 sheep, 135 goats and 163 cows were subjected to serological analysis. All sera were collected from females older than 6 months for sheep and goats or older than 12 months in the case of cattle. Serum was obtained by centrifugation and stored at $-20\ ^\circ$C until serological analysis.

Once the serological survey was completed, and with the aim of checking the presence of *C. burnetii* DNA in ruminant farms, 25 farmers, who did not know about the status of *C. burnetii* infection in their farms, agreed to participate voluntarily in the study. Seven dairy cattle farms, 5 goat farms, 2 sheep farms and 11 mixed flocks (with sheep, goats and/or cattle) were visited once between February and October 2019. In each farm, aerosols were taken inside the animal premises using the air sampler "MD8" Sartorius (Goettingen, Germany), performing an aspiration of 50 L/min air for 10 min. Dust samples were taken from different surfaces of animal premises with sterile swabs to detect the presence of *C. burnetii* DNA by real-time PCR.

2.2.2. Wildlife

The percentage of protected areas in the principality of Asturias amounts to almost 22 percent of its territory, which in terms of surface area represents 228,879 hectares. These areas harbor a high percentage of the continental vertebrate species present in Spain (67%). Thus, faunistic richness in Asturias is high. A total of 327 blood samples from wild ungulates were included in the study (83 Iberian red deer, 57 roe deer, 41 Cantabrian chamois, 73 fallow deer and 73 European wild boars). Roe deer predominate in the west of the territory, fallow deer in the east, and the remaining species (red deer, chamois and wild boar) are present throughout the territory. Blood samples were collected in hunting seasons between 2004 and 2018 in the frame of SERIDA's research projects related to wildlife populations. After blood centrifugation at the laboratory, sera were kept at$-20\ ^\circ$C until serological analyses.

2.2.3. Human Population Investigated

Blood samples were collected from patients who attended outpatient health services with compatible symptoms of Q fever (based on physicians' criteria) to investigate the presence of antibodies against *C. burnetii*. A total of 1312 samples were submitted throughout 2018 to the Microbiology Service of the Central University Hospital of Asturias (HUCA) from 6 of the 8 Health Areas (HA) of the region (Occidente, Suroccidente, Oviedo, Mieres, Langreo, Oriente). Data collected included age, gender, HA and month of sampling.

2.3. Serological Analyses

2.3.1. Animal Sera

An indirect ELISA test (PrioCHECK™ ruminant Q fever Ab plate kit, Thermo Fisher Scientific) was performed according to the manufacturer's instructions. This commercial kit uses protein G as a conjugate, valid to analyze sera of wild ungulates, as reported elsewhere [18]. Antibody results in animal sera were expressed by titers based on the calculation of the sample/positive ratio (S/P = OD sample − ODm NC/ODm PC − ODm NC). Titers equal to or greater than 1:40 were considered positive.

2.3.2. Human Sera

Sera from patients were first analyzed by indirect chemiluminescent immunoassay (CLIA) (*Coxiella burnetii* VirClia©, Vircell, Granada, Spain) to determine the presence of specific IgG antibodies against *C. burnetii* phase II in serum or plasma. Later, samples with a CLIA-positive result were titrated by indirect immunofluorescence assay (IFA) to detect antibodies anti-phase II (I + II IFA IgG/IgM/IgA, Vircell, Granada, Spain). Since a second sample was not taken 2–4 weeks apart to study seroconversion, a single IFA-positive convalescent serum IgG phase II $\geq$ 1:128 in a patient with compatible symptoms of Q fever of over 1 week's duration was considered a probable acute infection [19].

2.3.3. Molecular Analyses

Before DNA extraction, dust swabs were treated with 300 μL of TE buffer (Tris base 10 mM, EDTA 1 mM, pH 8) before being mixed with ATL and proteinase K for 1 h at 56 °C, and then, DNA extraction continued using the QIAmp DNA blood mini kit (Qiagen, Hilden, Germany). For aerosol samples, gelatine filters used in the air sampler device were treated with 2 mL of ATL buffer until gelatine was dissolved. This solution was mixed with 500 μL of buffer ATL and vortexed, centrifuged and heated at 56 °C. Then, two aliquots of 1 mL each were taken, and 50 μL of proteinase K (8 mg/mL) were added to each one, and the mixture was incubated for 1 h at 56 °C. The extraction process continued using QIAmp DNA blood mini kit (Qiagen, Hilden, Germany) following the manufacturer's instructions. Negative extraction controls were included every 10 samples to rule out DNA contamination. The presence of *C. burnetii* DNA was investigated by a real-time PCR amplification targeting the transposon-like repetitive region IS1111 of the *C. burnetii* genome [20]. A commercial internal amplification control (IAC) (TaqMan®® exogenous internal positive control, Thermo Fisher Scientific) was included in the assay to monitor for PCR inhibitors.

2.3.4. Statistical Analysis

Logistic regression was used to analyze the possible influence of the different potential risk factors studied over seroprevalence against *C. burnetii* in domestic ruminants, i.e., animal species (categorical; sheep, goats, cattle), period of sampling (categorical; spring, summer, autumn/winter), year (categorical; year of sampling, applicable to sheep and goats only), herd size (categorical; <50 animals, >50 animals), and geographical location (categorical; east, central and western Asturias). Similarly, for wild ungulates, the animal species (categorical; red deer, roe deer, fallow deer, chamois, wild boar), the year of sampling (categorical; 2004–2007, 2008–2018) and the geographical area (categorical; east, west, all territory) were considered for the analysis.

In addition, the influence of risk factors over the presence of *C. burnetii* DNA in aerosols or dust was analyzed in 25 livestock farms using logistic regression. The variables included in the model were animal species (categorical; sheep, goats, cattle), production (categorical; milk, meat), herd size (categorical; <50 animals, >50 animals), geographical location (categorical; east, central and western Asturias), the month of sampling (categorical; February, March, April, May, July, August, September, October), recent abortions (categorical; yes, no), biosecurity measures implemented in the farm (categorical; poor, moderate, good) and livestock housing (categorical; old, modern, adapted shed). The final model was selected as the one with the lowest Akaike's information criterion (AIC) value from all of the models performed. Odds ratio (OR) values were computed by raising "e" to the power of the logistic coefficient over the reference category. The seroprevalence against *C. burnetii* was calculated for each animal species. Statistical uncertainty was assessed by estimating the 95% confidence interval (CI) for each of the proportions according to the following formula: CI 95% = 1.96 [p (1 − p)/n]1/2, where p is the seroprevalence and n is the sample size.

To analyze human infection, patient data were grouped into categories, such as age (, 41–60, 61–80, 81–100 years old), sampling period (March–May, June–October, November–February), gender (male/female), and geographical location (east Asturias (HA of Langreo and Oriente), central Asturias (HA of Oviedo and Mieres), and western Asturias (HA of Occidente and Suroccidente). A hierarchical cluster analysis was performed to determine the natural groupings of variables regarding seropositivity to *C. burnetii*. Homogeneous clusters of these categorical variables were identified using the ClustOfVar R package [21]. FactoMine R package for Multiple correspondence analysis (MCA) [22] was used to explore for associations between categories of qualitative variables related to demographic (age, gender), temporal (month of sampling), and geographical variables with Q fever acute infection (patients with symptoms compatible with Q fever and with IFA serology positive). MCA is an analytical method used to detect and display the underlying structure of a set of nominal categorical data using Euclidean distances. MCA graphically displays data relationships. Data are converted to a K-by-K table of all pairwise tabulations and represented on a two-dimensional graph where more proximal variables show a more similar distribution. The human census was compiled by geographical location, and the incidence of acute Q fever per 100,000 inhabitants was calculated. All statistical analyses were performed using the statistical software R version 4.0.20 [23].

2.3.5. Ethics Statement

Anonymized animal and human data were provided by LSAPA and the Microbiology Service of HUCA, respectively. Blood sampling from domestic ruminants was carried out as part of the Livestock Official Sanitation Campaign, and human blood samples were taken in the course of disease diagnosis; therefore, written consent from farmers and patients was not required. The study protocol was approved by the Investigation Ethics Committee of the Principality of Asturias (N° 274/19 for the animal study and 125/17 for the human study).

3. Results

3.1. Seroprevalence in Livestock

Overall, *C. burnetii* seroprevalence was 8.4% (95% CI: 5–14) in sheep, 18.4% (95% CI: 13–25) in cattle, and 24.4% (95% CI: 18–32) in goats (Table 1 and Table S1). Geographically, seroprevalence in sheep was slightly higher in eastern Asturias (15.2%), whereas in goats, values were higher in eastern and central Asturias (29.1% and 28.0%, respectively). In general, higher prevalences were observed in areas where sheep and goat census are larger (Table 1). Conversely, the highest seroprevalence in cattle was found in western Asturias (24.4%), where the cattle census is slightly larger (Table 1).

Table 1. *C. burnetii* seroprevalence in domestic ruminants in the three geographical areas of Asturias.

AREA	Sheep				Goats				Cattle			
	Census (*n*)	Analyzed (*n*)	ELISA + (*n*)	Seropre-valence	Census (*n*)	Analyzed (*n*)	ELISA + (*n*)	Seropre-valence	Census (*n*)	Analyzed (*n*)	ELISA + (*n*)	Seropre-valence
West	7493	26	1	3.8	6609	24	1	4.2	150,250	45	11	24.4
Central	14,557	49	0	0.0	5676	25	7	28.0	132,841	39	6	15.4
East	23,954	79	12	15.2	18,738	86	25	29.1	109,698	79	13	16.5
Asturias	46,004	154	13	8.4	31,023	135	33	24.4	392,789	163	30	18.4

Logistic regression models identified sampling year and flock size as variables associated with seropositivity in sheep (Table 2). Prevalence was significantly higher in 2018 than in other years (p = 0.0339; OR 7.48), and Q fever infection was associated with flocks with more than 50 animals (p = 0.0017; OR 7.18). In goats, the geographical location of the herd explained the prevalence (Table 2). Hence, flocks in the eastern region had a significantly higher prevalence than flocks in the western region (p = 0.0309; OR 9.65), and those in central Asturias marginally higher than herds in the western region (p = 0.0866; OR 6.70). No explanatory variables were found in cattle.

Table 2. Logistic regression models for the seroprevalence against *C. burnetii* in sheep (A) and goats (B).

A—Sheep	Estimate	Z-Value	Pr (>|t|)	OR	CI 95%
Intercept	−3.9802	−4.978	0.0001	0.02	0.01–0.07
Sampling 2016 (ref.)					
Sampling 2017	1.0638	1.260	0.2060	2.90	0.64–20.60
Sampling 2018	2.0249	2.121	0.0339	7.58	1.25–62.66
Census 1–49 animals (ref.)					
Census 50–120	1.9716	3.139	0.0017	7.18	2.10–25.78

B—Goats	Estimate	Z-Value	Pr (>|t|)	OR	CI 95%
Intercept	−3.0910	−3.0236	0.0025	0.05	0.01–0.22
Western Asturias (ref.)					
Central Asturias	1.9014	1.7135	0.0866	6.70	1.07–130.56
Eastern Asturias	2.2669	2.1587	0.0309	9.65	1.86–177.50

3.2. Investigation of C. burnetii DNA in Animal Premises

Twenty of the 25 farms (80%) tested positive for the presence of *Coxiella* DNA in aerosols, dust or both (Table S1). Positive aerosols were detected in 5 farms (1/7 dairy cattle, 2/5 goat herds, and 2/11 mixed herds). The risk of detecting *C. burnetii* DNA in aerosols was associated with recent abortions (estimate 2.485; z value 1.937; p= 0.05275; OR = 12.00). Concerning the dust collected from surfaces in the animal premises, *C. burnetii* DNA was detected in 6/7 cattle herds, 2/5 goat herds, 2/2 sheep flocks and 6/11 mixed herds. The risk of detecting *C. burnetii* in dust was significantly associated with the productive aptitude of the herds, with dairy herds showing higher risk compared to meat-producing herds (estimate 1.132; z value −2.369; p = 0.0178; OR = 14.7). The remaining variables included in the models did not show a significant association.

3.3. Seroprevalence in Wildlife

Twenty-one of the 327 wild ungulates showed antibodies against *C. burnetii* (Table 3 and Table S1). The highest seroprevalence was observed in red deer (8.43%, 95% CI: 3–14) and the lowest in roe deer (3.51%, 95% CI: 1–8). Regarding the risk of exposure to *C. burnetii*, no significant associations were observed for any of the variables included in the model.

Table 3. *C. burnetii* seroprevalence in wild ungulates from Asturias.

Wildlife Species	Analyzed (*n*)	ELISA Positive (*n*)	Seropre-Valence
Iberian red deer (*Cervus elaphus hispanicus*)	83	7	8.43
Roe deer (*Capreolus capreolus*)	57	2	3.51
Cantabrian chamois (*Rupicapra rupicapra*)	41	3	7.32
Fallow deer (*Dama dama*)	73	5	6.85
European wild boar (*Sus scrofa*)	73	4	5.48
Total	327	21	6.42

3.4. Estimation of Q Fever Incidence in Humans

A total of 1312 patients from Health Centers and hospital admissions showing symptoms compatible with Q fever were included in the study (Table S1). Of them, 226 were CLIA-positive (17.2%, 226/1312), but only 208 (144 men and 64 women) had IFA titers $\geq$ 1/128 (Table 4) and were, therefore, considered probable acute Q fever cases.

Table 4. Distribution of human cases considered as probable Q fever by age and gender, season, and IFAT titer.

AGE	Men	Women	*n*	SEASON	*n*	IFAT Titer	*n*
1–40	17	13	30	Spring (Mar–May)	74	1:128	76
41–60	46	17	63	Summer (Jun–Oct)	68	1:256	56
61–80	62	23	85	Autumn–Winter	66	1:512	44
81–100	19	11	30	(Nov–Feb)		1:1024	31
						1:4096	1
Total	144	64	208	Total	208	Total	208

The number of cases mainly concentrated among the age groups 41–60 years (30.3%) and 61–80 years (40.9%). The distribution of cases was constant throughout the year, with 35% of cases occurring in March–May, 32.7% in June–October and 31.7% in November–February (Table 4). Geographically, the incidence was higher in western HA (54.2 cases/100,000 inhabitants) compared to central (29.3 cases/100,000 inhabitants) and eastern HA (32.6 cases/100,000 inhabitants).

Figure 1 shows a hierarchical representation of the analyzed variables showing an association between age and geographical area.

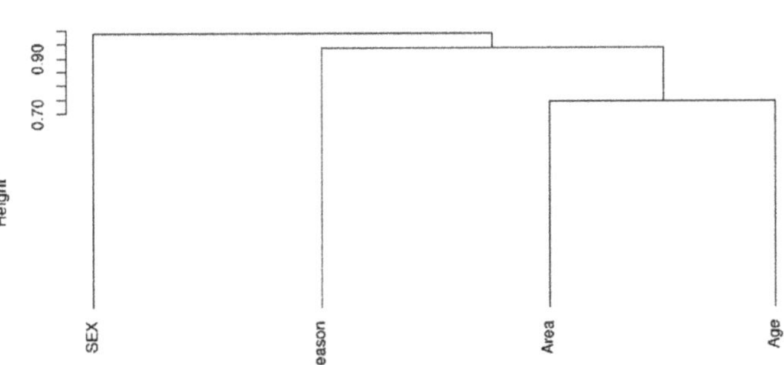

Figure 1. Dendrogram showing the relationship among variables affecting probable cases of Q fever.

MCA analysis identified the most important associations among the categorical variables. A graphic presentation constructed in a series of 2-dimensional spaces is shown in Figure 2.

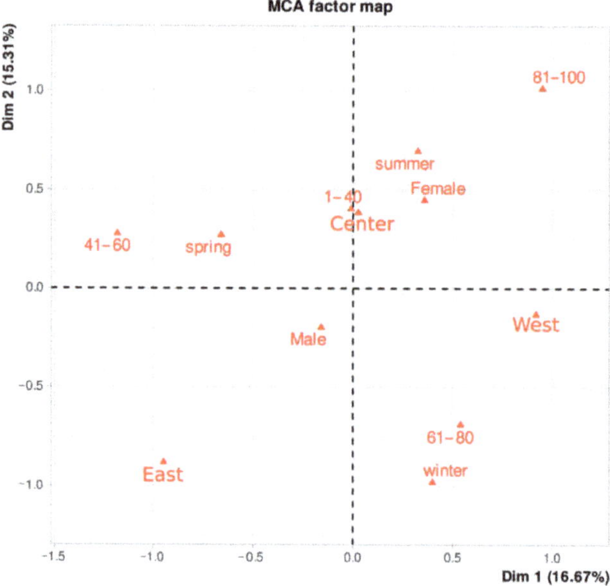

Figure 2. Multiple correspondence analysis describing associations between categories of age, gender, the month of sampling and geographical location of patients with Q fever.

The two first principal factors derived from the MCA analysis were retained to plot the coordinates of the studied variables and categories. Factorial axis1 (dimension 1) captured 16.1% of the variability and showed a geographical location gradient (east-central-western Asturias). The second axis (dimension 2) captured 15.3% of the variability and showed a gradient in the seasonal appearance of cases. The two dimensions 1 and 2 are sufficient to retain 32.0% of the total inertia (variation) contained in the data. As shown in Figure 2, Q fever cases among older patients in the western region mainly occurred in summer, cases among 60–80-year-old patients concentrated in winter, and 41–60-year-old cases in the eastern region were associated with spring. Q fever cases in younger people (<40 years old) were mainly found in the central region during the summer and spring months. No associations were found with patients' gender.

Incidence of probable human Q fever cases and seroprevalence against *C. burnetii* in domestic ruminants per geographical region are compiled in Figure 3.

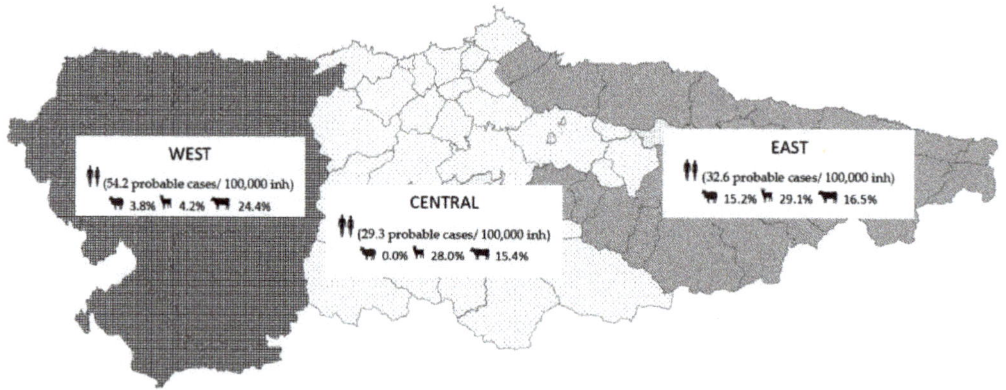

Figure 3. Map of Asturias indicating the number of probable cases of Q fever/100,000 inhabitants and the mean seroprevalence against *C. burnetii* in domestic ruminants in each of the three geographical zones.

4. Discussion

The incidence of human zoonotic infections, like Q fever, reflects the circulation of the bacteria in the animal reservoirs, i.e., domestic ruminants and several wildlife species [1,5]. Therefore, Q fever prevention strategies should incorporate professionals from human health, animal health, and environmental health integrated into a "One Health" approach [24]. In the study area, representing the Cantabrian coast regions, livestock production and hunting activities related to wild ungulates are very important, meeting the conditions for studying *C. burnetii* infection at the wildlife–livestock–human interface.

Contact with domestic ruminants is considered one of the most relevant risk factors in human *C. burnetii* infections [3]. To the best of our knowledge, no previous data on the status of Q fever in domestic ruminants was available in this area, except for a study conducted almost 20 years ago in sheep that showed 5.6% seroprevalence using the complement fixation test (CFT) [25]. ELISA, the technique currently used in most seroprevalence studies, is much more sensitive than CFT [26]. The results obtained in the current study showed that, in general, seroprevalence in domestic ruminants is higher compared to that observed in wild ungulates, suggesting that livestock, and particularly goats, may be the most important reservoir of infection in Asturias. In fact, the most important Q fever outbreak reported in Europe, which occurred in the Netherlands, was associated with goats [2,4], as were the most recent outbreaks reported in the Basque Country, a nearby region in northern Spain [27–29].

The seroprevalence values detected in sheep were similar to those observed in this species in other areas of the Iberian Peninsula (11.4%) [30] and comparable to those described in other regions in northern Spain (8.44% vs. 11.8%) [31]. Other Spanish areas, such as the Canary Islands, where the incidence of Q fever in humans is high [7,8], have shown higher seroprevalence in goats and sheep (60.4% and 31.7%, respectively) [32] compared with the studied area. Interestingly, the risk factors associated with increased exposure to *Coxiella* in sheep were the size of the herd and the year of sampling, with significantly higher prevalence in 2018 compared to previous years. In goats, the model found an association with geographical location. Although the risk of transmission of *C. burnetii* from small ruminants to humans seems to be higher than from cattle [3], the current study highlights that the role of cattle as *Coxiella* reservoir must not be underestimated.

Wildlife species are of paramount importance in Asturias due to their diversity, abundance, and interaction with domestic fauna. A large number of samples and species were analyzed in the current study, and the results indicated a different degree of exposure to *C. burnetii* infection among ungulates, with higher seroprevalence in red deer (8.4%) compared to other wild ungulates. Interestingly, seroprevalence in red deer was similar to that observed in previous studies [13] in the same region. Considering that red deer are widely distributed throughout the whole territory of Asturias, its impact as the reservoir of *C. burnetii* would be higher compared to other species with lower seroprevalences that occupied more restricted areas, like fallow deer (eastern Asturias) or roe deer (western Asturias). The rates of exposure of wild ungulates to *C. burnetii* determined here by ELISA were very similar to the infection rates obtained by PCR in roe deer (5.1%) or wild boar (4.3%) in neighboring regions [33].

Considering the results of seroprevalence, it seems that infection is more active within the domestic cycle than in the wild cycle. The fact that *C. burnetii* DNA was detected in 20 of the 25 farms where environmental samples were investigated demonstrates the importance of infection within the domestic cycle. Thus, 80% of the farms harbored *Coxiella* DNA in aerosols and/or environmental dust, indicating that a high percentage of ruminant herds may have had an active infection by *C. burnetii* (aerosol positive) at sampling [28], or have suffered it recently (dust positive) [34]. The recent occurrence of abortions is a risk factor for the presence of *C. burnetii* in aerosols, as has been observed in other studies [35]. The production system also appears to be a key factor for *C. burnetii* infection, with a higher risk in milking herds than meat herds. More intensive management systems where animals remain indoors favor contact between the animals, thus increasing the transmission of the bacteria, especially at lambing/calving time [36].

Considering the high prevalence observed in domestic ruminant farms and the challenging natural environment that facilitates contact with wild species, a high degree of exposure to *Coxiella* would be expected in the local population. This was confirmed by the IFAT results that showed that 15.9% (208/1312) of the people attending health centers with suspected symptoms of Q fever had IFAT titers ≥128 suggestive of probable acute infection. Unfortunately, no paired sera samples were collected 2–3 weeks apart to assess seroconversion, and therefore, it was not possible to confirm diagnosing Q fever [37]. A total of 208 probable cases of Q fever were detected along the study year (2018), representing 33.3 cases/100,000 inhabitants.

Age, geographical location and season were associated with Q fever exposure. In the eastern region, where ovine census is the largest and goats are also present at high numbers, the prevalence among 41–60-year-old patients was higher in spring. The abundance of small ruminants (the main reservoir of *C. burnetii*) in the area may pose a high risk of infection for humans [2–4]. Moreover, human Q fever cases in other Spanish regions have been associated with the months following the peak of the ovine lambing [38], which in general concentrated in spring. In the western region of Asturias, the number of farms is smaller, but herd size is larger, cattle predominates over small ruminants, and management is, in general, more professional [17]. In case of infection, many animals (herds with a large census) favors environmental contamination [39,40] and the risk of spread and transmission of *C. burnetii* to surrounding areas through the wind [41]. Q fever exposition in western Asturias occurred predominantly in summer in older patients (81–100 years old) and in winter in 61–80-year-old patients. The wider seasonal distribution might be because calvings occur all along the year in cattle herds. The association of probable cases of Q fever in the elderly with summer is difficult to explain, although summer months might be when older people spend more time in contact with nature. On the other hand, Q fever cases in younger people (<40 years old) were mainly found in the central region during the summer and spring months. Central Asturias is mostly urban, and summer and spring are the seasons when the population spends more leisure time outdoors in nature, thus increasing the possibility of exposure to rural infection sources. In addition, the incidence of Q fever is generally lower under the age of 30 years [7,42], in agreement with the results reported here. Similarly, the incidence is higher in men compared with women [1,7]. This was also the case in this study, though gender was not identified as an explanatory variable.

5. Conclusions

This work integrated the collaboration of groups working in animal and public health to obtain a global perspective of the Q fever situation in Asturias (northwestern Spain) in domestic ruminants, wildlife and the human population. Serological and molecular techniques provided an estimation of recent exposure to *Coxiella* and identified a wide distribution of *C. burnetii* infection among domestic ruminants. The results presented might help local authorities to set priorities when implementing control measures.

Supplementary Materials: The following are available online at https://www.mdpi.com/article/10.3390/ani11051395/s1; Table S1: Serological results and demographic, temporal, and geographical variables compiled from humans and domestic ruminants (sheep, cattle and goats) and wildlife. Furthermore, environmental research carried out in 25 farms is included.

Author Contributions: A.E. designed the study, collected the data from the different Laboratories and drafted the manuscript; M.R.-P. performed the human serological tests and compiled the data; A.d.C. performed the molecular analysis of the environmental samples and like L.D.R.-M., helped in the design and samples data collection and performed the animal serological tests; Á.O. carried out the work related to wildlife samples; C.M.L. conducted the statistical analysis; J.F.B., A.H. and A.L.G.-P. assisted with the interpretation of results and contributed to the discussion, revision and edition of the manuscript. All authors have read and agreed to the published version of the manuscript.

Funding: This work was funded by INIA—Spanish National Institute for Agricultural and Food Research and Technology (RTA2017-00055-C02-02), the European Regional Development Funds (ERDF), and PCTI 2018–2020 (GRUPIN: IDI2018-000237).

Institutional Review Board Statement: The study was conducted according to the guidelines of the Declaration of Helsinki, and the study protocol was approved by the Investigation Ethics Committee of the Principality of Asturias (no. 274/19 for the animal study and 125/17 for the human study).

Informed Consent Statement: As indicated in Section 2.3.5 anonymized animal and human data were provided by LSAPA and the Microbiology Service of HUCA therefore, written consent from farmers and patients was not required.

Data Availability Statement: Data supporting results can be found in Supplementary Material Table S1.

Acknowledgments: We are grateful to Roberto Agustín Lopez Acebo, Salvador Llorens García, Victor Manuel Álvarez González and other Veterinarians of the Government of the Principality of Asturias for their assistance in selecting and contacting the farms. We are grateful to farmers for their collaboration.

Conflicts of Interest: The authors declare no conflict of interest. The funders had no role in the design of the study; in the collection, analyses, or interpretation of data; in the writing of the manuscript, or in the decision to publish the results.

References

1. Eldin, C.; Mélenotte, C.; Mediannikov, O.; Ghigo, E.; Million, M.; Edouard, S.; Mege, J.-L.; Maurin, M.; Raoult, D. From Q Fever to Coxiella burnetii Infection: A Paradigm Change. *Clin. Microbiol. Rev.* **2017**, *30*, 115–190. [CrossRef] [PubMed]
2. Roest, H.I.; Bossers, A.; Van Zijderveld, F.G.; Rebel, J.M. Clinical microbiology ofCoxiella burnetiiand relevant aspects for the diagnosis and control of the zoonotic disease Q fever. *Vet. Q.* **2013**, *33*, 148–160. [CrossRef] [PubMed]
3. EFSA Panel on Animal Health and Welfare (AHAW). Scientific Opinion on Q fever. *EFSA J.* **2010**, *8*, 1595. [CrossRef]
4. Brom, R.D.; Engelen, E.; Roest, H.; der Hoek, W.; Vellema, P. Coxiella burnetii infections in sheep or goats: An opinionated review. *Vet. Microbiol.* **2015**, *181*, 119–129. [CrossRef] [PubMed]
5. González-Barrio, D.; Ruiz-Fons, F. Coxiella burnetiiin wild mammals: A systematic review. *Transbound. Emerg. Dis.* **2019**, *66*, 662–671. [CrossRef]
6. Gortazar, C.; Reperant, L.A.; Kuiken, T.; De La Fuente, J.; Boadella, M.; Martínez-Lopez, B.; Ruiz-Fons, F.; Estrada-Peña, A.; Drosten, C.; Medley, G.; et al. Crossing the Interspecies Barrier: Opening the Door to Zoonotic Pathogens. *PLoS Pathog.* **2014**, *10*, e1004129. [CrossRef]
7. Rodríguez-Alonso, B.; Almeida, H.; Alonso-Sardón, M.; López-Bernus, A.; Pardo-Lledias, J.; Velasco-Tirado, V.; Carranza-Rodríguez, C.; Pérez-Arellano, J.L.; Belhassen-García, M. Epidemiological scenario of Q fever hospitalized patients in the Spanish Health System: What's new. *Int. J. Infect. Dis.* **2020**, *90*, 226–233. [CrossRef]
8. Instituto de Salud Carlos III.Red Nacional de Vigilancia Epidemiológica. Informe anual del Sistema de Información Microbiológica 2016. Available online: https://repisalud.isciii.es/bitstream/20.500.12105/6975/7/InformeAnualDelSistema_2017.pdf (accessed on 15 September 2020).
9. Alende-Castro, V.; Macía-Rodríguez, C.; Novo-Veleiro, I.; García-Fernández, X.; Treviño-Castellano, M.; Rodríguez-Fernández, S.; González-Quintela, A. Q fever in Spain: Description of a new series, and systematic review. *PLoS Negl. Trop. Dis.* **2018**, *12*, e0006338. [CrossRef] [PubMed]
10. Clemente, M.M.G.; Seco-García, A.J.; Gutiérrez-Rodríguez, M.; Romero-Álvarez, P.; Fernández-Bustamante, J.; Rodríguez-Pérez, M. Brote epidémico de neumonía por Coxiella burnetii. *Enferm. Infecc. Microbiol. Clínica* **2007**, *25*, 184–186. [CrossRef]
11. Margolles, M.; Herrojo, A.; Ávarez, B.S.O.; González, M.; Redondo, M.A. *La Brucelosis y la Fiebre Q: Actuaciones Clínicas y EpiDemiológicas. Situación en un Colectivo de Alto Riesgo. Asturias 1993*; Direccion Regional de Salud Pública. Consejería de Sanidad. Principado de Asturias: Oviedo, Spain, 1994; pp. 1–33.
12. Nuño, F.J.; Noval, J.; Campoamor, M.T.; del Valle, A. Fiebre Q aguda en Asturias. *Rev. Clin. Esp.* **2002**, *202*, 569–573.
13. González-Barrio, D.; Ávila, A.L.V.; Boadella, M.; Beltrán-Beck, B.; Barasona, J.Á.; Santos, J.P.V.; Queirós, J.; García-Pérez, A.L.; Barral, M.; Ruiz-Fons, F. Host and Environmental Factors Modulate the Exposure of Free-Ranging and Farmed Red Deer (Cervus elaphus) to Coxiella burnetii. *Appl. Environ. Microbiol.* **2015**, *81*, 6223–6231. [CrossRef]
14. Ruiz-Fons, F.; Rodríguez, Ó.; Torina, A.; Naranjo, V.; Gortázar, C.; De La Fuente, J. Prevalence of Coxiella burnetti infection in wild and farmed ungulates. *Vet. Microbiol.* **2008**, *126*, 282–286. [CrossRef]
15. Instituto de Desarrollo Económico del Principado de Asturias. Demografía y población a fecha (2019). Available online: https://www.idepa.es/conocimiento/asturias-en-cifras/demografia (accessed on 15 September 2020).
16. Peel, M.C.; Finlayson, B.L.; McMahon, T.A. Updated world map of the Köppen-Geiger climate classification. *Hydrol. Earth Syst. Sci.* **2007**, *11*, 1633–1644. [CrossRef]

17. Consejería de Desarrollo Rural y Recursos Naturales: Sección de Prospectiva y Estadística. Declaración Anual Obligatoria—PACA—Estadísticas a fecha 15 de noviembre de 2017. Available online: https://www.asturias.es/portal/site/webasturias/template.PAGE/BuscadorOpenDataTransparencia (accessed on 15 September 2020).
18. Stöbel, K.; Schönberg, A.; Staak, C. A new non-species dependent ELISA for detection of antibodies to Borrelia burgdorferi s. l. in zoo animals. *Int. J. Med. Microbiol.* **2002**, *291*, 88–99. [CrossRef]
19. CDC. Q Fever (Coxiella burnetii) 2009 Case Definition. Available online: https://wwwn.cdc.gov/nndss/conditions/q-fever/case-definition/2009 (accessed on 15 September 2020).
20. Schets, F.; De Heer, L.; Husman, A.D.R. Coxiella burnetii in sewage water at sewage water treatment plants in a Q fever epidemic area. *Int. J. Hyg. Environ. Health* **2013**, *216*, 698–702. [CrossRef] [PubMed]
21. Chavent, M.; Kuentz, V.; Liquet, B.; Saracco, J. ClustOfVar: Clustering of variables. R package version 1.1.2017. Available online: https://CRAN.R-project.org/package=ClustOfVar (accessed on 15 September 2020).
22. Lê, S.; Josse, J.; Husson, F. FactoMineR: AnRPackage for Multivariate Analysis. *J. Stat. Softw.* **2008**, *25*, 1–18. [CrossRef]
23. R Core Team. R: A Language and Environment for Statistical Computing. R Foundation for Statistical Computing, Vienna, Austria. Available online: http://www.R-project.org/ (accessed on 11 March 2021).
24. Rahaman, R.; Milazzo, A.; Marshall, H.; Bi, P. Is a One Health Approach Utilized for Q Fever Control? A Comprehensive Literature Review. *Int. J. Environ. Res. Public Health* **2019**, *16*, 730. [CrossRef]
25. Espi, A.; Prieto, M.; Álvarez, M. *Situación sanitaria del ovino en Asturias: Seroprevalencia de la Enfermedad de la Frontera, Maedi-Visna, Aborto Enzoótico, Agalaxia Contagiosa, Fiebre Q y Leptospirosis*; Consejería de Agricultura y Pesca. Viceconsejería. Servicio de Publicaciones y Divulgación: Sevilla, Spain, 2001; pp. 708–713.
26. WHO (World Organization for Animal Health). Manual of diagnostic Tests and vaccines for terrestrial animals (mammals, birds and bees). Chapter 2.7.11 "Q fever". Available online: https://www.oie.int/doc/ged/D7709.PDF (accessed on 15 September 2020).
27. Alonso, E.; Eizaguirre, D.; Lopez-Etxaniz, I.; Olaizola, J.I.; Ocabo, B.; Barandika, J.F.; Jado, I.; Álvarez-Alonso, R.; Hurtado, A.; García-Pérez, A.L. A Q fever outbreak associated to courier transport of pets. *PLoS ONE* **2019**, *14*, e0225605. [CrossRef] [PubMed]
28. Álvarez-Alonso, R.; Basterretxea, M.; Barandika, J.F.; Hurtado, A.; Idiazabal, J.; Jado, I.; Beraza, X.; Montes, M.; Liendo, P.; García-Pérez, A.L. A Q Fever Outbreak with a High Rate of Abortions at a Dairy Goat Farm:Coxiella burnetiiShedding, Environmental Contamination, and Viability. *Appl. Environ. Microbiol.* **2018**, *84*, 01650–18. [CrossRef]
29. Hurtado, A.; Alonso, E.; Aspiritxaga, I.; Etxaniz, I.L.; Ocabo, B.; Barandika, J.F.; De Murúa, J.I.F.-O.; Urbaneja, F.; Álvarez-Alonso, R.; Jado, I.; et al. Environmental sampling coupled with real-time PCR and genotyping to investigate the source of a Q fever outbreak in a work setting. *Epidemiol. Infect.* **2017**, *145*, 1834–1842. [CrossRef]
30. Cruz, R.; Esteves, F.; Vasconcelos-Nóbrega, C.; Santos, C.; Ferreira, A.S.; Mega, C.; Coelho, A.C.; Vala, H.; Mesquita, J.R. A Nationwide Seroepidemiologic Study on Q Fever Antibodies in Sheep of Portugal. *Vector Borne Zoonotic Dis.* **2018**, *18*, 601–604. [CrossRef]
31. Ruiz-Fons, F.; Astobiza, I.; Barandika, J.F.; Hurtado, A.; Atxaerandio, R.; Juste, R.A.; Garcia-Perez, A.L. Seroepidemiological study of Q fever in domestic ruminants in semi-extensive grazing systems. *BMC Vet. Res.* **2010**, *6*, 3. [CrossRef] [PubMed]
32. Rodríguez, N.F.; Carranza, C.; Bolaños, M.; Pérez-Arellano, J.L.; Gutierrez, C. Seroprevalence of Coxiella burnetii in domestic ruminants in Gran Canaria island, Spain. *Transbound. Emerg. Dis.* **2010**, *57*, 66–67. [CrossRef]
33. Astobiza, I.; Barral, M.; Ruiz-Fons, F.; Barandika, J.; Gerrikagoitia, X.; Hurtado, A.; García-Pérez, A. Molecular investigation of the occurrence of Coxiella burnetii in wildlife and ticks in an endemic area. *Vet. Microbiol.* **2011**, *147*, 190–194. [CrossRef] [PubMed]
34. Álvarez-Alonso, R.; Zendoia, I.I.; Barandika, J.F.; Jado, I.; Hurtado, A.; López, C.M.; García-Pérez, A.L. Monitoring Coxiella burnetii Infection in Naturally Infected Dairy Sheep Flocks Throughout Four Lambing Seasons and Investigation of Viable Bacteria. *Front. Vet. Sci.* **2020**, *7*, 352. [CrossRef] [PubMed]
35. Roest, H.J.; van Gelderen, B.; Dinkla, A.; Frangoulidis, D.; van Zijderveld, F.G.; Rebel, J.; van Keulen, L. Q fever in pregnant goats: Pathogenesis and excretion of Coxiella burnetii. *PLoS ONE* **2012**, *7*, e48949. [CrossRef] [PubMed]
36. Mori, M.; Roest, H.-J. Farming Q fever and public health: Agricultural practices and beyond. *Arch. Public Health* **2018**, *76*, 2. [CrossRef]
37. Anderson, A.D.; Bijlmer, H.; Fournier, P.E.; Graves, S.; Hartzell, J.; Kersh, G.J.; Limonard, G.J.; Marrie, T.J.; Massung, R.F.; McQuiston, J.H.; et al. Diagnosis and management of Q fever—United States, 2013: Recommendations from CDC and the Q Fever Working Group. *MMWR* **2013**, *62*, 1–30.
38. Cilla, G.; Montes, M.; Pérez-Trallero, E. Q fever in the Netherlands—What matters is seriousness of disease rather than quantity. *Eurosurveillance* **2008**, *13*, 18975. [CrossRef]
39. Agger, J.F.; Paul, S. Increasing prevalence of Coxiella burnetii seropositive Danish dairy cattle herds. *Acta Vet. Scand.* **2014**, *56*, 46. [CrossRef]
40. Carrié, P.; Barry, S.; Rousset, E.; De Crémoux, R.; Sala, C.; Calavas, D.; Perrin, J.-B.; Bronner, A.; Gasqui, P.; Gilot-Fromont, E.; et al. Swab cloths as a tool for revealing environmental contamination by Q fever in ruminant farms. *Transbound. Emerg. Dis.* **2019**, *66*, 1202–1209. [CrossRef] [PubMed]
41. Tissot-Dupont, H.; Amadei, M.-A.; Nezri, M.; Raoult, D. Wind in November, Q Fever in December. *Emerg. Infect. Dis.* **2004**, *10*, 1264–1269. [CrossRef] [PubMed]
42. Maurin, M.; Raoult, D. Q Fever. *Clin. Microbiol. Rev.* **1999**, *12*, 518–553. [CrossRef] [PubMed]

MDPI
St. Alban-Anlage 66
4052 Basel
Switzerland
Tel. +41 61 683 77 34
Fax +41 61 302 89 18
www.mdpi.com

Animals Editorial Office
E-mail: animals@mdpi.com
www.mdpi.com/journal/animals

www.ingramcontent.com/pod-product-compliance
Lightning Source LLC
LaVergne TN
LVHW070913170726
843515LV00005B/767